THE ECOLOGY AND EVOLUTION OF INDUCIBLE DEFENSES

THE ECOLOGY AND EVOLUTION
OF INDUCIBLE DEFENSES

Edited by Ralph Tollrian and C. Drew Harvell

PRINCETON UNIVERSITY PRESS   PRINCETON, NEW JERSEY

Copyright © 1999 by Princeton University Press
Published by Princeton University Press, 41 William Street,
Princeton, New Jersey 08540
In the United Kingdom: Princeton University Press, Chichester, West Sussex
All Rights Reserved

Library of Congress Cataloging-in-Publication Data

The ecology and evolution of inducible defenses / edited by Ralph Tollrian and
C. Drew Harvell.
p.   cm.
Includes bibliographical references and index.
ISBN 0-691-01221-0 (cloth : alk. paper)
ISBN 0-691-00494-3 (pbk.)
1. Animal defenses. 2. Plant defenses. 3. Ecology. 4. Evolution (Biology)
I. Tollrian, Ralph, 1960–   . II. Harvell, C. Drew, 1954–   .
QL759.E335   1998
578.4'7—dc21   98-17182

This book has been composed in Times Roman

Princeton University Press books are printed on acid-free paper and
meet the guidelines for permanence and durability of the Committee
on Production Guidelines for Book Longevity of the Council on
Library Resources

http://pup.princeton.edu

Printed in the United States of America

10 9 8 7 6 5 4 3 2 1

10 9 8 7 6 5 4 3 2 1
(pbk.)

Contents

Acknowledgments	vii
List of Contributors	ix
Why Inducible Defenses? *C. Drew Harvell and Ralph Tollrian*	3
1. Coping with Life as a Menu Option: Inducible Defenses of the Wild Parsnip *May R. Berenbaum and Arthur R. Zangerl*	10
2. Adaptive Status of Localized and Systemic Defense Responses in Plants *Johannes Järemo, Juha Tuomi, and Patric Nilsson*	33
3. Why Induced Defenses May Be Favored over Constitutive Strategies in Plants *Anurag A. Agrawal and Richard Karban*	45
4. Evolution of Induced Indirect Defense of Plants *Marcel Dicke*	62
5. Consumer-Induced Changes in Phytoplankton: Inducibility, Costs, Benefits, and the Impact on Grazers *Ellen Van Donk, Miquel Lürling, and Winfried Lampert*	89
6. The Immune System as an Inducible Defense *Simon D. W. Frost*	104
7. Kairomone-Induced Morphological Defenses in Rotifers *John J. Gilbert*	127
8. Predator-Induced Defenses in Ciliated Protozoa *Hans-Werner Kuhlmann, Jürgen Kusch, and Klaus Heckmann*	142
9. Ecology and Evolution of Predator-Induced Behavior of Zooplankton: Depth Selection Behavior and Diel Vertical Migration *Luc De Meester, Piotr Dawidowicz, Erik van Gool, and Carsten J. Loose*	160

10. Inducible Defenses in Cladocera: Constraints, Costs, and Multipredator Environments — 177
Ralph Tollrian and Stanley I. Dodson

11. Predator-Induced Defense in Crucian Carp — 203
Christer Brönmark, Lars B. Pettersson, and P. Anders Nilsson

12. Density-Dependent Consequences of Induced Behavior — 218
Bradley R. Anholt and Earl E. Werner

13. Complex Biotic Environments, Coloniality, and Heritable Variation for Inducible Defenses — 231
C. Drew Harvell

14. Developmental Strategies in Spatially Variable Environments: Barnacle Shell Dimorphism and Strategic Models of Selection — 245
Curtis M. Lively

15. Evolution of Forager Responses to Inducible Defenses — 259
Frederick R. Adler and Daniel Grünbaum

16. Evolution of Reversible Plastic Responses: Inducible Defenses and Environmental Tolerance — 286
Wilfried Gabriel

17. The Evolution of Inducible Defenses: Current Ideas — 306
Ralph Tollrian and C. Drew Harvell

References — 323

Index — 377

Acknowledgments

WE EXTEND deep thanks to the many people who helped us produce this volume. First, we are grateful to Winfried Lampert, who suggested the idea of an edited book on inducible defenses. We appreciate the interest of the many colleagues at the Inducible Defense Symposium in Plön (1995) who gave freely of ideas and inspiration for this project and to the European Science Foundation for funding this special session. We are especially grateful to Simon Levin of Princeton University for his initial enthusiasm for publishing our book with Princeton University Press and for his ongoing support. We called on the services of a large number of external reviewers, and to them we extend a warm thank you. Members of our lab groups often helped with reviewing or discussion on short notice; particular thanks for this work goes to Kiho Kim, Andrea Graham, Erika Iyengar, and Wilfried Gabriel. The Graduate Core course at Cornell was also generous in critiquing sections of the book, and we appreciate the students' help. Finally, we are especially grateful to Bob Paine and an anonymous reviewer for reviewing the book in its entirety and providing useful suggestions for improvement. We are grateful also to the National Science Foundation Program in Ecological Physiology, which supported Harvell's research and also some of the time she devoted to this book; to the Section of Ecology and Systematics at Cornell University for providing some secretarial time; to the Ecology Department at the Ludwig-Maximilians-Universität Munich for support; and to Rosie Brainard for her help with countless details of correspondence. We appreciate the help from the staff at Princeton University Press, in particular Alice Calaprice, Sam Elworthy, and Emily Wilkinson. Last but not least, we thank all the contributors for their cooperation and for their patience in putting up with the editors' countless demands and unending deadlines.

List of Contributors

FREDERICK R. ADLER
Department of Mathematics and Department of Biology, University of Utah, Salt Lake City, UT 84112, USA

ANURAG A. AGRAWAL
Department of Entomology and Center for Population Biology, University of California, Davis, CA 95616-8584, USA

BRADLEY R. ANHOLT
Department of Biology, University of Victoria, Box 3020, Victoria, B.C., V8W 3N5, Canada

MAY R. BERENBAUM
Department of Entomology, 320 Morrill Hall, University of Illinois, 505 S. Goodwin, Urbana, IL 61801-3795, USA

CHRISTER BRÖNMARK
Department of Ecology, Ecology Building, Lund University, S-223 62 Lund, Sweden

PIOTR DAWIDOWICZ
Department of Hydrobiology, University of Warsaw, Nowy Swiat 67, 00-046 Warsaw, Poland

LUC DE MEESTER
Laboratory of Ecology and Aquaculture, Katholieke Universiteit Leuven, Naamsestraat 59, 3000 Louvaine, Belgium

MARCEL DICKE
Department of Entomology, Wageningen Agricultural University, P.O. Box 8031, NL-6700 EH Wageningen, The Netherlands

STANLEY I. DODSON
Department of Zoology, University of Wisconsin, 430 Lincoln Drive, Madison, Wisconsin 53706-1381, USA

SIMON D. W. FROST
Centre for HIV Research, Institute of Cell, Animal and Population Biology, University of Edinburgh, Waddington Building, Kings Buildings, West Mains Rd., Edinburgh, Scotland

WILFRIED GABRIEL
Zoologisches Institut, Ludwig-Maximilians-Universität München, Karlstr. 25, D-80333 Munich, Germany

JOHN J. GILBERT
Department of Biological Sciences, Dartmouth College, Hanover, New Hampshire 03755, USA

DANIEL GRÜNBAUM
Department of Mathematics, University of Utah, Salt Lake City, UT 84112, USA

C. DREW HARVELL
Section of Ecology and Systematics, Division of Biological Sciences, Cornell University, Ithaca, NY 14853, USA

KLAUS HECKMANN
Institut für Allgemeine Zoologie und Genetik, Universität Münster, Schlossplatz 5, D-48149 Münster, Germany

JOHANNES JÄREMO
Department of Theoretical Ecology, Ecology Building, Lund University, S-223 62 Lund, Sweden

RICHARD KARBAN
Department of Entomology and Center for Population Biology, University of California, Davis, CA 95616-8584, USA

HANS-WERNER KUHLMANN
Institut für Allgemeine Zoologie und Genetik, Universität Münster, Schlossplatz 5, D-48149 Münster, Germany

JÜRGEN KUSCH
Universität Kaiserslautern, Abteilung Ökologie, Erwin-Schrödinger-Straße 13/14, D-67663 Kaiserslautern, Germany

WINFRIED LAMPERT
Max Planck Institute for Limnology, Postfach 165, 24302 Plön, Germany

CURTIS M. LIVELY
Department of Biology, Indiana University, Bloomington IN 47405, USA

CARSTEN J. LOOSE
Alfred Wegener Institute for Polar and Marine Research, Columbusstraße, 27568 Bremerhaven, Germany

MIQUEL LÜRLING
Department of Water Quality Management and Aquatic Ecology, Agricultural University, P.O. Box 8080, 6700 DD Wageningen, The Netherlands

P. ANDERS NILSSON
Department of Ecology, Ecology Building, Lund University, S-223 62 Lund, Sweden

PATRIC NILSSON
Department of Theoretical Ecology, Ecology Building, Lund University, S-223 62 Lund, Sweden

LARS B. PETTERSSON
Department of Ecology, Ecology Building, Lund University, S-223 62 Lund, Sweden

RALPH TOLLRIAN
Zoologisches Institut, Ludwig-Maximilians-Universität München, Karlstraße 25, D-80333 Munich, Germany

LIST OF CONTRIBUTORS

JUHA TUOMI
Department of Biology, University of Oulu, Linnanmaa, FIN-90570, Oulu, Finland

ELLEN VAN DONK
Netherlands Institute of Ecology, Centre for Limnology (NIOO-CL), Rijksstraatweg 6, 3631 AC Nieuwersluis, The Netherlands

ERIK VAN GOOL
Department of Aquatic Ecology, University of Amsterdam; and Netherlands Institute of Ecology, Centre for Limnology, Rijksstraatweg 6, 3631 AC Nieuwersluis, The Netherlands

EARL E. WERNER
Department of Biology, University of Michigan, Ann Arbor, MI 48109, USA

ARTHUR R. ZANGERL
Department of Entomology, 320 Morrill Hall, University of Illinois, 505 S. Goodwin, Urbana, IL 61801-3795, USA

THE ECOLOGY AND EVOLUTION OF INDUCIBLE DEFENSES

Why Inducible Defenses?

C. DREW HARVELL AND RALPH TOLLRIAN

INDUCIBLE DEFENSES are phenotypic changes induced directly by cues associated with biotic agents; most can measurably diminish the effects of subsequent attacks by these agents. The study of these shifting phenotypes not only is the focus of the organismal and evolutionary biologist, but it may make a contribution by promoting major advances in the multidisciplinary study of ecology, development, evolution, chemistry, and theory that is the hallmark of this discipline. Issues surrounding the evolution of inducible defenses are fundamental to understanding biotic evolution: what conditions favor the evolution of adaptive, phenotypically variable responses to biotic agents? At its extreme, the answers give us insight into what is perhaps our greatest medical problem: the evolution of inducible resistance to disease and its sometimes disastrous failure. Recent tabulations indicate that inducible defenses are widespread and occur in many invertebrate groups, in vertebrates, and in plants. In invertebrates and vertebrates, defensive shifts in morphology, life history, and behavior are induced by proximity to predators and competitors. In insects, antibacterial cecropins are induced by pathogens. In vertebrates, selective induction of multiple mechanisms of resistance in the immune system is perhaps the most striking example in any organism of precision in tracking biotic agents. In plants, changes in chemistry and morphology are similarly induced by cues from herbivores or pathogens.

Inducible morphological changes in animals include production of spines (bryozoans, cladocerans, and rotifers), helmets (cladocerans), and protective variation in shell shape, body shape, and coloration (ciliates, cladocerans, barnacles, gastropods, fishes, and amphibians). The visibility of these phenotypic shifts and their experimental tractability has made them powerful study tools for understanding ecological consequences and evolutionary causes of inducible responses. Inducible shifts in life-history parameters such as size at maturity and offspring size allow reproduction in more predator-resistant size classes (ciliates, cladocerans, rotifers, gastropods) or higher offspring survival (ciliates, cladocerans).

Work with freshwater zooplankton (cladocerans, copepods, insect larvae) has revealed the importance of inducible shifts in stereotypic behaviors such as vertical migration. In this spectacular phenomenon, whole segments of the zooplankton community in marine and freshwater systems are changing their vertical position in a diel mode. Diel vertical migration can be induced by proximity of predators. Other induced behavioral defenses like changes in activity (amphibia, cladocera, fishes) and increased alertness (ciliates, cladocera) reduce predator encounter probabilities and increase the escape probability after predator attacks. In a broad sense nearly all escape reactions are inducible defenses, triggered by visual, mechanical, or chemical cues.

Research in aquatic and terrestrial habitats and with plants, animals, and parasitized hosts has all to some degree developed independently, often with different terminology to describe the same phenomena. The inducible mechanisms that increase fitness in the presence of predators are usually called "defenses"; the mechanisms resulting from exposure to herbivores or pathogens are usually termed "resistance." Whatever the terminological or procedural differences, each field can benefit from insights generated in other systems. All the diverse phenomena reported in this book are linked by being phenotypic changes, often triggered by chemical cues associated with the biotic environment.

The goal of this book is to provide a detailed overview of the well-studied systems of inducible defenses of plants and animals from terrestrial, marine, and freshwater habitats to pinpoint unifying factors favoring the evolution of inducible defenses. Finally, we advocate further generalization and conceptual brainstorming through presenting several theoretical studies that investigate the evolution of inducible defenses.

Since De Beauchamp (1952a) and Gilbert (1966) first discovered the inducible spines of rotifers and Haukioja (1977) suggested that population cycles of northern herbivores are a function of variation in plant resistance, new inducible defense systems have been discovered. With the examples of quite dramatic inducible defenses piling up in all types of organisms, it is clear now that inducible defenses are almost common and quite widespread. The obvious cases in well-studied systems have been found first; in animals where the inducible traits are visible and must only be linked to inducing agents, or in plants where the effect of the defense is visible (resistance to herbivores) while the defense itself (e.g., chemical compound) is not. However, as other systems are better studied, more cryptic defenses become visible. An induced helmet is clearly easier to detect than a change in reproductive effort in the same organism. More intensive efforts in even the well-studied organisms such as *Daphnia*, the protozoan *Euplotes*, and the marine bryozoan *Membranipora* are now yielding insights into genetic variability and oscillations of defensive traits that can be related to complex biotic regimes.

It is becoming increasingly clear that inducible defenses, beyond being striking examples of phenotypic plasticity, also evolve under specific ecological conditions. Examples of inducible defenses may be found in all organisms under selective pressure by predators, parasites, herbivores, pathogens, and even competitors. Four factors emerge as prerequisites for the evolution of inducible defenses (Havel 1987; Sih 1987a; Rhoades 1989; Dodson 1989b; Harvell 1990a; Adler and Harvell 1990):

1. The selective pressure of the inducing agent has to be variable and unpredictable, but sometimes strong. If the inducer is constantly present, permanent defenses should evolve.

2. A reliable cue is necessary to indicate the proximity of the threat and activate the defense.

3. The defense must be effective.

4. A major hypothesis about the advantage of an inducible defense is cost savings. If a defense is inducible, it could incur a cost that offsets the benefit of the defense. If there is no trade-off, it is widely postulated that the trait will be fixed in the genome.

Related to these four prerequisites for the evolution of inducible defense are four main issues that currently require investigation and form the conceptual underpinnings of the book.

1. Many of the responses described in this volume are induced by chemical factors released by biotic agents. Specific cues are important where prey species live in environments populated by predators with different selectivities. The role of cues and their reliability (Moran 1992) has been a difficult issue to evaluate experimentally due to a general lack of success in isolating and elucidating the chemical structures of the cues. After years of effort, the chemical structures of some of the kairomones that induce morphological shifts are being described. This will open a new frontier allowing investigation of the sensory structures detecting infochemicals and also the genetic mechanisms underlying inducible changes.

Although Price (1986) popularized the perspective that predator-prey (or herbivore-plant) interactions must be viewed in the context of their entire biotic environment, only recently have studies of tritrophic interactions revealed the chemical mechanisms by which plants could "signal" natural enemies of their consumers and thereby eliminate them. This indirect effect may work as effectively as an inducible defense in lowering the fitness of an herbivore (Dicke, chapter 4).

Recent theoretical work (Clark and Harvell 1992; Padilla and Adolph 1996) also suggests the importance of time lags in the response to cues: what are the rates of change in phenotypes relative to the biotic threats that challenge them? Is the distribution of inducible defenses among organisms limited by the rate at which a transformation can be accomplished?

2. The cost of defense issue is not resolved with a yes or no. When are costs most likely to be a constraint in the evolution of inducible defense? In what organisms are the magnitudes of cost large? How do we measure costs? Scientists working with plants, rotifers, cladocerans, bryozoans, barnacles, protozoans, and fish have examined the relative fitness of inducibly and non-

inducibly defended predators. The collective effort has revealed (1) costs in some organisms, (2) a definitive lack of costs in other organisms, and (3) a treasure trove of pitfalls and insights associated with assessments of fitness. The variable success in detecting costs of defense has to lead us to consider other hypotheses associated with the evolution of inducible defenses. The importance of this issue and the difficulties associated with it are indicated by most contributors to this book, including them as a main or partial theme in their papers. Cost is addressed in virtually every chapter of the book.

3. While allocation costs have been the traditional constraint thought to favor the evolution of inducible defenses, more recently it has become clear that other issues may be equally important. Recent studies point to a recognition of the importance of multiple biotic influences as selective agents for the evolution of multiple prey states. For example, work with *Daphnia* in multiple predator environments of both fish and invertebrates shows the integrated operation of changes in vertical migration, morphology, and body size and the associated life-history shifts as a means of keeping a clone alive (Tollrian, in prep.). Although many of the induced changes are associated with costs in growth or reproduction that undoubtedly affect population dynamics, the real importance of the inducibility of the character may well have more to do with the importance of responding without error to a changing predator field, than to cost savings. Do defenses against one type of consumer lead to a higher vulnerability to other types of consumers (Tallamy 1985; Tallamy and McCloud 1991; Taylor and Gabriel 1992)?

4. Finally, the coevolutionary and genetic aspects of inducible defenses remain a new frontier. Do inducible defense provide a "moving target" that slows counteradaptation of consumers (Adler and Karban 1994)? Is an inducible "dialogue" between predator and prey played out in ecological as well as evolutionary time? Irrespective of the selective factors favoring the evolution of inducible defenses, evolution will not occur in the absence of heritable variation in inducibility. An area of vigorous investigation is determining the heritability of the inducible characters, their phenotypic range, and the composition of natural populations. This is revealing heritable variation in inducible defenses and polymorphism in inducibility, leading to questions about what factors maintain the variation in these characters. Although such microevolutionary studies of inducibility are just now accumulating, assessments of macroevolutionary or phylogenetic patterns in inducibility are very rare. These are discussed in the concluding chapter, since the area is too scant to comprise a chapter.

This book is organized into organismally defined sections, with a section on plant inducible defenses, animal inducible defenses, and theoretical approaches to inducible defenses. Although we could have organized this volume by habitat or ecological discipline, we felt that the organismal distinctions were the strongest. We attempted to be nearly comprehensive in the animal section, since there are no recent reviews of this area. Since Karban and Baldwin (1997) recently evaluated inducible defenses of plants, we have introduced here selected major issues in plant inducible defenses and examined a few model systems. Berenbaum and Zangerl (chapter 1) provide an overview of the wild parsnip study system, which covers many of the historical and emerging issues in plant inducible defenses. In particular, they provide some insight into the difference between inducible resistance to pathogens and inducible defenses to herbivores. Järemo et al. (chapter 2) examine causes and consequences of different spatial scales in the inducible responses of plants and outline general conditions where selection should favor localized or systemic responses within plants. They develop this theme into an evaluation of the evolution of "interplant communication" in a game-theoretical context. Agrawal and Karban (chapter 3) examine directly the question of why inducible defenses are favored over constitutive defenses in some plants. In an attempt to shift attention away from the "allocation cost" model, they focus on several alternative hypotheses for the benefits of induction. Given that most plants interact with multiple specialist and generalist herbivores, various pathogens, microbes, and mutualists, they point out that a host of constraints arise maintaining constantly high levels of resistance. Dicke (chapter 4) takes on another emerging area in the study of inducible defenses with a review of studies showing indirect defenses of plants: the production of specific chemical cues by plants that attract parasitoid enemies of herbivores. Finally, Van Donk et al. (chapter 5) provide an overview highlighting their recent discovery that aquatic phytoplankton show an unusual inducible response: changing from single cells to colonies in response to cues from herbivores. Not only is this the first record of inducible defenses in phytoplankton, but it is a dramatic morphological shift. Although not reviewed here, studies also show that marine benthic algae and macrophytes are capable of inducible chemical responses to herbivores (Van Alstyne 1988; Paul and Van Alstyne 1992; Cronin and Hay 1996; Wolfe et al. 1997).

The animal section spans an organismal range from protozoans to vertebrates, but focuses (as do most inducible defense studies) on aquatic invertebrates. The animal section starts off with a consideration by Frost (chapter 6) of the vertebrate immune system as an inducible defense. This represents the most complex and highly inducible of all defense systems, where questions can be asked about specificity and time course of the different components of the defense. We are hopeful that a consideration of this complex, well-studied defense system will stimulate researchers in other areas to ask similar questions. And indeed we see this happening already, with plant (Karban and Adler 1996) and invertebrate (Harvell 1990a) researchers asking questions about the role of memory in inducible responses. Gilbert's (1966) work with rotifers stands as the seminal work in inducible defenses, and the rotifer system continues to break ground in showing new frontiers to be addressed in the evolution of inducible defenses. In particular, Gilbert (chapter 7) focuses on the nature of cues inducing the change and the technical problems in measuring the costs. Kuhlmann et al. (chapter 8) show the amazing insights gained in the study of protozoan inducible defenses and, perhaps alone among all these chapters, can take on the field in its entirety, evaluating effectiveness, costs, cues, and consequences of the dramatic phenotypic permutations accomplished by protozoans. De Meester et al. (chapter 9) review the emerging discipline of inducible stereotypic behaviors in induced vertical migration behaviors in zooplankton, emphasizing the role of clonal variation. Tollrian and Dodson (chapter 10) review the extensive literature on inducible changes in Cladocera. This group contains *Daphnia*, which has perhaps been the best-studied animal in this context and which has shown an amazing and complex array of predator-induced morphologies, behaviors, and life-history shifts. They emphasize the role of multipredator environments in shaping multiple inducible defenses. Brönmark et al. (chapter 11) summarize their work with predator-induced shifts in body allometry of crucian carp. This is the first example of predator-induced morphological changes in a vertebrate and should prompt examination of other cases where vertebrates might change their body allometry in different environments. Anholt and Werner (chapter 12) examine density-dependent consequences of induced behaviors in amphibians, and show the fitness consequences of predator and food-induced behavioral shifts. Their chapter also includes a consideration of the importance of indirect effects and emphasizes the complexity of simultaneously considering the predator and competitor environment. Harvell (chapter 13) summarizes her work with predator and competitor-induced structures in a marine bryozoan. Her emphasis is on the importance of considering multiple selective agents and on quantifying the range of genetically based defensive variants comprising natural populations.

The theoretical chapters treat an assortment of issues at the heart of evolving inducible defenses. Lively (chapter 14) analyzes the maintenance of polymorphism and uses his previous work on barnacle shell dimorphism as an example. He evaluates the conditions for evolution of genetic polymorphism of canalized morphs and mixtures of canalized and inducible morphs. Adler and Grünbaum (chapter 15) pick up a completely new topic in a theoretical examination of the coevolution of predators to inducible prey. They contend that a forager-based perspective is essential both to understanding the evolution of inducible defense systems and to assessing the community-level effects of inducible defenses. Similarly, Gabriel (chapter 16) raises a new theoretical issue in pointing out that previous theory deals mainly with irreversible phenotypes and that most inducible phenotypes are reversible. His models show the consequences of different selective environments and the importance of reversibility of response.

WHY INDUCIBLE DEFENSES?

We present this book now, emphasizing the factors favoring the evolution of inducible defenses and spanning a wide organismal range, because the field of inducible resistance is at a critical juncture and must be viewed as a whole to illuminate emerging issues and their importance to the evolution of these adaptive, phenotypically plastic characters.

1

Coping with Life as a Menu Option: Inducible Defenses of the Wild Parsnip

MAY R. BERENBAUM AND ARTHUR R. ZANGERL

Abstract

Chemical defenses offer plants great flexibility in terms of ecological responsiveness to stress, but they are not without disadvantages. Allocation costs of chemical defenses arise as a result of diversion of energy and materials from reproduction; genetic costs arise via negative pleiotropy; and ecological costs result from differential efficacy of particular chemicals against a wide range of enemies. These costs are thought to underlie the evolution of inducibility in chemical defense systems. Using the wild parsnip, *Pastinaca sativa*, as a paradigm, we examine the evidence for these costs. *P. sativa* produces an array of furanocoumarins toxic to a wide range of organisms. These compounds are inducible by damage inflicted by both generalist and specialist enemies in three kingdoms. Genetic costs exist in the form of negative genetic correlations between furanocoumarin content and fruit production, between the concentrations of furanocoumarins in different pathways, and between furanocoumarin content and size of storage organs. Ecological costs likely exist as a result of the demonstrably different capacity of enemies to detoxify and hence tolerate particular compounds. That induction occurs irrespective of resource availability suggests that opportunity costs—consequences of a failure to induce—are operative in this system. Molecular studies on the biosynthesis of furanocoumarins provide insights into the mechanisms underlying cost minimization in chemical defense systems. These include specificity of promoter elements to minimize ecological costs, gene duplication to minimize genetic costs, and transcriptional inhibition of cell division to minimize allocation costs. In contrast with abiotic stress agents, biotic agents may be particularly effective selective agents for the diversification of recognition systems and regulatory mechanisms in plant chemical defense due to their unique capacity to evolve resistance to defense.

Introduction

As the principal autotrophs in the majority of terrestrial ecosystems, angiosperm plants are at the bottom of most food webs and thus are in a position to serve as the ultimate energy and nutrient sources for heterotrophs of all descriptions. That the world remains green is a reflection in part of the fact that plants have evolved all manner of defenses to minimize the impact of herbivores.

Morphological defenses such as thorns, spines, and hairs serve as deterrents, primarily to larger (generally vertebrate) herbivores. Among the most widespread defenses against microherbivores, including insects, mites, and microbial pathogens, are chemical defenses. From the perspective of plants, which with very few exceptions have at best a limited capacity for movement and thus for running away from danger, chemical defenses offer significant advantages over other forms of defense. Because even slight alterations in chemical structure can bring about a change in biological activity, plants are capable of biosynthetically tailoring chemical defenses for particular environmental stresses; thus, chemical defenses offer greater potential for specificity than do morphological, phenological, or other forms of defense. As well, the production of chemical substances that have great biological activity at extremely low concentrations allows plants the option of diverting fewer energy and material resources away from growth and reproduction than would be required for morphological, phenological, or other forms of defense. Furthermore, chemical defenses also offer greater flexibility in terms of ecological responsiveness—the lability and interconnectedness of biosynthetic pathways, as well as the multiplicity of regulatory mechanisms, allow plants to switch chemical defense options more rapidly and more reversibly than many other forms of defense.

Chemical defenses, however, are not without disadvantages. Associated with the production of chemical defenses are various costs (Rausher 1996). Allocation costs are incurred as a result of investment of energy, materials, or other resources into resistance that might otherwise be invested into growth and reproduction. Genetic costs are incurred as a result of pleiotropy; the alteration of physiological functions that brings about resistance may at the same time decrease the efficiency with which other vital life processes are carried out. Finally, some forms of chemical resistance mechanisms may confer an ecological cost, in that plants are exposed to conflicting selection pressures from different mortality factors; resistance against one agent of mortality may increase susceptibility to other enemies.

In view of the potential costs associated with chemically mediated resistance, it is not altogether surprising that many plant defense systems are inducible. Inducibility, defined here as control of the expression of resistance by means of transcriptionally activated genes, provides a mechanism for delaying the costs associated with chemical defenses until such time as they are warranted; defense costs are not incurred in the absence of the stress agent. Evidence exists of the inducibility of a wide range of chemically based plant resistance factors in response to both biotic and abiotic stress agents. In the context of plant-pathogen interactions, Kombrink and Somssich (1995) recognize three classes of such inducible defense reactions to stress. Immediate early responses involve recognition and signaling processes, including changes in ion fluxes, initiation of oxidative reactions, hypersensitive cell death, callose formation, and intracellular rearrangement. After immediate early responses, locally initiated mechanisms are activated; responses at this stage involve de novo biosynthesis within the phenylpropanoid pathways, production of intracellular pathogenesis-related proteins, and induction of peroxidases and lipoxygenases. Finally, systemic reactions come into play; these tend to be broad-spectrum in nature and are associated with induction of extracellular pathogenesis-related proteins, 1, 2-beta-glucanases, chitinases, and the like. This classification may be broadly applicable to inducible defense responses to other biotic stress agents (such as herbivores) as well as to abiotic stress agents, albeit on a different time scale.

The notion that the inducibility of plant defenses arises as a result of trade-offs between chemically based resistance and fitness costs is a fundamental premise of theories of plant-herbivore interactions, despite the fact that these trade-offs have proved difficult to document experimentally (Simms 1992). Part of the problem in documenting these trade-offs has been defining them experimentally. Costs in defense are typically defined as negative correlations between fitness (generally represented as yield, growth rate, or seed production) and defense (generally measured either as the proportionate reduction in damage inflicted by enemies as compared to a susceptible phenotype, or as the quantity produced of secondary metabolites associated with resistance to enemies) in the absence of consumers. The basis for these costs may involve resource trade-offs or genetic constraints. Resource trade-offs are the direct result of competition for limiting resources within a plant; investments of energy or materials into defense are at the expense of equivalent investments in growth (Mole 1994). Such allocation costs may or may not be influenced by the genotype of individuals and, consequently, may or may not be subject to evolution. Just as resource trade-offs, in the form of phenotypic correlations, may not have a genetic component, negative genetic correlations between fitness and defense (e.g., Reznick 1985) may not involve allocation processes. Ecological costs, costs arising from conflicting biotic selection pressures, are evidenced as differences in the sign of correlations between defense and fitness in the presence of different consumers. As such, they are the most difficult to estimate, since to attempt to do so requires an intimate knowledge of the full range of prospective selective agents, as well as at least passing familiarity with the principal resistance factors associated with each.

As Mole (1994) points out, there are conceptual niceties associated with multiple approaches to studying costs of plant defense. On the one hand, working at the phenotypic level allows investigators to manipulate particular resources in order to determine experimentally precise constraints on defense investments. However, with such an approach, the evolutionary significance of findings is unclear; negative phenotypic correlations may not be genetically determined responses. On the other hand, working at the level of genotype and demonstrating a negative genetic correlation sheds no light on the physiological relationship between fitness and the defense trait in question; a genetic correlation per se provides little or no information on the physiological underpinnings of possible trade-offs. And, although discussions of trade-offs have traditionally been the province of ecologists and evolutionary biologists, there remains the inescapable fact that, because ecological and evolutionary phenomena arise as a result of biochemical and genetic processes, trade-offs can be defined at these levels as well.

No single approach will provide sufficient information to allow an accurate assessment of whether plant defense against herbivores exacts a cost in the form of reduced fitness. It is perhaps for this reason that debate on the cost of defense continues without resolution; evidence for (or against) such costs has been frustratingly equivocal. In the conclusion to his review of trade-offs and constraints, Mole (1994) states that "at present there is a critical need for physiological studies of trade-offs involving defensive traits, as only these can address current theory. In particular, such studies need to be performed with systems where there is also a genetic component to the trade-off. Only this will allow for an analysis of the relative impact of formal and genetic constraints upon defense related trade-offs observed at the phenotypic level." To interpret phenotypic responses requires knowledge of the resource trade-offs, the patterns of allocation of nutrients and energy; to understand genetic associations requires knowledge of the constraints—the biosynthetic pathways, genetic linkages, and other products of evolutionary history; and to understand both genetic and phenotypic correlations in an environmental context requires a thorough knowledge of the selective regime to which the plant is exposed. For few plants is sufficient information available, even about the chemical basis for defense, to allow a comprehensive estimate of costs and constraints associated with inducible defenses.

The Parsnip as Paradigm

Whatever its shortcomings as either an edible root crop or roadside weed, the parsnip, *Pastinaca sativa*, provides a useful model for examining induc-

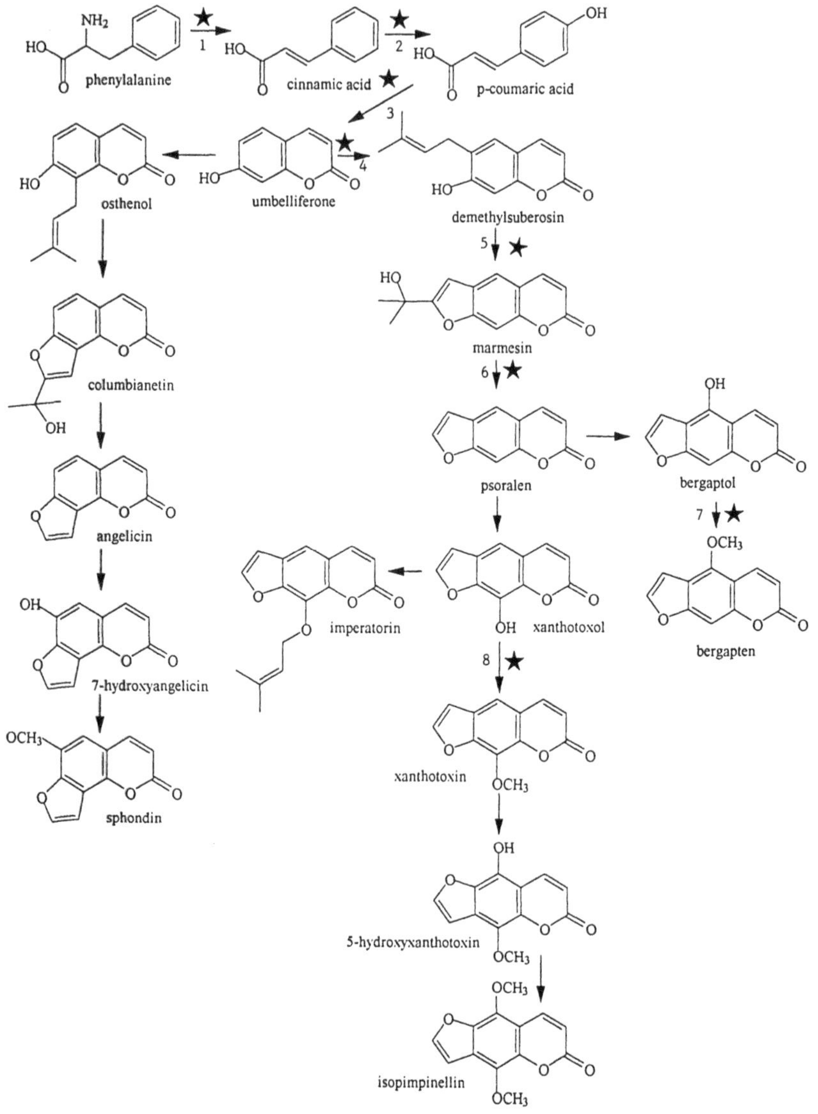

Fig. 1.1. Biosynthesis of furanocoumarins in *Petroselinum sativum*. 1. Phenylalanine ammoni lyase: encoded by four genes (Lois et al. 1989; Appert et al. 1994) and inducible by UV light an fungal elicitor (Hahlbrock and Scheel 1989). 2. Cinnamic acid hydroxylase. 3. Coumarate 4 co ligase: encoded by two genes and inducible by UV light and fungal elicitor (Douglas et al. 1987 4. Dimethylallylpyrophosphate: umbelliferone dimethylallyltransferase: inducible by fungal elicito (Ebel 1988). 5. Marmesin synthase: inducible by fungal elicitor (Hamerski et al. 1990). 6. Psorale synthase: inducible by fungal elicitor (Hamerski et al. 1990). 7. S-adenosylmethionine: bergaptol (methyltransferase: inducible by fungal elicitor (Hauffe et al. 1985). 8. S-adenosylmethionine: xan thotoxol-O-methyltransferase: inducible by fungal elicitor (Hauffe et al. 1985). *Inducible by fur gal elicitor Pmg.

ible defenses in plants. P. sativa is a facultative biennial native to Eurasia but extensively naturalized throughout North America, where it is found primarily along roadsides, in old fields, and in waste places. In its first year, the plant produces a rosette; if sufficient biomass is accumulated during its first year of growth, it will produce a flowering stalk its second year. As a monocarpic biennial, P. sativa flowers only once in its life. The majority of flowers are borne in the primary umbel, terminating the flowering stalk, although many other higher-order umbels can be produced. Flowers in the primary umbel are centrifugally protandrous and are insect-pollinated. The schizocarpic fruits are dispersed by wind or by gravity and require a chilling period prior to germination (Baskin and Baskin 1979).

Although little about its life history makes the plant uniquely well suited to studies of inducible defense, aspects of its chemistry certainly do. Like many other apioid umbellifers, P. sativa produces an array of furanocoumarins, tricyclic derivatives of the phenylpropanoid pathway (fig. 1.1). Plants that produce furanocoumarins as secondary compounds present a distinctive toxicological challenge to a wide range of organisms (Berenbaum 1991; Berenbaum and Zangerl 1996). Furanocoumarins are generally capable of absorbing ultraviolet light energy to form an excited triplet state; the highly reactive triplet state molecule can interact with duplexed DNA to form irreversible crosslinks, with amino acids to cause protein denaturation, with unsaturated fatty acids to form cycloadducts, and with ground state oxygen to form toxic oxyradicals that can damage a wide range of biomolecules. Thus, they are toxic to a broad range of organisms, including insects (review: Berenbaum 1991; Berenbaum 1995a; table 1.1).

Furanocoumarins fall into two major structural types—linear furanocoumarins, with the furan ring attached at the 6,7 positions of the benz-2-pyrone nucleus (e.g., psoralen, fig. 1.1), and angular furanocoumarins, with the furan ring attached at the 7,8 positions of the benz-2-pyrone nucleus (e.g., angelicin, fig. 1.1). These two groups, while sharing a common precursor, are distinguished biosynthetically by the action of site-specific prenylating enzymes that initiate the attachment of the furan ring (Berenbaum 1991). As a result of their configuration, angular furanocoumarins cannot form crosslinks between DNA base pairs, but they can form monoadducts and thus are toxic to insects as well as other organisms (Berenbaum and Zangerl 1996; table 1.1).

Although *Pastinaca sativa* produces a range of furanocoumarins, substances that are demonstrably toxic to potential consumers, it is nonetheless subject to attack, as are all green plants, by a diversity of enemies (table 1.2). This sizable community of enemies, including representatives from at least three kingdoms, is distinctive in that, across taxa, it is dominated by specialists—i.e., organisms that attack *P. sativa* and its close relatives, generally confamilials in the Apiaceae. Many of these specialists apparently are

TABLE 1.1
Relative Biological Activity of Furanocoumarins Found in Wild Parsnip as Measured in Bioassay Organisms

Organism	Effect	Species	Sensitivity
Bacteria	Mutagenesis	*Bacillus subtilis*	bol > xan = xol, ber
		Escherichia coli	pso > xan > ang
		Escherichia coli	pso > xan = ang
	Mortality	*Bacillus subtilis*	pso > xan > xol > bol
		Escherichia coli	pso > xan > ber > iso
	Growth inhibition	*Bacillus subtilis*	pso = xan > ber > iso
		Staphylococcus aureus	pso > xan > ber > iso
		Streptococcus faecalis	pso > xan > ber > iso
		Escherichia coli	pso > xan > ber, iso
Fungi	Growth inhibition	*Candida albicans*	ber > xan
		Curvularia lunata	pso > ber = pim = imp
		Curvularia lunata	xan > ang = xol
		Aspergillus niger	pim > ber
		Aspergillus niger	ang > xan, xol
	Mortality	*Candida albicans*	ber > xan
	Ribosomal RNA binding	Yeast	pso > xan > ber > xol
Plants	Growth inhibition	*Raphanus sativa*	xan = iso
		Lactuca sativa	pso = xan = ang, ber, iso, imp xol, bol
		Allium root tips	ber > pso > ang > xol > xan
	Germination inhibition	*Lactuca sativa*	xan > ang > pso, ber, iso, imp, xol, bol
Protozoans	Growth inhibition	*Proteus vulgaris*	pso > xan > ber > iso
		Pseudomonas aeruginosa	pso > xan, ber, iso
	Cytolysis	*Paramecium caudatum*	ber > xan
		Tetrahymena pyriformis	ber > xan
Insects	Growth inhibition	*Heliothis virescens*	xan = ber = ang = imp
		Depressaria pastinacella	ber > xan
		Papilio polyxenes	ang > xan
		Spodoptera exigua	xan > pso > ber
		Aedes aegypti	xan > sph
	Antifeedant activity	*Spodoptera litura*	iso > ber > xan
		Spodoptera litura	iso > ber > ang >, xan > pso
		Spodoptera litura	ber > xan
		Spodoptera litura	iso > xan > pso > ang
		Periplaneta americana	xan = iso > ber
		Blattella germanica	ber > xan > iso
		Leptinotarsa decemlineata	ber = iso > xan > imp
	Feeding efficiency	*Papilio polyxenes*	ang > xan
	Mortality	*Culex pipiens pallens*	ber > xan > iso
		Trichoplusia ni	pso > ber
	Cytochrome P450 inhibition	*Manduca sexta*	xan > imp > ang >, ber > pso > iso > xol
Mollusks		*Biomphalaria mansoni*	ber > iso > xan
Vertebrates	Mortality	*Lebistes reticulatus*	ber > imp > xan > iso
		Toad	xan > imp > ber
		Rat	xan > imp > ber

BLE 1.1 (*Continued*)

ganism	Effect	Species	Sensitivity
	Erythemal responses	Humans	pso > xan = ber > ang, iso, imp, xol, bol
		Albino guinea pigs	pso > xan > ber, iso, xol
		Albino rabbits	pso > xan > ber
	DNA-binding	Calf thymus	pso> xan > be
		Human cells	xan > ber > ang
		Mouse	xan > iso > ang

ource: Adapted from Berenbaum and Zangerl 1996.
lotes: ang = angelicin; ber = bergapten; bol = bergaptol; imp = imperatorin; iso = isopimpinellin; pso ɔsoralen; sph = sphondin; xan = xanthotoxin; xol = xanthotoxol.

able to utilize this plant as a host at least in part because they are tolerant of furanocoumarins. In most cases, the principal mechanism for tolerating furanocoumarins is via metabolic detoxification, which, in insects and vertebrates, is effected by cytochrome P450 monooxygenases (Berenbaum 1995a, b). Although the same basic enzyme system is involved in furanocoumarin detoxification even in generalists associated with *P. sativa* (e.g., *Trichoplusia ni*, Berenbaum 1995a), the levels of activity displayed by specialists tend to be considerably (often orders of magnitude) higher than corresponding levels in generalists.

Pastinaca sativa responds to injury inflicted by both generalist and specialist enemies by induction of furanocoumarins (table 1.3). In this respect the plant resembles many, if not most, other furanocoumarin-producing plant species (Fungi: *Apium graveolens*—Heath-Pagliuso et al. 1992; Afek et al. 1995; *Ruta graveolens*—Bohlmann et al. 1995; *Ammi majus*—Matern et al. 1988; Bacteria: *Apium graveolens*—Surico et al. 1987; Virus: *Apium graveolens*—Lord et al. 1988). Individual furanocoumarins vary in their degree of responsiveness to damage, but, in general, they are unique among the secondary metabolites of parsnip in the degree to which they respond to damage. Indeed, in a survey of parsnip secondary metabolite responses to damage, only the furanocoumarins and a single other phenylpropanoid (myristicin) were induced; monoterpenes, sesquiterpenes, and fatty acids were unaffected or declined in concentration as a result of damage (Zangerl et al. 1997).

Abiotic sources of stress also cause de novo biosynthesis of furanocoumarins to occur in parsnip plants. Mechanical damage increases furanocoumarin concentrations by as much as 219% in leaves (Zangerl and Berenbaum 1990) and by as much as 84% in roots (Zangerl and Rutledge 1996). UVA radiation (Zangerl and Berenbaum 1987) can cause up to 50% increases in foliar furanocoumarins, although roots are not responsive to this form of stress (Li 1993). In other umbellifers, acidic fog, ozone (Dercks et

TABLE 1.2
Enemies of *Pastinaca sativa*

Enemies	Reference	Attacked Part	Specialist/Generalist
Viruses			
Parsnip mosaic potyvirus	Murant 1972	Foliage	S
Parsnip yellow fleck sequivirus	Murant and Goold 1968	Foliage	S
Anthriscus yellows	Elnagar and Murant 1976	Foliage	S
Carrot thin leaf potyvirus	Brunt et al. 1996	Foliage	S
Celery mosaic potyvirus	Brunt et al. 1996	Foliage	S
Celery yellow spot luteovirus	Brunt et al. 1996	Foliage	S
Coriander feathery red vein nucleorhabdovirus	Brunt et al. 1996	Foliage	S
Dandelion yellow mosaic sequivirus	Brunt et al. 1996	Foliage	G
Heracleum latent trichovirus	Brunt et al. 1996	Foliage	S
Parsnip potexvirus	Brunt et al. 1996	Foliage	S
Parsnip leafcurl virus	Brunt et al. 1996	Foliage	S
Strawberry latent ringspot nepovirus	Brunt et al. 1996	Foliage	G
Tobacco ringspot nepovirus	Brunt et al. 1996	Foliage	G
Bacteria			
Erwinia carotovora	Li 1993	Roots	S
Agrobacterium tumefaciens	Li 1993	Roots	G
Fungi			
Centrospora acerina	Channon 1965	Roots	S
Cylindrocarpon destructans	Channon and Thompson 1981	Roots, foliage	S
Erysiphe sp.	http://cygnus.tamu.edu/Texlab/Vegetables/vegq.html	Foliage	G
Alternaria sp.		Foliage	G
Fusarium sambucinum*	Desjardins et al. 1989a	Roots	G
Fusarium sporotrichoides	Desjardins et al. 1989b	Roots	G
Fusarium oxysporum	Li 1993	Roots	G
Fusarium solani	Li 1993	Roots	G
Itersonilia pastinacae	Cerkauskas 1986	Roots	S
Phoma complanata	Cerkauskas 1986	Roots, leaves	S
Insects			
Homoptera			
Cavariella pastinacae	Berenbaum 1981	Vascular sap	S
Cavariella aegopodii	Dunn and Kirkley 1966	Vascular sap	S

TABLE 1.2 (Continued)

Enemies	Reference	Attacked Part	Specialist/ Generalist
...laenus spumarius	Berenbaum 1981	Vascular sap	G
...miptera			
...us lineolaris	Berenbaum 1981	Vascular sap	G
...hops scutellata	Berenbaum 1981	Vascular sap	S
...giognathus politus	Berenbaum 1981	Vascular sap	G
...giognathus obscurus	Berenbaum 1981	Vascular sap	G
...pidoptera			
...onopterix clemensella*	Berenbaum 1981	Leaves	S
...phipyra tragopogonis	Berenbaum 1981	Leaves	G
...hips purpurana	Berenbaum 1981	Leaves	G
...pressaria pastinacella*	Berenbaum 1981	Flowers and fruits	S
...icrisia virginica	Tietz 1972	Leaves	G
...oa tessellata	Tietz 1972	Leaves	G
...icania unipuncta	Tietz 1972	Leaves	G
...paipema marginidens	Berenbaum 1983	Stem	S
...ilio polyxenes*	Berenbaum 1983	Leaves and flowers	S
...ilio bairdii	Tietz 1972	Leaves	S
...ilio brevicauda*	Tietz 1972	Leaves	S
...ilio zelicaon	Tietz 1972	Leaves	S
...choplusia ni*	Zangerl 1990	Leaves	G
...ptera			
...leia fratria	Berenbaum 1981	Leaves	S
...ytomyza angelicae	Berenbaum 1983	Leaves	S
...ytomyza pastinacae*	Berenbaum 1983	Leaves	S
...la rosae		Roots	S
...rtebrates			
...mo sapiens*		Roots	G
...uus caballus	Grieve 1971	Roots	G
...s taurus	Grieve 1971	Roots	G
...: scrophula	Grieve 1971	Roots	G

Notes: S = specialist; G = generalist.
*Indicates species known to metabolize furanocoumarins (Berenbaum 1995a; Desjardins et al. 1989a).

al. 1990), cold temperatures, sodium hypochlorite, and copper sulfate (Beier and Oertli 1983) can act as elicitors of furanocoumarin production; these agents are as yet untested against parsnips. Overall, biotic stresses induce a response of greater magnitude in both roots and foliage than do abiotic stresses; in leaves, for example, bergapten and sphondin induction was up to 50% greater in response to insect damage than to a comparable amount of mechanical damage (Zangerl 1990). Presumably, these differences are due to the presence of elicitors of insect origin; although not characterized in this

TABLE 1.3
Agents Capable of Inducing Furanocoumarins in Parsnip

Agent	Reference	Attacked Part	Fold Increase
Biotic			
"Disease"	Ceska et al. 1986	Roots	25.6
Alternaria sp.	Johnson et al. 1973	Roots	
Ceratocystis fimbriata	Johnson et al. 1973	Roots	20
Colletotrichum lindemuthianum	Johnson et al. 1973	Roots	
Fusarium oxysporum	Li 1993	Roots	2.09–7.62
Fusarium solani	Li 1993	Roots	11.5–21.2
Fusarium sporotrichioides	Johnson et al. 1973	Roots	
Gibberella pulicaris	Desjardins et al. 1989a	Roots	~20
Fusarium sporotrichioides	Desjardins et al. 1989b	Roots	~13
Helminthosporium carbonum	Johnson et al. 1973	Roots	
Phoma complanata	Cerkauskas and Chiba 1991	Roots	21.7–1029
		Leaves	1.2–5
		Stems	1–5.2
Erwinia carotovora	Li 1993	Roots	1.3–1.6
Agrobacterium tumifasciens	Li 1993	Roots	2.39
Trichoplusia ni	Zangerl 1990	Leaves	1.38–2.68
Depressaria pastinacella	Li 1993	Roots	4.04–11.53
Abiotic			
Mechanical damage	Zangerl 1990	Leaves	1.10–2.19
Mechanical damage	Li 1993	Roots	1.05–4.02
UVA radiation	Zangerl and Berenbaum 1987	Leaves	1.46

system, such elicitors have been identified as important in other plant-insect interactions (Tallamy and Raupp 1991).

Adaptiveness of Furanocoumarin Induction as a Defense Response

An increase in furanocoumarin production after injury inflicted by an enemy tolerant of furanocoumarins is likely to be defensive only if the increase is of sufficient magnitude to overcome the tolerance mechanism of the enemy, of sufficient rapidity to avert lethal damage, and of sufficient efficacy that the cost of preventing the injury does not exceed the cost of sustaining it. In wild parsnip foliage, induction of xanthotoxin, a linear furanocoumarin, in response to mechanical damage is highly localized, involving only the damaged leaflet itself and certain neighboring leaflets (Zangerl and Berenbaum

1994/95). Moreover, it is rapid—mechanical damage induces an increase in xanthotoxin concentration that reaches a maximum level within 6 hours; concentrations decline to preinduction levels within 7 days. That this induction results from de novo biosynthesis rather than translocation of preformed furanocoumarins from elsewhere in the plant is indicated by the fact that detached leaves are able to respond to damage with a significant increase in furanocoumarin concentration.

Thus, xanthotoxin induction after simulated herbivory is both localized and rapid; both attributes suggest that such a response could serve to reduce the impact of the damaging herbivore. That the induction response is of sufficient magnitude to reduce further damage by the eliciting agent is suggested by Zangerl (1990), who demonstrated that the cabbage looper, *Trichoplusia ni*, an occasional feeder on parsnip foliage, experienced reduced growth (by 65%) and feeding (by 30%) on foliage with induced levels of furanocoumarins. Furthermore, the 20-fold increase in xanthotoxin concentrations following infection by several species of nonpathogenic fungi in parsnip roots results in levels that are demonstrably fungitoxic (Johnson et al. 1973), although these concentrations are not necessarily toxic to pathogens capable of metabolizing furanocoumarins (Desjardins et al. 1989a).

Allocation Costs of Inducibility

Variation in furanocoumarin content of both foliage and reproductive structures of wild parsnip is at least in part under genetic control (Berenbaum et al. 1986). Half-sib heritability estimates for furanocoumarins in fruits are uniformly high (0.54 to 0.98 under field conditions); heritability estimates for foliar furanocoumarins are more variable (ranging from 0.18 to 1.17 in the field) but are also in almost all cases significantly different from zero. Yet, despite the fact that genetic factors profoundly influence furanocoumarin production in the wild parsnip, environmental factors have an influence as well. Light availability and nutrient availability combine to affect constitutive foliar concentrations of four furanocoumarins—xanthotoxin, bergapten, imperatorin, and isopimpinellin (Zangerl and Berenbaum 1987); low levels of either resource limited the effects of variation in the other resource. Resource limitation similarly can alter induction response, contributing to environmental variation in furanocoumarin concentrations and composition.

In a series of experiments designed to estimate physiological costs of furanocoumarin biosynthesis, photosynthetic and respiration rates were measured directly and correlated with constitutive and induced furanocoumarin production (Zangerl et al. 1997). Leaf area was removed by mechanical damage; removal of 2% of leaf area resulted in an 8.6% decline in total biomass and a 14% decline in root biomass over a four-week period. Coincident with the accumulation of furanocoumarins following damage to foliage (on average 1.18 μg/mg) was a significant increase in respiration rates (0.56 μmol/m^2/sec,) and a significant decrease in photosynthesis rates (2.8 μmol CO_2 m^{-2} sec^{-1}), although the decrease in photosynthesis rates was short-lived and was not evident after 48 hours. These changes in respiration were localized to the leaflet experiencing damage. Expressed in glucose-equivalents, the energy cost of furanocoumarin induction, 12.6 μg glucose/cm^2, was virtually identical with the increase of respiration, 12.0 μg glucose/cm^2. Variation in damage-induced furanocoumarin production was correlated with damage-induced increases in respiration and was not significantly correlated with damage-induced decreases in photosynthesis. These results indicate that growth reduction after damage is due not solely to removal of photosynthetic tissue but rather to the energy and material costs of furanocoumarin biosynthesis.

There is compelling evidence, then, of a physiological cost of induction; plants that produce elevated levels of furanocoumarins experience reduced growth. In that induction has been suggested as a mechanism for conserving limited resources, this cost suggests that, in situations of severe resource limitation, induction of furanocoumarin production may be constrained. To determine effects of nutrient limitation on induction, forty plants germinated from seed collected from a natural population (University of Illinois Phillips Tract, 6 km northeast of Urbana, Champaign County, IL) were allowed to grow for 3 months, during which time they received nutrient supplements (20% solution of Miracle-Gro, 15N-30P-15 K; Stern's Miracle-Gro Products, Port Washington, NY), and forty plants were grown for the same amount of time without supplementation. During this time, plants that received no supplements ceased growth after 8 weeks, and plants receiving supplements had significantly more leaves at the end of the experimental period than did nutrient-deprived plants (mean 4.8 leaves vs. 3.2 leaves, p > 0.0001, df 1, 18, ANOVA). To induce furanocoumarin production, we mechanically damaged individual leaflets, increasing furanocoumarins to levels comparable to those observed in leaves damaged by insects (Zangerl 1990). After 24 hr, furanocoumarin content of the damage-induced leaflet and its "sister" leaflet, which is not induced, were compared. In a second experiment, to determine effects of light limitation on inducibility of furanocoumarins, nutrient-deprived plants from the first experiment were clipped back so that only the youngest leaf remained; cutting back the plants served to provide new undamaged leaves for sampling and to reduce stored carbon in the substantial tap root. Half of the plants were shaded under a frame that reduced ambient light levels by 90% (in contrast with an earlier study, Zangerl and Berenbaum 1987, in which light levels were reduced only 50%). After 3 weeks, during which time nutrient supplements were provided so as to insure mineral nutrition was not limiting, undamaged leaves from both control and

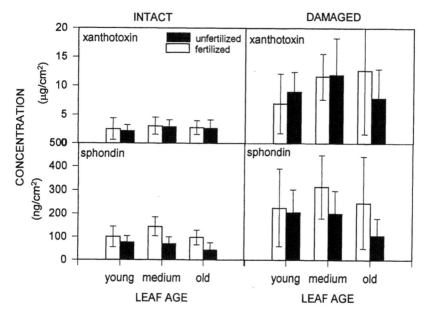

Fig. 1.2. Effects of nutrient limitation on constitutive and induced concentrations of furanocoumarins in young, intermediate, and old leaves of *Pastinaca sativa*. Error bars are standard deviations. All values in this figure and in figure 1.3 are expressed on the basis of leaf area rather than weight. By our definition of induction, the absolute amount of defense compounds must increase. Because some of the treatments that we imposed may affect leaf mass without changing the amount of furanocoumarins, a false impression that induction occurred could result if furanocoumarin concentration is expressed in terms of leaf mass. Since furanocoumarins are localized within oil tubes in leaf veins and leaf vein architecture is correlated with leaf area, leaf area provides a better basis for estimation of induction in these experiments.

experimental plants were sampled to compare total furanocoumarin content. That light limitation was achieved in the experimental treatment was evidenced by the significantly greater level of etiolation displayed by experimental plants (Zangerl and Berenbaum 1994/95).

In the first experiment, in which nutrient levels were varied to determine effects on constitutive levels of furanocoumarins in wild parsnip foliage, instrument problems precluded quantitation of three furanocoumarins known to occur (bergapten, imperatorin, and isopimpinellin); however, clear, repeatable, and reliable separations were obtained for two furanocoumarins (xanthotoxin and sphondin). Furanocoumarin analysis revealed that severe deficiency of mineral nutrients did not limit constitutive xanthotoxin production in wild parsnip foliage (fig. 1.2). In contrast, constitutive production of the angular furanocoumarin sphondin was significantly reduced by nutrient limitation in all age classes of foliage ($p = 0.0012$; fig. 1.2). In neither case,

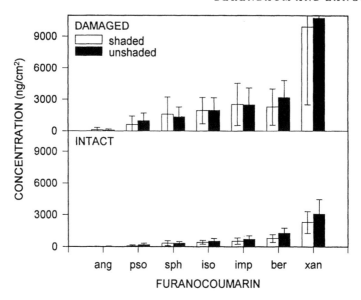

Fig. 1.3. Effects of shading on constitutive and induced concentrations of seven furanocoumarins in leaves of wild parsnip. Error bars are standard deviations.

however, was the inducibility of the furanocoumarin affected by treatment; xanthotoxin and sphondin concentrations increased approximately fivefold and twofold, respectively, after damage. In the second experiment, in which light levels were varied in order to determine the consequences of carbon limitation on furanocoumarin production, production of seven furanocoumarins was quantified. Light limitation resulted in constitutive production of lower levels of linear furanocoumarins ($p = 0.014$) but did not affect production of the angular furanocoumarins ($p = 0.76$) (fig. 1.3). As was the case with nutrient limitation, light limitation had no influence on the inducibility of furanocoumarins; under extreme shading, all five furanocoumarins increased in concentration between 3- and 4-fold.

Thus, even under conditions that severely restrict growth, parsnip plants respond to leaf damage by induction of furanocoumarins to substantially higher levels. For the wild parsnip, costs of defense, as measured by carbon or nutrient expenditures, are clearly subordinate factors in determining patterns of response to damage once it is inflicted. Presumably, the benefit of damage-induced defense is of far greater importance than are additional costs (in terms of nutrient or energy investment) associated with increasing levels of defense. Benefit (protection from enemies) appears to be the driving force behind induction in this system. Furanocoumarins are induced to the same concentrations after damage irrespective of resource supply, suggesting that the value of induced responses is of the same magnitude under resource-poor as it is under resource-rich conditions.

Resource-based theories of defense allocation (e.g., Coley et al. 1985) were by and large generated based on studies of constitutive defense and are difficult to apply to inducible defenses. Fundamental to such theories is the notion that material costs of defense constrain allocations to defense. At least for wild parsnips, material cost does appear to constrain production of constitutive defenses, but impacts of material costs on induction are more difficult to classify. Defense investment is not necessarily fueled by allocation away from growth; plants that have ceased growing can nonetheless respond to damage by increasing furanocoumarin concentrations several-fold. In this case, materials may be reallocated from some other plant function, incurring an as yet unknown functional cost.

Underlying most current theories of defense allocation is the notion that defense chemicals are produced prophylactically, even in the absence of a stress factor. Such a concept limits the usefulness of these theories in predicting patterns of allocations to induced defenses, inasmuch as these allocations are made only after damage is incurred. More important under these circumstances than material or energy costs are opportunity costs—the consequences to the plant if the investment is not made. Given that parsnip is monocarpic and only has one opportunity to reproduce during its lifetime, a failure to respond to attack due to resource limitation may result in reductions in fitness far greater than those that would have resulted from lost opportunities for growth.

Genetic Costs of Induction Response

Suggestive evidence of genetic control of inducibility of furanocoumarin production is found in comparisons among edible parsnip cultivars to infection by the fungus *Phoma complanata* (Cerkauskas and Chiba 1990). Across ten different cultivars, inducibility of bergapten in roots ranged from 1.3-fold to 2.2-fold; inducibility of xanthotoxin ranged from 1.8-fold to 5-fold, and inducibility of psoralen ranged from 1.8 to 4.4-fold. In only one of the cultivars was angelicin, an angular furanocoumarin, inducible. Inducibility in leaves also appears to be partly under genetic control. In *P. sativa* foliage, the inducibility of at least two furanocoumarins has a heritability significantly different from zero, suggesting that genetic variation is available for selection on this trait (Zangerl and Berenbaum 1990). That selection has acted to modify furanocoumarin inducibility is indicated by patterns of inducibility between populations experiencing different levels of herbivory. In east-central Illinois, the furanocoumarin chemistry of the parsnip plant reflects its local associations with herbivores. A population of parsnip with historically high levels of damage to foliage had higher constitutive levels of furanocoumarins than another population with historically lower levels of damage (Zangerl and Berenbaum 1990); furthermore, levels of at least one furanocoumarin, bergapten, are inducible to higher levels after damage in foliage of plants from this population.

Within the plant as well, patterns of inducibility reflect probabilities of experiencing insect damage. Those plant parts hardest (and most predictably) hit by insects also contain the highest constitutive level of furanocoumarins. While furanocoumarin production is low and inducible in leaves and roots, it is constitutively high and uninducible in reproductive parts, which are almost invariably damaged in east-central Illinois by the parsnip webworm, *Depressaria pastinacella* (Zangerl and Rutledge 1996).

Although there is evidence that furanocoumarin inducibility is a trait capable of evolving in natural populations of parsnip, there is little information on the extent to which this evolution is constrained genetically. Such genetic constraints apparently affect constitutive production of furanocoumarins. Bergapten levels in fruits, for example, are negatively genetically correlated with potential fruit production (Berenbaum et al. 1986), indicating a pleiotropic fitness cost associated with increasing bergapten concentrations. Similarly, production of the angular furanocoumarin sphondin is negatively genetically correlated with production of the linear furanocoumarin xanthotoxin in fruits and with the linear furanocoumarin bergapten in leaves, suggesting the possibility that the two biosynthetic pathways compete for precursors in limiting supply (Berenbaum and Zangerl 1988). Morphological traits may also constrain furanocoumarin responsiveness. In the fruits, seed morphology is a factor contributing to variation in furanocoumarin content; in particular, a significant positive family mean correlation was found between the area of the oil tube and the content of the angular furanocoumarin sphondin (Zangerl et al. 1989).

Ecological Costs of Furanocoumarin Induction

To begin to estimate ecological costs of defense induction requires a thorough knowledge not only of the chemistry of defense induction but also of the toxicology of the target enemies. The information in table 1.1 demonstrates dramatically that no single furanocoumarin is predictably toxic against all bioassay organisms; relative rank is not even consistent within any given phylum. The toxicological information available for the principal enemies of parsnip (table 1.2) is sketchy but even so is suggestive of possible trade-offs (table 1.1). The parsnip webworm, *Depressaria pastinacella*, which feeds primarily on reproductive parts, is capable of very high levels of xanthotoxin metabolism but can metabolize bergapten and sphondin only at substantially lower levels (Berenbaum and Zangerl 1992). Levels of bergapten and sphondin are in fact positively correlated with fruit production in the field in the presence of webworms (Berenbaum et al. 1986) and plants with higher levels of sphondin are more likely to escape webworm attack in the field than are plants with lower levels of sphondin (Zangerl and Berenbaum 1993). Administered in artificial diets, bergapten significantly reduces growth and development of webworms (Berenbaum et al. 1989). Bergapten and sphondin, then, appear to be important resistance factors for the plant against webworms. In contrast, the black swallowtail, *Papilio polyxenes*, which feeds on both foliage and flowers and which can reach appreciable numbers on parsnip in banner years, can metabolize bergapten and xanthotoxin with equal facility (Berenbaum and Zangerl 1993). Bergapten induction in response to webworm attack may thus have no impact on susceptibility to black swallowtails or may even have the negative consequence of rendering plants more prone to attack by these insects.

Ecological constraints on furanocoumarin defense systems appear to be less than dramatic for the wild parsnip, at least against insects. This absence of striking trade-offs may be a reflection of the fact that, as an introduced species, the plant may have left some portion of its ecological constraints behind; its insect fauna is relatively depauperate in comparison with some native plants that grow sympatrically with wild parsnip (e.g., goldenrod; Maddox and Root 1990). The nature of the biochemical resistance mechanisms utilized by specialists can also affect the likelihood that certain kinds of ecological constraints arise. As far as is known, none of the insect herbivores of parsnip sequester particular furanocoumarins for their own defense, in the way that consumers of asclepiadaceous plants sequester cardenolides (Malcolm and Zalucki 1996); such adaptations set up the paradoxical situation whereby increasing concentrations of defensive chemicals in response to damage may actually benefit the herbivore by increasing its resistance to its own enemies. Webworms are, however, capable of sequestering unmetabolized xanthotoxin and bergapten into their silk (Nitao 1990); if the role of sequestered furanocoumarins is defensive, as speculated (e.g., to protect against microbial overgrowth of the silk or to deter cocoon-penetrating parasitoids), then a paradoxical situation may apply here as well, although to a more limited extent.

Molecular Basis for Costs of Inducibility

Understanding the molecular mechanisms underlying the induction response can provide insights into how plants can minimize allocation, genetic, and ecological costs. Allocation costs may be reduced by the evolution of biosynthetic pathways that allow "crosstalk" and shared use of intermediates or enzymes (Jones and Firn 1991). Genetic costs reflecting negative pleiotropic effects of resistance may be minimized by the evolution of multiple forms of critical enzymes, each with a particular dedicated function. In maize, for example, entire segments of chromosomes are duplicated, and within these regions are located multiple resistance genes that may have arisen from duplication events. Duplication may provide a mechanism for acquiring new alleles and new functions in the face of new selection pressures without sacrificing other functions (McMullen and Simcox 1995). Ecological costs to some degree may be minimized by regulatory mechanisms that provide specificity to the induced response.

The majority of studies on mechanisms of furanocoumarin inducibility have been conducted not on parsnip but on cell cultures of *Petroselinum sativum*, another apioid umbellifer. For the purposes of discussion, it can be assumed that the fundamental characteristics of the biosynthesis of furanocoumarins in parsnip are similar; however, the degree to which studies conducted on parsley apply to other plants, even relatively closely related ones, is a matter of speculation (as is, for that matter, the applicability of studies on cell cultures to intact plants). In these studies, transcriptional activation by biotic elicitors has been demonstrated for virtually all of the enzymes that participate in the biosynthesis of furanocoumarins (fig. 1.1).

Induction of the enzymes involved in furanocoumarin synthesis depends to a large degree on the nature of the stimulus that elicits it. Tietjen et al. (1983) demonstrated specificity in the spectrum of furanocoumarins produced in response to particular fungal elicitors; such specificity in nature may act to reduce ecological costs of induction. Whereas application of the elicitor from the nonpathogenic fungus *Phythophthora megasperma* (Pmg) results in production of psoralen and xanthotoxin, application of another nonpathogenic fungus, *Alternaria carthami*, leads to the production of bergapten and isopimpinellin. At the molecular level, such specificity may be determined by interactions between receptors and elicitors. Elicitor activity of the nonpathogenic fungus *Phytophthora megasperma* f. sp. *glycinea* is associated with a 42kDa glycoprotein that binds to a putative receptor in the parsley plasma membrane. Interactions between this same fungus and soybean, on which it is pathogenic, are mediated by a glucan elicitor that binds to a 70kDa protein receptor in soybean plasma membranes. Parsley cells do not recognize the glucan elicitor, and soybean cells do not recognize the glycoprotein elicitor (Nürnberger et al. 1994).

Although ultraviolet light, like Pms fungal elicitor, induces several of the enzymes involved early in furanocoumarin biosynthesis, as well as several others along the flavonoid pathway (Kuhn et al. 1984), it is ineffective at inducing 5-adenosylmethionine:bergaptol O-methyl transferase (BMT), the final enzyme involved in bergapten synthesis. This response pattern parallels those in intact plants, in which UV irradiation induces flavonoids but not furanocoumarins in the absence of fungal elicitors (Lozoya et al. 1991). Moreover, exposure to the fungal elicitor blocks light-induced flavonoid production, and exposure to light reduces elicitor-induced furanocoumarin production. This arrangement suggests that "cross-talk" to reduce allocation costs may be minimal. DNA footprinting studies demonstrate that two promoter motifs are involved in both UV and elicitor induction for phenylalanine ammonia lyase 1 (PAL1) and coumarate-4-CoA-ligase (4CL) (Lois et al. 1989). Of the three inducible footprints in PAL1, one is brought about in response to elicitor but not to UV light, suggesting that different transacting factors may be involved in induction. Forms of stress other than fungal elicitor or UV irradiation can override typical responses. The administration of simultaneous heat-shock treatment led to the loss of inducibility of PAL and chalcone synthase in parsley cell cultures (Walter 1989). Taken together, studies of stimulus-specificity of furanocoumarin induction indicate that trade-offs occur at the molecular level; as Walter (1989) states, "The plant response to several competing stresses is not organized on a first come, first served basis. . . . It rather appears that a plant, while limited in its adaptation potential, is able to preferentially respond to a new and stronger stress at the expense of defense against the weaker stress" (p. 7).

Furanocoumarin induction responses are tissue-specific as well as stimulus-specific in parsley. Leaf wounding causes an increase in flavonoid-initiating chalcone synthase (CHS) mRNA and an initial decline, followed by a rise, in S-adenosyl-L-methionine: bergaptol methyltransferase (BMT) mRNA (Lois and Hahlbrock 1991). Root wounding, in contrast, causes CHS mRNA to decline while BMT displays a U-shaped response curve. In multigene families such as PAL, in which-tissue specific response is controlled at the gene level, selection may have favored multiple genes to minimize genetic costs arising from pleiotropy. Whereas induction in response to fungal elicitor is associated with PAL3 in leaves, in contrast, constitutive expression of PAL activity is associated with PAL2 in roots. These enzymes may retain specificity by their localization as well as their structure. In young leaves PAL and 4CL are in epidermal cells of xylem of vascular bundles and in epithelial cells of oil ducts (Lois and Hahlbrock 1991), consistent with their function in central phenylpropanoid reactions, while BMT is expressed exclusively in the epithelial cells of oil ducts (Wu and Hahlbrock 1992) and accumulates locally at infection sites, consistent with its singular role in late stages of furanocoumarin synthesis. Specificity of induction response, then, may be facilitated by the existence of multiple enzymes, each with potentially different activities, tissue specificity, and multiple response elements in promoter regions.

Examining furanocoumarin biosynthesis at the molecular level also reveals evidence for induction costs. Logemann et al. (1995) showed that UV exposure or application of fungal elicitor stops cell growth in parsley suspension cultures, probably as a result of arrest of cell division. Such cessation of cell division may be "a prerequisite for full commitment of the cells to transcriptional activation of pathways involved in UV protection or pathogen defense." Thus, histone H3 mRNA decreases near the fungal infection sites. In order for the cell to produce biosynthetic enzymes necessary to respond to attack, histone protein synthesis, and thus normal cell function, declines inversely to the rate at which PAL mRNA increases. Light inhibition of histone production lasts 24 hours, and elicitor inhibition of histone production much longer. That histone synthesis is transcriptionally inhibited in concert with (and in proportion to) PAL transcriptional induction by light or elicitor, coupled with the fact that repression of histone genes is associated with repression of cell division, indicates that allocation costs of induction may begin at the molecular level.

Additional evidence that the biosynthetic machinery for furanocoumarin metabolism is turned on at the cost of normal cell function comes from studies of Haudenschild and Hartmann (1995), in which the addition of *Phytophthora megasperma* f sp. *glycinea* elicitor to parsley cell cultures results in the inhibition of sterol biosynthesis within 7 hours; this inhibition immediately precedes the accumulation of furanocoumarins in the culture medium. The isoprenoid pathway is involved in synthesis of both sterols and furanocoumarins, contributing the sterol nucleus in its entirety and contributing the prenyl group to form the furan ring attached to the coumarin nucleus. Evidently, in the event of exposure to potential pathogens, commitment to a defense pathway precludes production along a competing "housekeeping" pathway.

Conclusions

As sedentary organisms, plants are subject to an extraordinary variety of environmental insults; such sources of stress include not only biotic agents but also a wide range of abiotic phenomena, including freezing, desiccation, chilling, salinity, flooding, and heavy metals, among others. Similarities in physiological responses to such abiotic stresses have led to the suggestion that there exists in plants a generalized adaptation syndrome (Leshem and Kuiper 1996) akin to that in mammals. Many different forms of stress elicit an identical reaction, and exposure to one form of stress can induce tolerance to a form of stress not yet experienced. These stress responses can be classified into two groups: responses that allow the organism to tolerate the stress, and responses that act to neutralize or eliminate the causes of physiological stress. To cope with oxidative stress, for example, which can be generated by both abiotic stressors (such as ultraviolet light, ozone exposure, or drought) and by biotic stressors (such as fungal pathogens or insect herbivores that damage tissues), plants produce a series of antioxidative enzymes, including catalase and superoxide dismutase, and free radical or singlet oxygen quenchers, including glutathione, ascorbic acid, carotenoids, and flavonoids.

By neutralizing toxic oxygen species, these defenses reduce the destructive consequences of oxidative stress. Characteristic of such a generalized adaptation syndrome are mechanisms that prevent the stress that has been inflicted from causing irreparable harm as well as mechanisms that enhance or accelerate post-stress repair processes.

To the extent that, irrespective of its origin, stress can cause damage to cells and tissues, it is likely that a generalized adaptation syndrome may exist in connection with both biotic and abiotic stress agents. There are, after all, only so many ways that a cell can die and there are presumably only a limited number of ways of preventing that death from taking place. However, biotic agents of stress offer plants an evolutionary opportunity for diversification that simply does not exist with respect to abiotic stress; plants can exercise control over the source of stress to prevent further damage. By no extraordinary physiological mechanism can a plant make the sun stop shining or the wind stop blowing, nor can plants influence rates of precipitation or ambient temperatures; thus, there is a premium on evolving mechanisms of coping with these forces and preventing the damage they can cause. By producing toxins, however, plants do have a mechanism for eliminating the stress agent altogether or at least profoundly interfering with its population dynamics. Biotic stress, then, is unique as a selective agent and is likely to select not only for variety (not uniformity) in chemical structures but also for variety in mechanisms regulating biosynthesis. It is highly likely, then, that at least some components of the responses of plants to biotic stress are highly specific and result from long evolutionary associations with the stress agents. There are indications of such specificity in the pattern of induced furanocoumarin response in parsnip. Microbes elicit a response distinct from that elicited by insect herbivores, nonpathogenic fungi elicit a response distinct from that elicited by pathogenic fungi, and that response in turn is distinct from that elicited by abiotic stressors such as ultraviolet light (Li 1993; table 1.3; fig. 1.4). It follows that "adaptive hormones" such as absicic acid or jasmonates (Leshem and Kuiper 1996) may be insufficient to initiate induction in response to attack by potential consumers; since each consumer is differentially susceptible to defense compounds, a "generalized" response is unlikely to be optimal. Indeed, pretreatment of parsley cell cultures with salicylic acid or dichloroisonicotinic acid, putative stress hormones, leads to increased furanocoumarin production only when fungal elicitor is present (Kauss 1994).

Although they may be menu options, plants are, by virtue of chemical-defense compounds, capable of inflicting more than a few bites themselves on their erstwhile consumers. An individual plant must be able to repair damage inflicted by herbivores and pathogens to survive or resist attack (Kuc 1995); while it surely benefits by possessing physiological mechanisms for preventing further damage, its opportunities for high lifetime fitness will

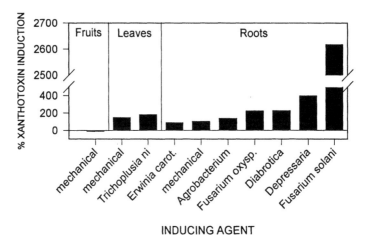

Fig. 1.4. Induction of furanocoumarins in response to biotic and abiotic stress in roots, foliage, and reproductive parts of *Pastinaca sativa*. Full names for inducing agents are *Erwinia carotovora*, *Agrobacterium tumefaciens*, *Fusarium oxysporum*, *Diabrotica undecimpunctata howardi*, and *Depressaria pastinacella*. (Data from Li 1993 for roots, from Zangerl and Rutledge 1996 for fruits, and from Zangerl 1990 for leaves.)

be enhanced if more of its offspring survive by virtue of the fact that they encounter fewer enemies. It remains to be seen whether the tremendous diversity of chemical defenses displayed by plants is rivaled by a diversity of recognition systems and regulatory mechanisms designed to insure that the defense armamenterium is utilized to best effect.

2

Adaptive Status of Localized and Systemic Defense Responses in Plants

JOHANNES JÄREMO, JUHA TUOMI, AND PATRIC NILSSON

Abstract

Compared to constitutive resistance, inducible defenses in plants imply lower potential costs of defense but presumably also a higher cost of herbivory, since an initial damage is required for induction. Consequently, inducible defenses are most likely to evolve when the risk of herbivory is relatively low and the initial damage made by the herbivore is small. The herbivore damage pattern also determines the magnitude of induction. We expect to find localized responses when herbivores are small in comparison to the plant and when defense compounds are effective in small dosages. Systemic responses, on the other hand, are economical when herbivores are free to move around in the foliage and when an initial damage is a reliable cue for a high future cost of herbivory. As an extreme case, systemic defense responses might also spread from an attacked plant to an unattacked neighbor, i.e., the "talking-tree" hypothesis. However, as far as "interplant communication" implies benefits for both the attacked and the unattacked plant, it would evolve only under rather specific conditions, e.g., when the plants obtain synergistic benefits from each other's defenses.

Introduction

Inducible defense responses in plants vary both in time and magnitude of induction. Rapidly inducible responses start immediately or within hours after damage (e.g., Green and Ryan 1972; Baldwin and Schmelz 1996), whereas defoliation of deciduous trees sometimes increases resistance to herbivores in the following year (delayed inducible resistance; Haukioja 1980). Inducible responses also vary in spatial extent in that they can be localized to the nearest surroundings of damaged areas (localized responses), or they may spread within the plant (systemic responses) (e.g., Berryman 1988). As an extreme case, it has been proposed that inducible responses may even spread from a damaged plant to undamaged neighbors (Baldwin and Schultz

1983; Rhoades 1983; Farmer and Ryan 1990), i.e., the talking-tree hypothesis, criticized by Fowler and Lawton (1985). We discuss the evolutionary background of these patterns of inducible defenses. First, we define potential costs and benefits of inducible defenses, and briefly discuss their evolution in relation to the risk of herbivore attack. Second, we outline general conditions where selection should favor localized or, alternatively, systemic responses within plants. Finally, we discuss the evolution of "interplant communication" in a game-theoretical context. Our analyses are based on the economic principles of plant defense theory, which intends to explain adaptive patterns in plant resistance in terms of fitness costs and benefits (e.g., Zangerl and Bazzaz 1992; Herms and Mattson 1992).

Optimal Defense and Risk of Herbivory

Tissue losses to herbivores are usually costly in terms of plant fitness (Crawley 1983). As plant defenses reduce these losses, they imply a decrease in costs of herbivory, indicating the benefits of defenses to the plant. The costs of defenses, on the other hand, include the metabolic energy expended to construct and maintain defenses (direct costs) as well as the lost photosynthetic gain when resources are reallocated to defense instead of other plant functions (indirect costs; Givnish 1986). The costs may be expressed as a decrease in growth, competitive ability, and/or fecundity, since a part of limited resources is used for defense (Coley et al. 1985; Gulmon and Mooney 1986; Bazzaz et al. 1987; Fagerström 1989).

Let us envisage a highly simplified cost-benefit relation (for more extensive analyses, see, e.g., Simms and Rausher 1987; Clark and Harvell 1992; Adler and Karban 1994). Assign $W(D)$ as a measure of plant fitness that is a function of the proportion of resources allocated to defense, D. There is a cost of allocation to defense in terms of reduced growth and reproduction, $C(D)$, and a cost of herbivory, $H(D)$, when the plant is attacked. The cost of defense increases, whereas the cost of herbivory is a decreasing function of D. Thus, the benefits of defense for an attacked plant can be expressed in terms of a reduction in the costs of herbivory, or as $H(0) - H(D)$, where $H(0)$ is the cost of herbivory for an attacked undefended plant ($D = 0$). In a population where some plants are attacked and some are not, the expected benefit of defense should be weighted by the risk of herbivory, say m. Accordingly, m times $H(D)$ represents the expected cost of herbivory under a given risk of attack which is here, for simplicity, allowed to vary independently from the defense level. Hence, a simple expression for plant fitness as a function of defense allocation would be

$$W(D) = W_0 - C(D) - H(D)m, \qquad (1)$$

where W_0 is fitness for undefended plants in the absence of herbivores. This fitness function summarizes the situation for a plant with a constitutive defense (Tuomi and Augner 1993). In the case of inducible defenses, it may be useful to separate two kinds of costs since this strategy includes two states: the initial level before damage, and the induced level after damage. Without any loss of generality, we assume an initial defense level of $D = 0$. Let $C_c(D)$ now indicate a cost for the plant to maintain a capacity to increase a defense to a level D, while $C_i(D)$ denotes a cost of actually increasing the defense to a level D when a plant is attacked by herbivores. Then the fitness of a plant with a potential defense level D will be

$$W(D) = W_0 - C_c(D) - [H_i(D) + C_i(D)]m, \qquad (2)$$

which, after differentiation and making the derivative equal to zero, gives the optimality condition

$$-m\frac{dH_i(D)}{dD} = \frac{dC_c(D)}{dD} + m\frac{dC_i(D)}{dD}, \qquad (3)$$

where the expression on the left-hand side denotes marginal benefit, and the sum on the right-hand side denotes marginal costs of inducible defenses. Hence, the optimal defense investment is determined by marginal costs and benefits weighted by the risk of herbivory. A prerequisite for the evolution of inducible defenses is that they bring some kind of advantage to its possessor as compared to an ordinary constitutive defense. One of these advantages is generally assumed to be a reduction in the costs of defense. Assuming that $C_c(D) + C_i(D) \le C(D)$, we can deduce that $C_c(D) + [C_i(D)]m < C(D)$ for $m < 1$. In other words, the inducible defender will pay less costs than a constitutive defender. On the other hand, since an initial damage is necessary to mobilize inducible defenses, the cost of herbivory, $H_i(D)$, should be somewhat higher than the grazing costs for a constitutive defender, i.e., $H(D)$ in eq. (1). Therefore, the plant should choose inducible defenses when

$$[H_i(D) - H(D)]m < C(D) - [C_c(D) + C_i(D)m], \qquad (4)$$

or, nonmathematically speaking, when the plant will save more in the costs of defenses than it loses in terms of a higher initial cost of herbivory. Since the left-hand side of the inequality increases and the right-hand side decreases with m, there will be a critical value of m above which selection will favor constitutive defense and below which inducible defense is more favorable. Still, some risk of herbivory is always needed in order to favor inducible defenses ($D > 0$) over the nondefensive strategy ($D = 0$). These results are consistent with other, more extensive theoretical studies of plant defenses (e.g., Zangerl and Bazzaz 1992; Adler and Karban 1994; for a critical review and further discussion, see Agrawal and Karban, chapter 3).

While hard data on the evolution of inducible defenses are scarce, it is not difficult to envisage situations where an inducible defense would be more economical than a constitutive one. First, inducibility will reduce the cost of defense if the plant is dealing with herbivores that turn up occasionally and do not subject the plant to continuous grazing. Insect defoliators in temperate areas (Haukioja 1980), migrating hoards of ungulate herbivores (Georgiadis and McNaughton 1988) and cyclic rodent populations (Seldahl et al. 1994; Lundberg et al. 1994) fit into this pattern of herbivore appearance. We do not, however, consider temporal variation in grazing pressure as a necessary condition for the evolution of inducible defenses (cf. Adler and Karban 1994). A plant population can experience a constant risk of herbivory, but the damage can be locally concentrated on some plants only. Mass attacks by bark beetles on individual trees (Raffa and Berryman 1983) as well as dense colonies of insect pests on single host plants (Karban and Carey 1984) fit a spatial pattern of damage that could favor inducible resistance. Second, the cost of initial damage should be relatively low. This may be the case when an insect herbivore attacks a larger plant, but is far less evident when it comes to large mammalian grazers and browsers. Chemical and mechanical constitutive defenses might often be more economical against large herbivores (cf. Georgiadis and McNaughton 1988; Palo and Robbins 1991; Vicari and Bazely 1993).

Inducible Responses within Plants

On the presumption that natural selection will favor the evolution of inducible defenses, we extend our analysis to the optimal spatial extent of defense responses within individual plants. Therefore we redefine D to reflect the spatial distribution of the defense response. Specifically, we envisage two major alternatives, either a localized defense response in which a small proportion (D) of the foliage becomes induced, or a systemic response that includes almost the entire plant. Note that we do not define a natural breakpoint where a localized response becomes a systemic response and vice versa. We simply denote an induction of a large proportion of plant foliage (large D) as a systemic response, whereas a localized response involves an inducing of a relatively small proportion of the foliage (small D). Furthermore, since plants are not able to predict the position of a herbivore attack, the mechanism that enables inducibility should be represented over the whole plant. It is thus plausible to assume that the cost of maintaining this system, $C_c(D)$, is independent of the magnitude of the defense response, D. This being the case, the term $dC_c(D)/dD$ will vanish and the optimality condition (3) reduces to an especially simple and useful form,

LOCALIZED AND SYSTEMIC PLANT DEFENSES

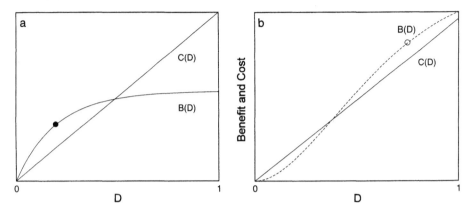

Fig. 2.1. The costs and benefits of defense, C(D) and B(D), as functions of the fraction of defended foliage (D). (a) Linear cost function and concave benefit function, and (b) linear cost function and sigmoid benefit function. The dots represent optimal strategies (see fig. 2.2).

$$-\frac{dH_i(D)}{dD} = \frac{dC_i(D)}{dD}, \qquad (5)$$

which defines the optimal defense level (D^*) for an attacked plant. Below we will follow this simplified form and consider $C(D) = C_i(D)$ as the cost of defense, while the benefit of defense is $B(D) = H(0) - H(D)$, where $H(D) = H_i(D)$ is the expected cost of herbivory for a plant defending a fraction D of its foliage.

Figure 2.1 shows an example of benefit and cost functions that can lead to two possibilities. The optimality condition $dB(D)/dD = -dH(D)/dD = dC(D)/dD$ is graphically obtained when the slope of the benefit curve is equal to that of the cost curve (fig. 2.2), and when $B(D^*) > C(D^*)$ (fig. 2.1). This point on the benefit curve is denoted in figure 2.1 by a dot which thus corresponds to the optimal strategy. We have, for simplicity, assumed a linear cost function, which makes the graphical interpretation much easier. First, we expect a localized defense response if the benefit curve first increases steeply with small values of D and then grows more slowly with larger D-values (fig. 2.1a). Second, the model generates a systemic defense response when the benefit curve first increases slowly with small values of D and more steeply with intermediate values of D (fig. 2.1b). We can thus go a step further and ask under what conditions can we expect the benefit curves to take the shapes presented in figure 2.1?

We assume first that the inducible responses occur rapidly and that the initial damage is a reliable cue of the expected degree of damage over the entire foliage. Then a possible scenario for figure 2.1a could be an insect herbivore attacking a larger plant. Because the expected degree of damage by a single insect herbivore is likely to be very limited, the benefit curve will

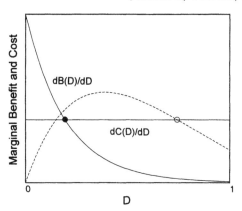

Fig. 2.2. The marginal costs, dC(D)/dD, and benefits, dB(D)/dD = −dH(D)/dD, in relation to the percentage of defended foliage (D). The optimal strategies are denoted by the dots. The filled dot corresponds to a localized defense response given by the cost and benefit curves in figure 2.1a. The open dot corresponds to a systemic response given by the curves in figure 2.1b.

increase fastest when the plant induces the defenses locally close to the initial damage, only (e.g., Fernandes 1990). It is often noted that these kinds of localized rapid responses by the host plant will force insect herbivores to move around in the foliage, with the consequence that the total damage caused by a herbivore is distributed as small local damages over the entire foliage (Edwards and Wratten 1983). A second possibility is that the attacking herbivore(s) will cause extensive damage if the plant does not defend itself at all. However, the mobilized defenses can be highly effective in small quantities so that the herbivore(s) will die, emigrate, or stop feeding even when the plant only defends a small fraction of the foliage. This reasoning directly leads to figure 2.1b where, in contrast, the benefits of defense investments only increases rapidly after a substantial proportion of the plant foliage has been induced (large D). Consequently, a local responses cannot stop the herbivores. Therefore, it is reasonable to assume that the benefit curve in figure 2.1b is most easily obtained when (1) the expected degree of damage is high, and (2) the plant defenses are relatively inefficient in small quantities. A possible example could be a sufficiently large colony of insect herbivores that are free to move around within the plant, or periodical mass attacks of insect defoliators on trees.

It is naturally possible that plants combine the strategies described in figure 2.1. It may be useful for the plant to defend the immediate surroundings of the damage site (fig. 2.1a). However, at some damage level it may become advantageous to induce defenses over the entire foliage (fig. 2.1b). Berryman (1988) has presented this kind of view of rapidly inducible plant defenses. Wounding is supposed to cause intensive local responses, while less intensive systemic responses can spread within the plant. We expect that systemic responses are most probable when the grazing will cause high potential costs (the plant is relatively small), and when the defenses are relatively ineffective in small dosages. In contrast, if the adaptive advantage of inducible defenses rests on avoiding the maintenance costs of high defense levels, large plants should be relatively conservative in triggering systemic responses. If the plant is very sensitive and reacts, for instance, to a wounding of a single leaf, the costs of systemic responses can often be higher than the benefits, as shown in figure 2.1a. In such a case, systemic responses would be economical for the plant only if the single wound can cause extensive damage. This is unlikely in relation to folivory as such, but it might be true in relation to pathogens that could invade from the wound site and then spread to the rest of the plant (for discussion, see Schultz et al. 1992). Berryman (1988) suggested that the induction proceeds in two phases. First, the controlled biosynthesis of relatively small quantities of defensive chemicals is induced by damage in living cells close to and even distant from the site of injury. Second, exogenous elicitors (e.g., microbial fragments) induce uncontrolled biosynthesis of large quantities of defensive chemicals close to damage. In this way, systemic responses would be initiated in cells ahead of the invasion front, while hyperactive, uncontrolled metabolism continues locally just in advance of the invader (Berryman 1988).

So far we have only discussed rapidly inducible plant defenses that we expect to involve localized responses if the plants are large and if they are primarily directed against solitary insect herbivores. In the case of delayed inducible responses to previous defoliation (Haukioja 1980), we expect that the responses should be systemic if the previous defoliation indicates a high risk of future damage over the entire foliage, as shown in figure 2.1b. Localized responses in a part of the plant, e.g., in a single branch (Tuomi at al. 1988), would be economical only when (1) herbivores predictably attack the same branches from one year to the next, or (2) the expected cost of an attack is higher for the previously defoliated branches, e.g., due to reduced photosynthetic leaf area and smaller resource pools, as compared to the previously undefoliated branches. The second alternative is especially interesting because available data indicate an association between delayed inducible accumulation of foliar phenolics and the mineral nutrition of previously defoliated trees and branches (Tuomi et al. 1988; Bryant et al. 1991a). There is also evidence that these responses can even be manipulated by an application of limiting mineral nutrients (Bryant et al. 1993; see also Hunter and Schultz 1995). Foliage removal can directly lead to a nutrient loss for a defoliated branch, and indirectly due to partial rootlet mortality or to a reduced sink strength of the defoliated branch relative to undefoliated branches of the same tree. From this perspective, it would be interesting to know whether the spatial patterns of delayed inducible responses are directly or indirectly determined by the nutritional status of plants, branches, and shoots, or have they in fact evolved to match the spatial and temporal patterns of herbivory.

Neighbor Interactions and Talking Trees

Rhoades (1983) suggested that systemic defense responses might spread not only within individual plants but likewise from an attacked plant to unattacked neighbors. Baldwin and Schultz (1983) presented evidence from controlled laboratory conditions that rapidly inducible responses might spread this way. So far, perhaps the best indication of possible "interplant communication" comes from three trophic-level interactions where plants attacked by insect herbivores elicit signal substances (synomones) that attract predatory insects or parasitoids (e.g., Dicke and Sabelis 1989). Synomone production is not only systemically induced in the attacked plants, but the effects may also spread to neighboring unattacked plants (Dicke et al. 1990b; Takabayashi et al. 1991b; Bruin et al. 1992).

Bruin et al. (1995) as well as Shoule and Bergelson (1995) have pointed out that the spread of inducible responses from one plant to another may not, however, require active communication among the plants. Instead, the unattacked neighbors may become passively contaminated by airborne signal substances that trigger their secondary metabolism. In practice, it is very difficult to decide whether interactions among plants involve active communication or not. From a theoretical point of view, it might be useful to ask (1) whether it is beneficial for an unattacked plant to increase its defense level if its neighbor is attacked, and (2) whether it is beneficial for the attacked plant to allow the unattacked neighbor to increase its defense level. We may then restrict communication among the plants into situations where it is beneficial for an attacked plant to signal its neighbors about an increased risk of damage and for an unattacked plant to respond if its neighbors will be attacked.

A major obstacle to the evolution of interplant communication may be that it is not beneficial to a plant to respond to attacks on its neighbors. Figure 2.3 presents a highly simplified example where plants are allowed either to defend (D) or not defend (N), and where the expected cost of herbivory ($H_{p,self,neighbor}$) for a plant (p) depends not only on its own strategy choice but also on the neighbor's response. The cost of defense is C when the plant chooses D instead of N. There are two plants growing close to each other. Plant 1 is attacked and $H_{1,self,neighbor}$ indicates the expected cost of the attack for the plant itself. Plant 2 is not attacked at the first place but an attack on plant 1 indicates an expected cost of herbivory $H_{2,self,neighbor}$ for plant 2. Plant 1, for instance, can be colonized by insect herbivores and there

LOCALIZED AND SYSTEMIC PLANT DEFENSES

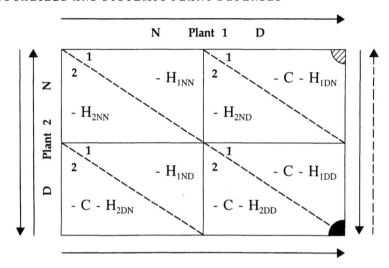

Fig. 2.3. Asymmetric payoff matrix for two plants when plant 1 is attacked and allowed to defend (D) or not (N). Plant 2 is a neighbor to the attacked plant and it has the same strategy choices (D or N). For each strategy combination, the upper payoffs are for plant 1 and the lower ones for plant 2. C = cost of defense, H = the expected cost of herbivory due to the initial attack on plant 1. The arrows indicate the dynamic conditions discussed in the text, and the colored corners denote the final outcomes; black corner = both plants choose D; shaded corner = plant 1 chooses D and plant 2 chooses N. These outcomes correspond to solid and broken arrows, respectively.

is a risk that they will move from plant 1 to plant 2. These kinds of asymmetrical situations have a number of game-theoretical solutions (Maynard Smith 1982). In order to restrict the number of possibilities, we specifically assume that it is always beneficial for plant 1 to defend itself when being attacked (see arrows in fig. 2.3). This will be so if $C < (H_{1NN} - H_{1DN})$ when plant 2 chooses N, and $C < (H_{1ND} - H_{2DD})$ when plant 2 chooses D. In such a case, the choice of plant 2 when plant 1 has chosen N is more or less irrelevant for the final outcome of the game. Instead, the solution of the game will depend on the strategy choice of plant 2 when plant 1 has chosen D. If $C < (H_{2ND} - H_{2DD})$, plant 2 should increase its defense level when plant 1 defends itself. If, on the other hand, $C > (H_{2ND} - H_{2DD})$, plant 2 should not respond even though plant 1 defends itself.

There are several reasons to expect that the last case often is a biologically more feasible situation. First, an attack on a neighbor may not indicate any increase in the expected herbivory load of the unattacked plant, or this increase may often be so small that it does not influence the strategy choice of the unattacked plant. Second, the attack may imply a significant increase in the expected herbivory load of the unattacked plant when the attacked neighbor does not defend itself, but not when the attacked neighbor defends itself

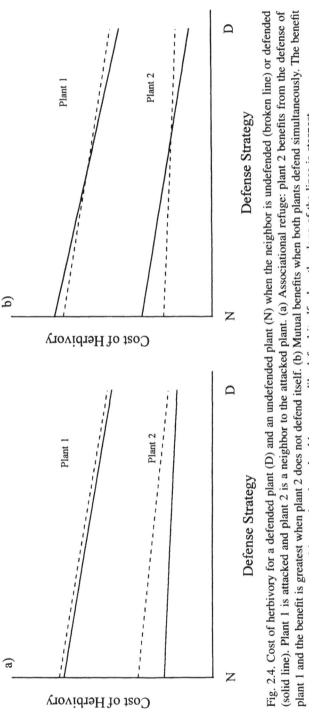

Fig. 2.4. Cost of herbivory for a defended plant (D) and an undefended plant (N) when the neighbor is undefended (broken line) or defended (solid line). Plant 1 is attacked and plant 2 is a neighbor to the attacked plant. (a) Associational refuge: plant 2 benefits from the defense of plant 1 and the benefit is greatest when plant 2 does not defend itself. (b) Mutual benefits when both plants defend simultaneously. The benefit of own defense is greatest, and hence the plant should most readily defend itself when the slope of the lines is steepest.

effectively. The herbivores may die, stop feeding, or emigrate from the patch with the consequence that a palatable plant growing close to an unpalatable neighbor experiences smaller losses to herbivores as compared to situations where the neighbor is palatable (e.g., Sabelis and de Jong 1988; Tuomi et al. 1994). These kinds of associational refuges (sensu Pfister and Hay 1988) would decrease the benefit of defense, $H_{2ND} - H_{2DD}$, for plant 2 (fig. 2.4a), and, hence, plant 2 should choose N rather than D when plant 1 has chosen D (broken-line arrow, fig. 2.3).

We do not, however, exclude the possibility that it may sometimes be beneficial to respond to the attacks on neighbors (Shoule and Bergelson 1995). Still, it requires that we can overcome the above obstacles. Figure 2.4b presents a hypothetical situation where the benefit of defense is larger when the neighbor defends than when the neighbor does not defend. In figure 2.4b, we have specifically assumed that $H_{2NN} < H_{2ND}$. In other words, undefended plant 2 experiences a higher cost of herbivory when plant 1 chooses D rather than N. This is possible if herbivores recognize that plant 1 increases its defense level and as a consequence move to the nearest undefended plant (cf. Dicke and Dijkman 1992). We have further assumed that $H_{2DD} < H_{2DN}$, indicating that the joint defensive response of both plants gives a more effective protection for plant 2 than the defense response of plant 2 alone. For instance, the herbivores may have already learned to recognize inducible defenses on plant 1 and leave plant 2 more readily if it chooses to defend itself. Alternatively, the plants may elicit volatiles that repel herbivores or that attract predators, and the protective effects are much stronger when two or more plants produce volatiles simultaneously. These synergistic, nonadditive effects are often assumed to favor mutual cooperation even though the interacting individuals were not genetically related to each other (Maynard Smith 1989; for plant defenses, see Tuomi and Augner 1993).

If the costs of herbivory for both plants 1 and 2 were qualitatively similar to figure 2.4b, this would satisfy our criteria of interplant communication. As far as $C < (H_{2ND} - H_{2DD})$, it will be beneficial for plant 2 to respond defensively when plant 1 triggers its inducible defenses. Moreover, plant 1 will benefit from plant 2 if $H_{1DD} < H_{1DN}$; that is, plant 1 suffers less damage when both plants defend than when plant 1 defends alone. Note also that an increase in defense level provides a reliable signal of an increased risk of damage in figure 2.4b, where defense is beneficial only when the expected cost of herbivory is sufficiently high. If the signal is false (plant 1 is not actually attacked), plant 1 itself gains nothing, but instead has to pay the cost, C. Similarly, it does not pay for plant 2 to cheat (i.e., to choose N if plant 1 chooses D) because it will be punished by a higher cost of herbivory ($H_{2DD} < H_{2ND}$).

Consequently, interplant communication might in principle work in situations as in figure 2.4b, but less likely so if palatable plants tend to gain associational protection from their unpalatable neighbors (fig. 2.4a). The herbivores should also have a highly patchy distribution over the plant population so that neighboring plants are not attacked simultaneously. As a whole, it seems to us that the conditions for the evolution of interplant communication can be too stringent to allow a common defense strategy against herbivores in general.

Conclusions

We have treated inducible resistance as a way of reducing defense costs as compared to constitutive defenses. Inducible resistance is expected to evolve under conditions where either herbivores appear occasionally, or the attacks are locally concentrated on some individuals only. The spatial extent of inducible defenses can vary from localized to systemic responses within a plant. Localized responses should be most profitable for the plant when the expected degree of damage will be low or when the mobilized defenses are effective in small dosages. Systemic responses are most probable when plant defenses are relatively ineffective in small quantities and when an initially small damage provides a reliable cue of high final costs of herbivory. Induction may even spread from an attacked plant to an unattacked neighbor. Such responses may or may not involve mutually beneficial "interplant communication." We present a scenario where a spread of induction from an attacked plant to an unrelated and unattacked neighbor would be beneficial for both of them. However, we suspect that, in most cases, it may not be advantageous for an unattacked plant to respond to attacks on its neighbor, or, if it responds, this may not be beneficial to the initially attacked plant. Consequently, our major conclusion is that plants should be somewhat conservative in mobilizing inducible defenses systemically if they are costly and primarily directed against herbivores. Therefore, we expect that localized responses within plants should more readily evolve than systemic defense responses, which in turn should be more common than interplant communication.

3

Why Induced Defenses May Be Favored Over Constitutive Strategies in Plants

ANURAG A. AGRAWAL AND RICHARD KARBAN

Abstract

Although induced resistance has been documented in over one hundred species of plants, why plants employ facultative defense strategies is not well understood. Although it has been widely accepted that induction may be a means of reducing resource allocations to defense when not needed, this explanation is not exclusive of a wide array of hypotheses for the advantages of induced defenses. In an attempt to shift attention away from the "allocation cost" model, here we focus on several alternative hypotheses for the benefits of induction. Given that most plants interact with multiple specialist and generalist herbivores, various pathogens, microbes, and mutualists, a host of constraints on maintaining constantly high levels of resistance arise. The temporal and spatial variability in food quality for phytophages created by induction may be especially important. Variability may hinder herbivore performance in ecological time as well as the evolutionary ability of herbivores to adapt to host plants. Research on the various benefits of induced defenses seems to be the obvious next step in understanding why so many plants employ induction.

Introduction

This chapter examines the potential of induced responses to maximize plant defense and focuses on factors that may have favored induced over constitutive defenses. Induced responses to herbivory are common and well documented in plants (reviewed by Karban and Baldwin, 1997). We define induced resistance to herbivory as a plant response that leads to a reduction in the performance or preference of an herbivore that feeds on a damaged plant. We reserve the term "induced defense" for responses that confer resistance against herbivores and increase plant fitness. The study of induced responses is rich with examples of induced resistance across many taxa of plants and affecting various herbivores (e.g., Karban and Carey 1984; Haukioja and

Neuvonen 1985; Zangerl 1990; Baldwin 1991; Bryant et al. 1991b, Rausher et al. 1993). However, there are almost no clear examples of induced responses increasing the *fitness* of induced plants compared to uninduced controls in the presence of herbivores (but see Raffa and Berryman 1982). Demonstration of such effects on fitness components is necessary (but not sufficient) to infer that induced responses are the result of adaptations to herbivory.

In addition to the lack of evidence for induced responses affecting plant fitness (i.e., induced defense), there has been surprisingly little exploration of the various constraints and selection pressures that could favor an inducible over a constitutive strategy of defense (see Rhoades 1979; Karban and Myers 1989; Tuomi et al. 1991; Parker 1992; Adler and Karban 1994; Åström and Lundberg 1994; Padilla and Adolph 1996). If induced responses are effective at deterring and/or killing herbivores, why are they not on all of the time? Measurement of increased plant fitness of damage-induced plants compared to damaged but uninduced plants can demonstrate that induced responses are defensive, but not that inducibility would necessarily be favored over a constitutive strategy. Nearly all previous explanations for the evolution of induced defenses have implicated saving of allocation costs as the key factor. Other benefits have been less well explored. In this chapter we outline several potential benefits of induced responses to herbivory. Measuring the benefits of induced responses will allow us to (1) evaluate the ecological consequences of induction, and (2) begin to identify the selection pressures that may have resulted in inducible rather than constitutive defenses.

There are many reasons why inducible defenses may be favored over constitutive ones (table 3.1A). Induced defenses may be favored or maintained by selection because they are more effective than constitutive strategies. Inducibility can be viewed as an effective strategy if it minimizes costs (in the broadest sense) or if it maximizes benefits to the plant for a given level of investment in defense. These potential benefits do not demonstrate that induced responses are defensive, but rather that induced defenses may be favored over constitutive defenses. In table 3.1B we indicate mechanisms of induced responses that may benefit plant fitness and provide evidence for defense. Below we focus on the benefits described in table 3.1A and henceforth only consider defensive responses.

The Role of Allocation Costs

The hypothesis for the evolution of induced defenses that has been widely accepted posits that defense is maintained as a facultative trait because expression of defense traits diverts essential resources from growth and repro-

TABLE 3.1
Factors Favoring Induced Defenses and (B) Effects of Induced Responses

Benefit	Mechanism	Evidence	References
A. Factors Favoring an Induced Defense			
Reduces allocation costs	Resource allocation to growth and reproduction is saved when not induced.	Mixed	Simms 1992; Karban 1993b
Reduces host finding by specialists	Specialist herbivores that are attracted to resistance related compounds are avoided when the plant is not in the induced state.	Weak	Apriyanto & Potter 1990; Giamoustaris & Mithen 1995
Reduces susceptibility to pathogens	Herbivore resistance can inhibit pathogen resistance and vice versa.	Weak	Westphal et al. 1992; Doares et al. 1995a,b
Creates beneficial variability	Herbivores cannot deal with temporally and/or spatially variable food quality.	Weak	Brattsten et al. 1983; Lindroth 1991
Increases effectiveness of resistance	If resistance chemicals have exponentially increasing effects on herbivore performance, variation around a mean level provides more resistance than the mean level constitutively.	Theoretical	Karban et al. 1997
Slows evolution of herbivore adaptation	There is weak (nondirectional) selection pressure on herbivores to adapt to plants.	Theoretical	Whitham 1983
Disperses herbivory	Concentrated damage has a higher fitness cost than dispersed damage.	Moderate	Marquis 1992; Mauricio et al. 1993
Increases herbivore movement/predation	Moving herbivores are more visible to natural enemies.	Moderate	Bergelson & Lawton 1988
Reduces autotoxicity	Induction reduces plant exposure to autotoxic effects of secondary compounds.	Moderate	Baldwin & Callahan 1993
Reduces deleterious effects on predators and parasites	Plant resistance may negatively impact preference, provide a physical deterence, or poison natural enemies of herbivores.	Mixed	Hare 1992; Duffey et al. 1995
Reduces pollinator deterrence	Defense chemicals leak into flower parts and deter pollinators.	None	Baker & Baker 1975; Stephenson 1982

TABLE 3.1 (*Continued*)

Benefit	Mechanism	Evidence	References
B. Effects of Induced Responses on Plants			
Less damage to plants	Herbivores avoid or are poisoned by induced plants.	Strong	Reviewed in Karban & Baldwin 1997
Increases competitive ability	Allelopathic inhibition of competitors occurs.	Weak	Oleszek 1987; Sadras 1996
More tolerant to damage	Subsequent herbivory has less of a negative effect on plant fitness.	Theoretical/ weak	Lehtilä 1996; Wittmann & Schoenbeck 199
Increases predation and parasitism	Natural enemies are attracted by volatiles released from induced plants.	Moderate	Reviewed in Takabayashi & Dicke 1996

Notes: Inducibility may be favored as a defense strategy over constitutive defenses if inducibility is mo beneficial to plants (A). If induced responses benefit plants in environments with herbivores, induced plar should have higher fitness than uninduced plants. Although evidence for induced resistance is strong, effec on plant fitness have rarely been demonstrated (B).

duction (Zangerl and Bazzaz 1992). In the absence of herbivory, plants that do not allocate resources to defense traits are predicted to have higher levels of resources available for growth and reproduction (Herms and Mattson 1992; Rausher 1996). Experimental attempts to find such allocation costs of induction have met with limited support (Brown 1988; Simms 1992; Karban 1993b; A. Agrawal et al., unpublished manuscript; but see Baldwin et al. 1990). Although there is limited support for the "savings of allocation cost" argument, research on other benefits of induced defenses is currently lacking. Furthermore, many researchers continue to interpret patterns and develop mathematical models and verbal defense theories based largely on the notion that defense as an induced trait is a strategy to reduce allocation costs (Herms and Mattson 1992; Zangerl and Bazzaz 1992; Rausher et al. 1993; Zangerl and Rutledge 1996). Below we suggest that this interpretation may not be correct, or at least that it may not be complete, and that alternative factors favoring the evolution of induced defenses should be considered. Realistically, why might allocation costs be minimal or undetectable?

Selection can minimize initial costs of resistance. Numerous examples from taxa other than plants have demonstrated this phenomena and are reviewed by Simms (1992) and McKenzie (1996). In a recent study, Schrag and Perrot (1996) worked with bacteria that evolved resistance to an antibiotic. Two lines of resistant bacteria initially showed fitness reductions (costs) of 14% and 19%, respectively, in the absence of the antibiotic. Quite surprisingly, in bacterial cultures without the antibiotics but with both resistant and wild types (not resistant), the wild-type bacteria did not dominate the cultures. To the contrary, after 135 generations, all of the colonies assayed were resistant. After an additional 45 generations (in the absence of antibiotics), the resistant bacterial strains evolved further and minimized the cost of resistance until fitness was indistinguishable from wild types. In our own studies of costs of induction, we are finding that plants exhibit variation for costs among half-sibs grown in a common environment, suggesting that cost may be subject to selection (A. Agrawal et al., unpubl. data). Our interpretation of this is that costs of resistance are not universal and can be minimized by selection if they exist early in the evolutionary process.

In an extensive survey of costs of plant resistance, Bergelson and Purrington (1996) found that only 33% of the studies reviewed demonstrated costs of resistance to herbivores. Other reviews of costs of plant resistance, discussions of additional explanations for "no cost defenses," and problems with detecting costs can be found elsewhere (Parker 1992; Simms 1992; Karban 1993b; Antonovics and Thrall 1994; Mole 1994) and are not discussed further in this chapter.

Does the absence of allocation costs defy logic, or is it just a paradigm? In the late 1970s, ecologists studying species interactions held competition as a ubiquitous phenomenon, a paradigm of ecology (reviewed in Salt 1984). The view that competition was so important led some experimentalists who did not find competition among species to reject their experiments before rejecting the hypothesis that competition was occurring (Strong 1980). Allocation costs of plant resistance may be a similar paradigm. We view repeated failure to experimentally find significant allocation costs of plant resistance as a genuine contribution to plant-herbivore studies and as a window for new avenues of research on constraints and selection pressures on plant defense.

Even where detectable or undetectable allocation costs exist, they may not necessarily be the most important constraint favoring the evolution of induced defenses (Parker 1992; Simms 1992; Karban 1993b; Mole 1994). The importance of allocation costs for the evolution of induced defenses will be determined not only by attempts to detect such costs, but by comparing costs of maintaining induced versus constitutive defenses. If costs of induced and constitutive defenses are comparable, their relevance is dubious. Given our current state of knowledge, it seems shortsighted to overlook nonexclusive alternative hypotheses. In subsequent sections we consider less explored hypotheses for the evolutionary origin and/or maintenance of induced defenses in plants.

Induced Defenses Reduce Host Finding by Specialist Herbivores

Here we focus on situations where induced responses to herbivory lead to both resistance and susceptibility and might be a mixed blessing for the plant. Evidence from several systems suggests that damaged plants may sometimes become more attractive or of better quality to herbivores (Danell and Huss-Danell 1985; Apriyanto and Potter 1990; Haukioja et al. 1990; Karban and Niiho 1995; Baur et al. 1996). This is termed "induced susceptibility." For example, some phytochemicals that are implicated in resistance against generalist herbivores also serve as attractants for specialized herbivores. Thus, a potential benefit of employing induced defenses is that defense chemicals are used when needed, but specialist herbivores are not as attracted to plants in the uninduced state. Below we discuss a system where this may be occurring.

Induced responses and resistance to generalist herbivores have been studied in many brassicaceous plants including *Brassica napus*, *B. rapa*, *B. nigra*, *B. juncea*, *Raphanus raphanistrum*, *R. sativus*, *Lepidium virginicum*, and *Sinapis alba* (Koritsas et al. 1991; Bodnaryk 1992; Doughty et al. 1995; R. Mithen, per. comm.; A. Agrawal, unpubl. data). A general pattern of increased indole glucosinolates following damage and increased resistance against generalist herbivores has been found. There are also many studies demonstrating that glucosinolates are important stimulants of oviposition and larval feeding in pierid butterflies and other specialist herbivores (e.g., flea beetles and diamondback moths) of the Brassicaceae (David and Gardiner 1966; Feeny 1977; Blau et al. 1978; Chew 1980; Lamb 1989; Huang and Renwick 1993; Chew and Renwick 1995; Giamoustaris and Mithen 1995).

The question then arises, do plants with induced responses that result in elevated levels of glucosinolates suffer more damage from specialist herbivores and less from generalized feeders? Evidence comes from Baur et al. (1996), who demonstrated that damage to two different *Brassica* host plant species resulted in increased preference of the hosts to cabbage root flies (*Delia radicum*). In simultaneous-choice tests, root flies laid more eggs on plants that had experienced previous feeding. Other studies have shown that root fly feeding and mechanical damage resulted in the induction of glucosinolates (Griffiths et al. 1994; Birch et al. 1996) and that such glucosinolates are important oviposition cues (Roessingh et al. 1992). Taken together, these studies indicate that maintaining high levels of glucosinolates all of the time may increase the susceptibility of *Brassica* hosts to some specialist herbivores.

In an excellent study, Giamoustaris and Mithen (1995) bred twenty-eight lines of *Brassica napus* with continuous variation from low to high levels of glucosinolates. In field experiments, strong negative relationships were found between glucosinolates and herbivory by generalist slugs and pigeons. However, strong positive relationships were found between total glucosinolates and damage by specialist flea beetles and incidence of *Pieris rapae* larvae. This study demonstrates how plant resistance chemicals can functionally serve as "double agents." Because many brassicaceous plants exhibit an accumulation of glucosinolates following damage, individual plants can naturally vary in their levels of glucosinolates over time. A benefit of employing an inducible strategy may be to use resistance chemicals when necessary, without attracting specialist herbivores when the plant is not under attack. The same chemical strategies may mediate interactions between plants and pathogens (see Giamoustaris and Mithen 1997).

This phenomenon is not likely to be a special case for brassicaceous plants that employ inducible glucosinolates. For example, plants in the family Cucurbitaceae, which also employ suites of induced responses, are likely to be more resistant to generalists and more susceptible to specialists in the induced state (Da Costa and Jones 1971; Metcalf and Lampman 1989; Tallamy 1985; Apriyanto and Potter 1990; Tallamy and McCloud 1991). Carroll and Hoffman's (1980) classic paper demonstrated this double-edged cost of induction in *Cucurbita moschata*. Additional examples of this phenomena from other plant families can be found in Alados et al. (1996), Landau et al. (1994), Matsuki and MacLean (1994), Pasteels and Rowell-Rahier (1992), and Wink (1988).

Induced Defenses Improve Resistance against Different Enemies

Most plants are subject to attack, not only by multiple herbivores, but also by suites of damaging nematodes, fungi, bacteria, and viruses (Barbosa et al.1991b, Hatcher 1995). There are often different defense pathways of production and defense products that are activated in response to the various sets of plant attackers (Bennett and Wallsgrove 1994; Doares et al. 1995a,b; Benhamou 1996; Kunkel 1996). For example, the salicylic acid pathway which induces production of defense responses against some pathogens in brassicaceous plants produces different suites of glucosinolates than does the pathway that responds to insect damage (Doughty et al. 1995). Similarly, it is thought that there are multiple pathways of induction in Solanaceous plants such as tomato, and accumulation of induction products following attack by arthropods and fungal pathogens can be strikingly divergent (M. Stout, pers. comm.). It has recently been discovered that some of these pathways inhibit each other, although the physiological, ecological, and evolutionary reasons have not yet been explored (Doherty et al. 1988; Pena-Cortes et al. 1993; Doares et al. 1995a,b). For example, in both tomato and tobacco plants, exogenous application of salicylic acid inhibits induced defenses against herbivores but is likely to provide defense against fungal pathogens (Ward et al. 1991; Baldwin et al. 1997; M. Stout, pers. comm.).

A striking case of the potential incompatibility of plant resistance to pathogens and herbivores is the *Solanum dulcamara*–eriophyid mite system. Westphal and colleagues (1991) have demonstrated that gall induction by

Aceria cladophthirus mites induced resistance to subsequent attack by *A. cladophthirus* as well the rust mite, *Thamnacus solani*. Resistance to these mites in the *Solanum* system, however, is mediated by a hypersensitive plant response (Fernandes 1990), which leads to decreased leaf tissue damage and/ or growth of the mites (Westphal et al. 1991). The hypersensitive response was associated with increased peroxidase activity and other pathogenesis-related proteins in *S. dulcamara* (Bronner et al. 1991a,b). Such plant responses are most often associated with resistance against pathogens (i.e., Kuc 1987). If the plant pathways responding to pathogens and herbivores inhibit each other (Doherty et al. 1988; Pena-Cortes et al. 1993; Doares et al. 1995a,b), we predict that *S. dulcamara* plants infested with galling mites may become more susceptible to other herbivores. Indeed, Westphal et al. (1992) found that leaves of *S. dulcamara* damaged by the galling mites became more susceptible to the generalist two-spotted spider mite, *Tetranychus urticae*. *T. urticae* females were stimulated to oviposit and had significantly higher fecundity on leaves expressing the hypersensitive response. These results suggest that when plants respond to attack by producing pathogenesis-related defenses, they may become more susceptible to attack by other insect herbivores.

There have also been a considerable number of studies on the effects of arthropod and pathogen induced responses, their interaction, and their biological effects in the cucumber system. Infection of cucumber by anthracnose fungus induced systemic resistance to at least thirteen pathogens, including various fungi, viruses, and bacteria (Kuc et al. 1975; Kuc 1982, 1983, 1987). However, induced responses by anthracnose did not provide any resistance against spider mites, armyworms, or aphids (Ajlan and Potter 1991). In the reverse scenario, induction following spider mite and armyworm feeding did not provide resistance against anthracnose. Although there was no evidence of pathogen attack increasing plant susceptibility to herbivores, these studies provide additional evidence that there may be alternative pathways and different effects of induction by pathogens and herbivores. Similar results have been found for interactions between pathogens and arthropods on tobacco (Ajlan and Potter 1992).

Apriyanto and Potter (1990) report on another study on cucumber in which tobacco necrosis virus (TNV) was found to induce resistance against anthracnose. Consistent with the idea of multiple induction pathways, it was found that induction by TNV affected neither performance of spider mites and armyworms, nor preference of whiteflies and armyworms. Although cucurbitacins were not found to be induced in infected plants, leaves of infected plants were more susceptible to feeding by striped cucumber beetles, specialists on the Cucurbitaceae. In summary, cucurbits seem to have multiple induction pathways. Pathogens induce resistance against a diverse array of pathogens but not against arthropods and arthropod induction can cause resistance against arthropods, (Tallamy 1985; A. Agrawal, unpublished), but does not seem to cause resistance to pathogens (Ajlan and Potter 1991). Finally, induction following both pathogen and arthropod attack appears to result in susceptibility to some specialist arthropod herbivores (Carroll and Hoffman 1980; Apriyanto and Potter 1990; Tallamy and McCloud 1991). From the plant's perspective, such divergent results may be unavoidable and therefore may favor the facultative expression of resistance.

Changes in plant quality caused by herbivores or pathogens do not always result in such trade-offs in effects on different plant parasites. For example, there are a great number of examples of generalized induced responses that appear to provide resistance against many herbivores and pathogens (see extensive reviews in Barbosa et al. (1991b) and by Hatcher 1995). Especially within different guilds of plant enemies, there appear to be some generalized responses that confer resistance against several attackers. However, where multiple pathways do play a role, plant physiological constraints may favor inducible defenses against multiple attackers because alternative defense strategies may not be compatible. Adler and Karban (1994) theorize that inducible defense phenotypes would be favored over constitutive strategies when defenses against some attackers are ineffective against others. In summary, if defense-related products make plants more susceptible to specialist herbivores or if particular defense pathways inhibit other pathways, induced strategies may be favored.

Induced Defenses Create Variability in Food Quality

Variability, such as that imposed by induced defenses, could pose difficulties for herbivores. Induced plant responses can create variability in many ways (table 3.2). Plant defenses can be variable because they are induced and then relaxed over different temporal and spatial scales. In addition, biotic and abiotic factors can strongly affect the expression of induced defenses within and among individual plants. Factors listed in section A of table 3.2 make individual plants and populations more variable, while those in section B only have the potential to make populations of plants more variable. Below we consider the potential benefits of variability created by induced defenses.

Stockhoff (1993) provides the strongest evidence that performance of an herbivore cannot be simply predicted by the average food quality that it experiences in a heterogeneous environment. Extremely polyphagous gypsy moth larvae were unable to compensate for variability in food quality, resulting in decreased pupal mass and slowed development time when fed increasingly variable diets with the same mean nitrogen level. Stockhoff (1993) pointed out that unselective foragers would suffer the direct fitness decrement of natural variation in food quality, while selective foragers may expe-

TABLE 3.2
How Induced Responses Create Variability in Plant Quality Faced by Herbivores

Cause of Variation	Nature of Variation	References
	A. Within and among Plants	
Induction itself	Induction and relaxation create temporal variation.	Baldwin 1991, 1994; Malcolm and Zalucki 1996
Induction itself	Spatial: local vs. systemic responses.	Jones et al. 1993; Stout et al. 1994, 1996a
Induction itself	Different structures are more or less inducible.	Zangerl and Rutledge 1996; Baldwin and Karb 1995
Unpredictable induction	Highly variable change in quality following herbivory.	Stout et al. 1996; Van Dam and Vrieling 1994
Developmental state	Inducibility may change with plant age.	Alarcon and Malone 1995; Stout et al., 1996b
Fungal infection	Fungal infection may inhibit induced defenses to herbivory.	Westphal et al. 1992; Doares et al. 1995a,b; Hatcher 1995
Other herbivores	Other guilds may induce resistance or susceptibility.	Stout et al. 1996a; Karban and Niho 1995; Hatcher 1995
Plant genotype	Induction threshold and strength of inducibility.	Zangerl and Berenbaum 1990
Type of feeding	Induction products and the intensity of induction.	Stout et al. 1994, 1996a
	B. Among Plants	
Nutrient regime	Fertilized or shaded plants may be more or less inducible.	Haukioja and Neuvonen 1985; Hunter and Schultz 1995; Ruohomaki et al. 1996
Plant density	Crowded plants may be less inducible.	Karban et al. 1989; Karban 1993a; Hjalten et al. 1994
Light regime	UV light may affect expression of defenses.	Reviewed in Downum 1992

Notes: Induced responses can make individuals and populations of plants more variable (A). Abiotic and neighborhood effects can make induced responses in populations of plants more variable (B).

rience other costs associated with increased movement (see below). In either case, more variable plants should be better defended and thus favored by selection.

In addition to only partially compensating for food quality variation, herbivores often cannot respond well to large shifts in plant quality (Brattsten et al. 1983; Lindroth 1991). This is due in part to the enzymes used to detoxify plant defense chemicals which themselves can be inducible in herbivore guts (e.g., Berenbaum and Zangerl 1994); variation in plant defense chemicals may present herbivores with food environments not well matched with their current enzymatic capabilities. Furthermore, Brattsten et al. (1983) demonstrated that some herbivores may be able to compensate for heavy doses of defense chemicals, as long as they are increased in small increments; larger shifts in defense chemicals can poison the herbivore. Finally, many herbivores self-select appropriate diets (Waldbauer and Friedman 1991) and are able to learn to avoid lower-quality food. Variability created by induced responses may reduce the ability of herbivores to use such learned self-selection information and has the potential to maintain herbivores on suboptimal diets (Jones and Ramnani 1985). Where plant variability hinders herbivore performance in ecological time, we predict that variable plants will outperform more homogeneous genotypes.

Variability Itself Improves Resistance

By definition, induced defenses create temporal variability in the food quality experienced by herbivores (table 3.2). Below we consider a simplified model of plant defense to demonstrate how variability per se, as created by induced responses, could maximize defense. Consider the common case where levels of foliage toxins increase following herbivory (Baldwin 1994). If the level of secondary compounds is analyzed on a scale of 0 to 10, where 0 is no chemicals and 10 is a high dose of chemicals, then an herbivore on a plant with induced defenses may experience level 2 and level 8, before and after induction, respectively. If these compounds have a linearly negative effect on herbivores (i.e., from 0 to 10), then a herbivore exposed to these levels for equal times experiences a realized food quality at the mean toxin level, 5. An herbivore on a plant constitutively expressing chemical level 5 should have the same fitness as an herbivore faced with equal time eating variable levels 2 and 8.

However, consider food quality as a declining concave function of plant toxins, where a small amount of a chemical has little impact on the herbivore and higher chemical doses have exponentially increasing deleterious impacts on herbivore performance. A relationship of increasing effectiveness of toxins with higher doses has been reported from several plant-herbivore systems (reviewed by Karban, Agrawal, and Mangel, 1997). In this case, an herbivore experiencing chemical levels 2 and 8 experiences quantitatively poorer food quality than an herbivore feeding constantly at level 5. This inequality stems from the fact that higher chemical levels result in exponentially worse food quality. This model developed by Karban et al. (in press) demonstrates that variability, such as that created by induced defenses, could be favored because it increases the effectiveness of defense chemicals. In this model, as in all others considered in this chapter, plants must still be constrained to some level of defense; a plant expressing toxin level 10 would obviously be the most defended. However, no matter how the plant might be constrained, variation in defense could maximize the toxic effects of a given level of defense chemicals.

In conclusion, if the variability created by induced responses maximizes defense, then variable plants may be selected for by two mechanisms: (1) variable plants will have less damage and higher fitness because of direct negative effects on herbivores; and (2) variable plants will be avoided by herbivores and therefore have higher fitness. In addition to the potentially negative biological effects of variability (previous section), variability in itself may increase the quantitative effectiveness of defenses.

Induced Defenses Slow Adaptation of Herbivores

"Variation as a defense may place plant pests in an evolutionary 'squeeze play' that even rapidly evolving pests cannot easily surmount" (Whitham 1983). In the early 1980s, variation in plant defense was suggested as an important component of plant-herbivore interactions (Denno and McClure 1983). It was hypothesized and later mathematically modeled that variability in plant genotype, defense, or *Bacillus thuringiensis* expression at the individual or population levels could slow the evolutionary adaptation of herbivorous insects (Whitham and Slobodchikoff 1981; Whitham 1983; Whitham et al. 1984; Gould 1986a,b, 1995; Gould and Anderson 1991; Gould et al. 1991; Liu and Tabashnik 1997). Yet the idea that induced defenses create natural variation in plant defense is still in its infancy.

Variability in defense created by induction could potentially weaken the selection pressure for herbivores to adapt to plant defenses. Because selection will not always be strong, individuals not adapted to the defenses will continue to reproduce. Recombination and interbreeding between more and less adapted herbivores could thwart population level adaptation (Gould 1995). On the other hand, constitutively defended plants may impose a consistent directional selection pressure causing herbivores to adapt rapidly (e.g., Gould 1979; Fry 1989). However, it is not clear whether more slowly evolving pests on variable plants could in itself explain the evolution of induced defenses. For example, in order to avoid a group selection argument, inducibility must have arisen because of other selection pressures in short-lived plants. If, on the other hand, plants live long enough to have pests adapted to individuals (e.g., inducible *Quercus* spp.; see Faeth 1991 and Mopper et al. 1995), individuals with inducible defenses may have higher lifetime fitness than conspecifics with constitutively expressed defenses because they have fewer adapted herbivores.

Induced Defenses Disperse Damage and Increase Herbivore Movement

Induced responses may force herbivores to move following a bout of herbivory. If herbivores move as a result of induction, dispersing their damage, and dispersed damage reduces plant fitness less than concentrated damage, then induction may be favored indirectly. This benefit is especially plausible in cases with localized induced responses such as some of those in birch (Tuomi et al. 1988), wild parsnip (Zangerl and Berenbaum 1994/95), sunflower (Olson and Roseland 1991), and tomato (Stout et al. 1996a). Because methods to inhibit induced defenses are still being developed and have not been extensively used in ecological studies (see Baldwin 1988; Baldwin et al. 1990; Hartley 1988; Karban et al. 1989; Doares et al. 1995a,b), it has been difficult to test how localized responses influence herbivore feeding patterns because the null model has not been obvious. Studies that have measured herbivore feeding patterns on previously induced foliage (e.g., Wratten et al. 1990) did not measure the reaction of herbivores to induction following a bout of their own herbivory, but rather the reaction of herbivores to systemically induced leaves (i.e., the herbivore simply experiences a highly resistant plant). However, Bergelson et al. (1986) did find that herbivores were likely to move away from damage sites on leaves. Other demonstrations of damage-induced movement are reviewed in Edwards et al. (1991). If induction is hypothesized to be favored over a constitutive strategy because it forces herbivores to move, experiments will have to demonstrate that inhibition of induced responses causes herbivores to feed in a more concentrated fashion.

Experimental and theoretical evidence does suggest that concentrated damage can be more detrimental to plant fitness than the same amount of damage spread more widely (Lowman 1982; Marquis 1992, 1996; Mauricio et al. 1993; Lehtilä 1996). This finding is potentially due to the ability of plants to compensate locally for small amounts of damage. Although it is still not clear if this is a general result, others have suggested that the relationship between concentrated and dispersed damage will not always be as described above, but rather that the relationship will simply not be linear (P. Kareiva, pers. comm.). As long as different leaves or even different parts of individual leaves have different resource pools or productivity values for the plant, concentrated damage may decrease or increase plant fitness compared to plants with dispersed damage.

Increased herbivore movement promoted by induced responses may not only disperse damage, but also may lead to higher rates of predation and parasitism of herbivores (Edwards and Wratten 1983, Schultz 1983). Again, this is a difficult hypothesis to test because one must demonstrate: (1) that herbivores are moving more following induction than herbivores on uninduced controls, and (2) that this movement (and not other cues associated with induction) is attracting predators and parasites. Although Bergelson and Lawton (1988) found that herbivores only moved away from the immediate sites of damage, caterpillars experimentally forced to move suffered higher predation by ants. Other cues associated with induced responses may benefit the plant via increased attraction of natural enemies (table 3.1.B; Turlings et al. 1990, Whitman and Norlund 1994; Drukker et al. 1995; Takabayashi and Dicke 1996); however, it is not obvious that this would benefit plants more than a constitutive strategy that was always attractive.

In summary, induced defenses may be favored over constitutive defenses if (1) patterns of herbivore feeding/movement change following induction, and either (2a) the subsequent effect of herbivory on the plant or (2b) the subsequent level of predation and parasitism on the herbivore is influenced by the pattern of feeding. Presently, all of the individual links appear to have been satisfied in various systems; however, there have been no complete demonstrations of this potential benefit of induced defenses for any one system.

Induced Defenses Reduce Autotoxicity

Autotoxicity of plant secondary metabolites has been discussed as a possible physiological cost of plant defense since the 1970s (Chew and Rodman 1979; Fowden and Lea 1979). Karban and Myers (1989) suggested that one benefit of an induced defense over a constitutive defense strategy is that autotoxicity is avoided when the plants are not under attack and do not need the defenses. Some plant defense chemicals that are thought to be autotoxic, such as hydrogen cyanide (HCN) in clover, are stored as precursors and separated from catalyzing enzymes in vacuoles (reviewed by Duffey and Felton 1989). Feeding damage to such plants induces the defense by puncturing vacuoles; enzymes mix with cyanogenic glucosides to form HCN. Jones (1972) has hypothesized that noncyanogenic morphs of clover may be favored to reduce autotoxicity in environments where cell disruption and subsequent production of HCN frequently occur by abiotic processes.

Minimizing autotoxicity by inducing defenses may also be a strategy in other plant species where defense compounds are produced de novo following damage. Baldwin and Callahan (1993) provide evidence that nicotine, which is produced in response to herbivore feeding in wild tobaccos *Nicotiana sylvestris* and *N. glauca*, can reduce the photosynthetic capacity of the plants. These experiments used uninduced plants that were hydroponically fed realistic levels of nicotine to isolate the specific effects of autotoxicity. Indirect evidence also comes from Rasmussen et al. (1991), who demonstrated that salicylic acid (a known autotoxin) is produced following bacterial infection; natural heavy salicylic acid production or experimental endogenous injection is followed by veinal chlorosis and stunting of the leaves. Although not all plant defense compounds cause autotoxicity, the presence of compounds that do may favor inducible defenses.

Induced Defenses Reduce Deleterious Effects on Natural Enemies

Compatibility of biological control and plant resistance has been of longstanding interest to entomologists (reviewed in Boethel and Eikenbary 1986; Hare 1992; Duffey et al. 1995). In addition to the agricultural importance, a negative interaction between plant resistance to herbivores and predation and parasitism of herbivores could act as an evolutionary constraint favoring inducible defenses over constitutive strategies. Plant resistance traits have the potential to deter predators and parasites, interfere with their movement, and poison them when they consume plant parts or herbivores. Although several positive and even synergistic interactions between plant resistance and natural enemies of herbivores have been found (table 3.1.b.; Boethel and Eikenbary 1986, Hare 1992, Duffey et al. 1995), here we focus on the negative interactions to highlight a potential constraint on constitutive plant defenses.

Since many important predators are to some degree omnivorous, they may have reduced preference for herbivores on resistant plants (Stoner 1970; Kiman and Yeargan 1985). Development in plant tissue is fairly common among predaceous bugs, making the predators potentially very sensitive to plant defenses (Coll 1996). Many omnivores, such as the western flower thrips, often feed heavily on plants and they are also important predators of herbivores (Agrawal and Karban 1997). In such cases, generalized plant defenses may have a negative effect on preference of predators and parasites.

Several studies have highlighted the direct and indirect impacts of plant resistance on interference with predators and parasites. For example, Kauffman and Kennedy (1989) found that parasitoids suffered high mortality due to plant trichomes on wild tomato plants. In addition, plant resistance in the tomato system has been reported to negatively affect parasitoids indirectly through the herbivore (Campbell and Duffey 1979; Barbour et al. 1993). Even for herbivores that do not sequester chemicals from their host plants, consumption of herbivores feeding on more resistant plants led to decreased survival and fecundity and increased the development time of predators and parasites (Barbosa et al. 1991a; Stamp 1993; Stamp et al. 1997). Pathogens of herbivores may also be affected by plant resistance traits (Duffey et al. 1995). For example, Hunter and Schultz (1993) have found that herbivore-induced responses in oak trees negatively affected the ability of a nuclear polyhedrosis virus, a pathogen of gypsy moths, to kill moth larvae. Induction may be favored as a strategy to reduce such negative impacts of plant resistance on natural enemies of herbivores.

Induced Defenses Reduce Pollinator Deterrence

Secondary metabolites from some plants may be found in floral parts and nectar (Baker and Baker 1975; Stephenson 1982; Landolt and Lenczewski 1993; Langenheim 1994; Giamoustaris and Mithen 1996; F. Stermitz, pers. comm.). Such chemicals have been known to decrease floral visitation by certain pollinators (Baker and Baker 1975; Giamoustaris and Mithen 1996). For example, alkaloids can be found in the nectar and flower parts of many plants, and may even be toxic to some honeybees and lepidopteran visitors (Baker and Baker 1975). Levels of glucosinolates in the inflorescences of *Brassica napus* had strong negative effects on incidence of a pollen beetle (Giamoustaris and Mithen 1996). In addition, other volatile or surface compounds in fruits and foliage have the potential to deter particular pollinators. Chambliss and Jones (1966) found that cucumbers that did not express the bitter gene, and subsequent production of cucurbitacins, were significantly more attractive to bees and wasps than bitter controls. If such effects of secondary metabolites in vegetative plant parts decrease pollinator visitation and subsequent plant fitness, inducibility may be favored as a strategy to minimize pollinator deterrence.

Recent research has suggested that leaf damage may affect the pollination biology of plants (Quesada et al. 1995; Mutikainen and Delph 1996; Strauss et al. 1996). Although some of these effects on pollination caused by leaf damage are the likely result of changes in floral number and architecture, other effects may be due to changes in plant defense chemicals in foliage, flowers, nectar, and pollen (e.g., Baldwin and Karb 1995). Because of constraints on vascular architecture, plants may be unable to control where certain plant defense chemicals go (e.g., C. Jones et al. 1993). Lehtilä and Strauss (unpubl. ms.) have recently demonstrated that damaged wild radish plants suffered reduced visitation by pollinating syrphid flies even when differences in flower number and size were controlled. If such results are general and are caused by changes in plant chemistry, and if pollinator visitation affects plant fitness, then the evolution of plant defenses may be constrained by pollinator deterrence and induction may be favored.

Conclusions

Research on induced responses has broadened our knowledge of plant-herbivore interactions: plants are more dynamic than once thought and they employ complex and variable defensive phenotypes that herbivores must deal with. Induced defenses provide variability within and among individual plants. Therefore, we must first understand the factors that create and influence this variability. The listing in table 3.2 only begins to detail how and why plants are variable with regard to what herbivores perceive. The benefits of such variability are likely to be manifold. Given the constraints outlined in this chapter, we expect that in various systems, each of these many benefits may have contributed to favoring the induced defense strategy.

At this stage it would be difficult to assess the relative importance of these factors in promoting the evolutionary origin and maintenance of induced defenses, since the evidence for each of the benefits is circumstantial at best. It is likely that "reduced host finding" is at least important in certain plant families with highly specialized herbivores such as the Brassicaceae and Cucurbitaceae. Similarly, induction favored as a strategy to reduce large "autotoxic effects" of secondary compounds may be restricted to particular plant families with classes of chemicals such as cyanogenic glucosides, glucosinolates, and certain alkaloids. Maximizing defense against multiple plant enemies, dispersing herbivory, and taking advantage of the effects of variability may, on the other hand, be more general and widespread factors influencing the deployment of induced defenses. Although induction may have been favored as a strategy to "minimize herbivore adaptation" in long-lived plants with intimately associated herbivores, this benefit may only be important in maintaining induced defenses (and not necessarily in its evolutionary origin) in populations of shorter-lived plants. In systems where natural enemies are important regulating factors of herbivores, selection for reduced or inducible defenses may be common (see also Strong and Larsson 1994). Finally, "avoidance of poisoning the pollinator" may be a common constraint on plant defense, especially in plants which cannot control where particular compounds flow (i.e., C. Jones et al. 1993; and some parasitic plants—L. S. Adler, pers. comm.). In this chapter we have emphasized alternatives to the widely accepted explanation of induced defenses as a means of minimizing allocation costs. Although support for the role of allocation costs is surprisingly poor, we believe that further attempts to evaluate this hypothesized selective agent are important. However, we speculate that studying other potential benefits and constraints may be required to truly understand induced defenses in plants.

4

Evolution of Induced Indirect Defense of Plants

MARCEL DICKE

Abstract

Plants can defend themselves against herbivores indirectly by enhancing the effectiveness of carnivorous enemies of the herbivores. Plants respond to herbivory with the induced production of volatiles that attract carnivores. This can be done with specific volatiles that are produced de novo in response to herbivory and not in response to mechanical damage. Other plant species respond with nonspecific volatiles that are produced in response to herbivory as well as mechanical damage. Both in the case of specific and nonspecific volatiles, the composition of the blend and the amounts emitted can be used by carnivores to discriminate between herbivore-damaged and mechanically damaged plants. Moreover, damage by different herbivore species, or even different herbivore instars, can be distinguished on the basis of herbivore-induced plant volatiles. The information content of the volatile blend emitted by herbivore-infested plants is the most specific in the case of production of novel compounds in response to herbivory.

Research on induced indirect defense is firmly anchored in studies of population dynamics. The first demonstration of carnivore attraction to herbivore-infested plant parts was an investigation with the aim of understanding the mechanism of local extermination of herbivore populations by carnivores. This illustrates that induced indirect defense can be an important plant defense.

Herbivore-induced plant volatiles are of great importance to carnivorous arthropods. Carnivores that forage for herbivores are faced with a problem relating to reliability and detectability of available cues from herbivores and uninfested food plants of the herbivores. The herbivore-induced plant volatiles emitted by infested plants are a solution to this problem, since they can be both reliable (specific for plant and herbivore) and detectable. The degree of specificity clearly distinguishes induced indirect defense from induced direct defense. Specificity is likely to be vital to effective attraction of the right carnivore species and to the evolution of induced indirect defense, because the carnivores can learn to discriminate among different odor blends. Furthermore, the induced production of carnivore attractants occurs systemically. This results in an odor source that is larger than the infested plant tissue alone, which increases the chance of interception of the volatiles by downwind carnivores. The specificity and detectability of herbivore-induced plant volatiles are discussed in the context of evolution of indirect plant defense through plant-carnivore mutualism.

Introduction

Plants have evolved defenses that affect the behavior and physiology of herbivores such that herbivores avoid the plant, feed less from it, are intoxicated, or develop at a slower rate. As a consequence, plant defense affects the population dynamics of herbivores. In addition, herbivore population dynamics are influenced by carnivorous enemies of the herbivores. To understand how and why herbivore populations fluctuate, a wealth of studies has concentrated on the effects of resource availability, competition, parasitization, predation, and abiotic factors. Long discussions on the relative importance of bottom-up and top-down effects on herbivore population dynamics have not been resolved except for the hardly satisfying notion that the pattern is idiosyncratic (Hairston et al. 1960; Murdoch 1966; Bernays and Graham 1988; Thompson 1988; Jermy 1988; Rausher 1988; Price 1992; Schultz 1992). However, bottom-up and top-down effects are often dependent on each other. For instance, plant characteristics may affect herbivore development so that effects of the herbivore's predators are augmented because the herbivore remains longer in a vulnerable stage (Loader and Damman 1991). Or a predator may eliminate a competitive herbivore species to such an extent that the herbivore's resource availability is improved (Schoener 1993). Therefore, to understand the effect of plant traits on herbivore population dynamics and thus the contribution of the trait to plant defense, the effects on organisms at more than two trophic levels should be considered (Feeny 1976; Moran and Hamilton 1980; Price 1991, 1992; Sabelis et al. 1991; Schultz 1992; van Baalen and Sabelis 1993; Dicke 1995). In this chapter I will concentrate on the induction of plant defenses that affect herbivores indirectly, i.e., through interactions with carnivores.

Induced defense is a phenomenon that is highly influenced by information-conveying chemicals ("infochemicals," sensu Dicke and Sabelis 1988a). The induction process is dependent on information about the presence of an attacker, and the resultant defense may consist of chemical information that is used by the attacker as well as its natural enemies. Induced defense can lead to important variation in population fluctuations (Haukioja 1980). In this chapter I will focus on chemical information conveyance in induced defense in a multitrophic context. In doing so, I will review the current knowledge of this type of plant defense and discuss its evolution and its consequences for herbivore population dynamics.

Induced Defense: Two Options

Defense has traditionally been studied in a bitrophic setting. A predator or herbivore attacks, and the prey or plant defends itself through characteristics that prevent attack success or that minimize the effects of the attack. An additional defense option has been recognized in plant-arthropod interactions. Plants may promote the effectiveness of the carnivorous enemies of their own enemies and in this way reduce the effect of the herbivore (for reviews, see Price et al. 1980; Dicke and Sabelis 1988b). The bitrophic defense is referred to as "direct" or "intrinsic" defense and the defense that employs carnivores is called "indirect" or "extrinsic" defense (Price 1986; Dicke and Sabelis 1988b). The two defense options can be highly interrelated: a study of genetic variation of alkaloid concentration in *Senecio jacobea* revealed that a strong direct defense (high alkaloid concentration) was correlated with a weak indirect defense (predation by ants), due to the absence of aphids, which produce honeydew, on high-alkaloid plants (Vrieling et al. 1991).

In addition to being constitutively present, each of these two defense options may be induced, i.e., activated in response to enemy presence or attack: plants may induce direct defense after herbivores start feeding on the plant, resulting in the production of toxic or deterrent plant chemicals (Tallamy and Raupp 1991). Similarly, *indirect* defense of plants may also be induced by herbivory: feeding may result in the production of specific volatiles that attract predators of the herbivore (Dicke 1994).

In order to induce defense mechanisms, the attacked organism needs information about the attacker. This information may come from mechanical damage caused by the attacker, but also from chemical information. Plant defense can even be induced without mechanical damage: mere presence of herbivores on the plant (A. Shapiro and De Vay 1987; Blaakmeer et al. 1994b) or even volatiles from other plants can induce defensive changes (Farmer and Ryan 1990; Bruin et al. 1995). Furthermore, infochemicals can also be part of the actual defense: herbivore-damaged plants produce volatile infochemicals that attract carnivorous enemies of the herbivores (Dicke and Sabelis 1988b; Turlings et al. 1990; Vet and Dicke 1992).

Indirect defense of a plant against a herbivore involves the interaction of the herbivore's food and its enemies. Thus, indirect defense is a combination of a bottom-up and a top-down effect on herbivores. This exemplifies the interrelatedness of these two effects. Yet it remains interesting to investigate the relative importance of bottom-up and top-down effects. This relates to the question whether the plant plays an active role and decisively manipulates the enemies of the herbivore or whether the plant is more passively involved, with the herbivore's enemies taking advantage of infochemicals that are related to herbivore activities. This will be discussed below.

Induced Indirect Defense of Plants

The position of herbivores was once labeled to be "between the devil and the deep blue sea," represented by natural enemies and plants, respectively (Lawton and McNeill 1979). Miserable as this may be, the situation may get worse if the two collaborate, which occurs among plants and carnivorous arthropods. Carnivorous arthropods are faced with the difficult task of finding herbivores in complex vegetation. The first steps in the foraging process, i.e., host-habitat location and long-range host location, considerably limit the area to be searched. The information on which the carnivores can base their decisions seems to be limited. Herbivores are usually difficult to perceive from a distance, while information on the food plant of the herbivore, though more easy to detect, is usually not a reliable indicator of the herbivore's presence (see Vet and Dicke 1992 for discussion on the "reliability-detectability" problem of foraging carnivores). Carnivores can be attracted by volatiles from undamaged plants (e.g., Read et al. 1970), but volatiles from plants infested by their herbivorous host or prey are strongly preferred over volatiles from uninfested plants (Vet and Dicke 1992). The volatiles emitted from herbivore-damaged plants that attract carnivorous arthropods are produced by the plant in response to herbivore damage (Dicke and Sabelis 1988b; Dicke et al. 1990b; Turlings et al. 1990, 1993a). This plant response represents an induced *in*direct defense. The phenomenon of plant-carnivore mutualism through herbivore-induced carnivore attractants has now been studied for a substantial number of plant-herbivore-carnivore systems (table 4.1). In all tritrophic systems mentioned in table 4.1 herbivore damage leads to attraction of carnivorous enemies of the herbivores. Moreover, in all systems listed in table 4.1 for which the origin of the attractive infochemicals has been studied, the plant was identified as the producer. The plant response has been found in twelve plant families comprising monocotyledons and dicotyledons. The herbivores causing the response represent species in twelve families, and four insect orders as well as mites. The carnivores that react to the attractants include six insect families in the Hymenoptera, a family within the Hemiptera, and a family of mites. The information disseminated by the attacked plant appears to be highly detectable and often a reliable predictor of the presence and identity of the herbivore.

TABLE 4.1
Tritrophic Systems for Which Evidence for Carnivore Attraction by Infested Plants Has Been Demonstrated

Plant Species	Herbivore Species	Carnivore Species	Evidence for Herbivore-Induced Carnivore Attraction		
			Herbivory Leads to Carnivore Attraction	Behavioral Evidence for Plant Being Producer of Carnivore Attractants[a]	Chemical Evidence for Herbivore-Induced Plant Volatiles
Lima bean (*Phaseolus lunatus*) Fabaceae	Two-spotted spider mite (*Tetranychus urticae*) (Acari: Tetranychidae)	*Phytoseiulus persimilis* (Acari: Phytoseiidae)	Sabelis and van de Baan 1983; Dicke et al. 1990a	Sabelis et al. 1984a; Dicke et al. 1990a	Dicke et al. 1990a
Lima bean (*Phaseolus lunatus*) Fabaceae	Two-spotted spider mite (*Tetranychus urticae*) (Acari: Tetranychidae)	*Amblyseius andersoni* (=*A. potentillae*) (Acari: Phytoseiidae)	Sabelis and van de Baan 1983; Dicke and Groeneveld 1986	Dicke et al. 1990a	Dicke et al. 1990a
Lima bean (*Phaseolus lunatus*) Fabaceae	Two-spotted spider mite (*Tetranychus urticae*) (Acari: Tetranychidae)	*Amblyseius swirskii* (Acari: Phytoseiidae)	Dicke et al. 1989		
Lima bean (*Phaseolus lunatus*) Fabaceae	Western flower thrips (*Frankliniella occidentalis*)[b] (Thysanoptera: Thripidae)	*Amblyseius andersoni* (=*A. potentillae*) (Acari: Phytoseiidae)	Dicke and Groeneveld 1986		M. Dicke and M. A. Posthumus, unpubl. data
Lima bean (*Phaseolus lunatus*) Fabaceae	*Lyriomyza sativae* (Diptera: Agromyzidae)	*Opius dissitus* (Hymenoptera: Braconidae)	Petit et al. 1992		

Plant	Herbivore	Natural enemy			
Lima bean (*Phaseolus lunatus*) Fabaceae	*Tetranychus urticae* (Acari: Tetranychidae)	*Scolothrips takahashii* (Thysanoptera: Thripidae)	Shimoda et al. 1997	Shimoda et al. 1997	Dicke et al. 1990a
Bean (*Phaseolus vulgaris*) Fabaceae	*Tetranychus urticae* (Acari: Tetranychidae)	*Amblyseius andersoni* (Acari: Phytoseiidae)	Koveos et al. 1995	Koveos et al. 1995	
Bean (*Phaseolus vulgaris*) Fabaceae	*Liriomyza trifolii* (Diptera: Agromyzidae)	*Diglyphus isaea* (Hymenoptera: Eulophidae)	Finidori-Logli et al. 1996	Finidori-Logli et al. 1996	Finidori-Logli et al. 1996
Cucumber (*Cucumis sativus*) Cucurbitaceae	*Tetranychus urticae* (Acari: Tetranychidae)	*Phytoseiulus persimilis* (Acari: Phytoseiidae)	Dicke and Sabelis 1988b; Dicke et al. 1990b; Takabayashi et al. 1994a		Dicke et al. 1990b; Takabayashi et al. 1994a
Cucumber (*Cucumis sativus*) Cucurbitaceae	*Aphis gossypii* (Homoptera: Aphididae)	*Aphidius colemani* (Hymenoptera: Aphidiidae)	van Steenis 1995		
Tomato (*Lycopersicon esculentum*) Solanaceae	*Tetranychus urticae* (Acari: Tetranychidae)	*Phytoseiulus persimilis* (Acari: Phytoseiidae)	Dicke and Sabelis 1988b		Dicke et al. 1998
Tomato (*Lycopersicon esculentum*) Solanaceae	*Liriomyza bryoniae* (Diptera: Agromyzidae)	*Dacnusa sibirica* (Hymenoptera: Braconidae)	Dicke and Minkenberg 1991		
Rose (*Rosa X hybrida*) Rosaceae	*Tetranychus urticae* (Acari: Tetranychidae)	*Phytoseiulus persimilis* (Acari: Phytoseiidae)	Dicke and Sabelis 1988b		
Gerbera (*Gerbera jamesonii*) Asteraceae	*Tetranychus urticae* (Acari: Tetranychidae)	*Phytoseiulus persimilis* (Acari: Phytoseiidae)	Krips et al. 1996	Krips et al. 1996	O. E. Krips, P.E.L. Willems, and M. A. Posthumus, unpubl. data

TABLE 4.1 (*Continued*)

			Evidence for Herbivore-Induced Carnivore Attraction		
Plant Species	Herbivore Species	Carnivore Species	Herbivory Leads to Carnivore Attraction	Behavioral Evidence for Plant Being Producer of Carnivore Attractants[a]	Chemical Evidence for Herbivore-Induced Plant Volatiles
Pear (*Pyrus communis*) Rosaceae	*Tetranychus urticae* (Acari: Tetranychidae)	*Phytoseiulus persimilis* (Acari: Phytoseiidae)	Dicke and Sabelis 1988b		
Pear (*Pyrus communis*) Rosaceae	*Panonychus ulmi* (Acari: Tetranychidae)	*Amblyseius andersoni* (=*A. potentillae*) (Acari: Phytoseiidae)	Dicke and Groeneveld 1986		
Pear (*Pyrus communis*) Rosaceae	Pear psyllid (*Psylla pyricola*) (Homoptera: Psyllidae)	*Anthocoris* spp., *Orius* spp. (Hemiptera: Anthocoridae)	Drukker et al. 1995		
Ground ivy (*Glechoma hederacea*) Lamineae	*Tetranychus urticae* (Acari: Tetranychidae)	*Phytoseiulus persimilis* (Acari: Phytoseiidae)	Dicke and Sabelis 1988b		
Cassava (*Manihot esculenta*) Euphorbiaceae	Cassava green mite (*Mononychellus progresivus*) (Acari: Tetranychidae)	*Amblyseius anonymus*, *Amblyseius californicus*, *Amblyseius limonicus*, *Cydnodromella pilosa* (Acari: Phytoseiidae)	Janssen et al. 1990		
Cassava (*Manihot esculenta*) Euphorbiaceae	Cassava mealybug (*Phenacoccus manihoti*) (Homoptera: Pseudococcidae)	*Apoanagyrus lopezi* (=*Epidinocarsis lopezi*) (Hymenoptera: Encyrtidae)	Nadel and van Alphen 1987	Nadel and van Alphen 1987	

Plant	Herbivore	Natural enemy	References	
Apple (*Malus domestica*) Rosaceae	*Tetranychus urticae* (Acari: Tetranychidae)	*Phytoseiulus persimilis* (Acari: Phytoseiidae)	Sabelis and van de Baan 1983; Dicke and Sabelis 1988b	Takabayashi et al. 1991a
Apple (*Malus domestica*) Rosaceae	European red mite (*Panonychus ulmi*) (Acari: Tetranychidae)	*Typhlodromus pyri, Amblyseius andersoni* (=*A. potentillae*), *Amblyseius finlandicus* (Acari: Phytoseiidae)	Sabelis and van de Baan 1983; Dicke 1988; Dicke et al. 1986; Dicke et al. 1988	Takabayashi et al. 1991a
Apple (*Malus domestica*) Rosaceae	Apple rust mite (*Aculus schlechtendali*) (Acari: Eriophyidae)	*Typhlodromus pyri, Amblyseius andersoni* (=*A. potentillae*), *Amblyseius finlandicus* (Acari: Phytoseiidae)	Dicke 1988; Dicke et al. 1986, 1988	
Apple (*Malus domestica*) Rosaceae	*Phyllonorycter blancardella* (Lepidoptera: Gracillariidae)	*Apanteles* c.f. *circumscriptus* (Hymenoptera: Braconidae)	Lengwiler et al. 1994	
Corn (*Zea mays*) Poaceae	Beet army worm (*Spodoptera exigua*) (Lepidoptera: Noctuidae)	*Cotesia marginiventris* (Hymenoptera: Braconidae)	Turlings et al. 1990, 1991a	Turlings et al. 1990, 1991b
Corn (*Zea mays*) Poaceae	Common army worm (*Pseudaletia separata*) (Lepidoptera: Noctuidae)	*Cotesia kariyai* (Hymenoptera: Braconidae)	Takabayashi et al. 1995	Takabayashi et al. 1995
Corn (*Zea mays*) Poaceae	Corn stem borer *Chilo partellus* (Lepidoptera: Pyralidae)	*Cotesia flavipes* (Hymenoptera: Braconidae)	Potting et al. 1995; Ngi-Song et al. 1996	Potting et al. 1995

TABLE 4.1 (*Continued*)

			Evidence for Herbivore-Induced Carnivore Attraction		
Plant Species	Herbivore Species	Carnivore Species	Herbivory Leads to Carnivore Attraction	Behavioral Evidence for Plant Being Producer of Carnivore Attractants[a]	Chemical Evidence for Herbivore-Induced Plant Volatiles
Corn (*Zea mays*) Poaceae	*Chilo partellus* *C. orichalcociliellus* (Lepidoptera: Pyralidae) *Sesamia calamistis* (Lepidoptera: Noctuidae)	*Cotesia flavipes* *Cotesia sesamiae* (Hymenoptera: Braconidae)	Ngi-Song et al. 1996		
Wheat (*Triticum aestivum*) Poaceae	Greenbug (*Schizaphis graminum*) (Homoptera: Aphididae)	*Lysiphlebus testaceipes* (Hymenoptera: Aphidiidae)	Grasswitz and Paine 1993		
Cotton (*Gossypium hirsutum*) Malvaceae	Beet army worm (*Spodoptera exigua*) (Lepidoptera: Noctuidae)	*Cotesia marginiventris* (Hymenoptera: Braconidae)	McCall et al. 1994		Turlings et al. 1993a; Loughrin et al. 1994, 1995a
Cotton (*Gossypium hirsutum*) Malvaceae	Corn earworm (*Helicoverpa zea*) (Lepidoptera: Noctuidae)	*Microplitis croceipes* (Hymenoptera: Braconidae)	McCall et al. 1993	McCall et al. 1993	McCall et al. 1994
Cotton (*Gossypium hirsutum*) Malvaceae	*Heliothis virescens* (Lepidoptera: Noctuidae)	*Campoletis sonorensis* (Hymenoptera: Ichneumonidae)	McAuslane et al. 1991a	McAuslane et al. 1991b	
Cotton (*Gossypium hirsutum*) Malvaceae	Two-spotted spider mite (*Tetranychus urticae*) (Acari: Tetranychidae)	*Phytoseiulus persimilis* (Acari: Phytoseiidae)	Bruin et al. 1992		

Plant	Herbivore	Parasitoid	References		
Cowpea (*Vigna unguiculata*) Fabaceae	Corn earworm (*Helicoverpa zea*) (Lepidoptera: Noctuidae)	*Microplitis croceipes* (Hymenoptera: Braconidae)	McCall et al. 1993	McCall et al. 1993; Whitman and Eller 1990	
Cabbage (*Brassica oleracea*), various cultivars Brassicaceae	Large cabbage white (*Pieris brassicae*) (Lepidoptera: Pieridae)	*Cotesia glomerata* (Hymenoptera: Braconidae)	Steinberg et al. 1992; Geervliet et al. 1996	Steinberg et al. 1993	Blaakmeer et al. 1994a; Mattiacci et al. 1994
Cabbage (*Brassica oleracea*), various cultivars Brassicaceae	Small cabbage white (*Pieris rapae*) (Lepidoptera: Pieridae)	*Cotesia glomerata* (Hymenoptera: Braconidae)	Geervliet et al. 1996		
Cabbage (*Brassica oleracea*) Brassicaceae	Small veined white (*Pieris napi*) (Lepidoptera: Pieridae)	*Cotesia glomerata* (Hymenoptera: Braconidae)	Geervliet et al. 1996		
Hawthorn (*Crataegus* sp.) Rosaceae	Large veined white (*Aporia crataegi*) (Lepidoptera: Pieridae)	*Cotesia glomerata* (Hymenoptera: Braconidae)	Geervliet et al. 1996		
Cabbage (*Brassica oleracea*) Brassicaceae	*Plutella xylostella* (Lepidoptera: Plutellidae)	*Cotesia glomerata* (Hymenoptera: Braconidae)	Geervliet et al. 1996		
Cabbage (*Brassica oleracea*) Brassicaceae	*Plutella xylostella* (Lepidoptera: Plutellidae)	*Cotesia plutellae* (Hymenoptera: Braconidae)	Bogahawatte and van Emden 1996		
Cabbage (*Brassica oleracea*), various cultivars Brassicaceae	Large cabbage white (*Pieris brassicae*) (Lepidoptera: Pieridae)	*Cotesia rubecula* (Hymenoptera: Braconidae)	Geervliet et al. 1996		
Cabbage (*Brassica oleracea*), various cultivars Brassicaceae	Small cabbage white (*Pieris rapae*) (Lepidoptera: Pieridae)	*Cotesia rubecula* (Hymenoptera: Braconidae)	Geervliet et al. 1994, 1996; Agelopoulos and Keller 1994a	Geervliet et al. 1994; Agelopoulos and Keller 1994a	Blaakmeer et al. 1994a; Agelopoulos and Keller 1994b

TABLE 4.1 (*Continued*)

			Evidence for Herbivore-Induced Carnivore Attraction		
Plant Species	Herbivore Species	Carnivore Species	Herbivory Leads to Carnivore Attraction	Behavioral Evidence for Plant Being Producer of Carnivore Attractants[a]	Chemical Evidence for Herbivore-Induced Plant Volatiles
Cabbage (*Brassica oleracea*), various cultivars Brassicaceae	Small veined white (*Pieris napi*) (Lepidoptera: Pieridae)	*Cotesia rubecula* (Hymenoptera: Braconidae)	Geervliet et al. 1996		
Cabbage (*Brassica oleracea*) Brassicaceae	*Plutella xylostella* (Lepidoptera: Plutellidae)	*Cotesia rubecula* (Hymenoptera: Braconidae)	Agelopoulos and Keller 1994a; Geervliet et al. 1996	Agelopoulos and Keller 1994a	Agelopoulos and Keller 1994b
Cabbage (*Brassica oleracea*) Brassicaceae	*Mamestra brassicae* (Lepidoptera: Noctuidae)	*Cotesia glomerata, Cotesia rubecula* (Hymenoptera: Braconidae)	Geervliet et al. 1996		
Yellow cress (*Rorippa indica*) Brassicaceae	*Pieris rapae crucivora* (Lepidoptera: Pieridae)	*Cotesia glomerata* (= *Apanteles glomeratus*) (Hymenoptera: Braconidae)	Sato 1979	Sato 1979	
Potato (*Solanum tuberosum*) Solanaceae	Potato moth (*Phthorimaea operculella*) (Lepidoptera: Gelechiidae)	*Orgilus lepidus* (Hymenoptera: Braconidae)	Keller and Horne 1993	Keller and Horne 1993	
Potato (*Solanum tuberosum*) Solanaceae	*Heliothis virescens* (Lepidoptera: Noctuidae)	*Campoletis sonorensis* (Hymenoptera: Ichneumonidae)	McAuslane et al. 1991a		

Plant	Herbivore	Parasitoid	Reference
Sesame (*Sesamum indicum*) Scrophulariaceae	*Heliothis virescens* (Lepidoptera: Noctuidae)	*Campoletis sonorensis* (Hymenoptera: Ichneumonidae)	McAuslane et al. 1991a
Lettuce (*Lactuca sativa*) Lactucaceae and Chrysanthemum (*Chrysanthemum* sp.) Asteraceae	*Chromatomyia syngenesiae* (Diptera: Agromyzidae)	*Diglyphus isaea* (Hymenoptera: Eulophidae)	Cheah and Coaker 1992
Nasturtium (*Tropaeolum majus*) Tropaeolaceae	*Pieris brassicae, P. rapae, P. napi* (Lepidoptera: Pieridae)	*Cotesia glomerata, Cotesia rubecula* (Hymenoptera: Braconidae)	Geervliet et al. 1996
Broad bean (*Vicia faba*) Fabaceae	Pea aphid (*Acyrthosiphon pisum*) (Homoptera: Aphididae)	*Aphidius ervi* (Hymenoptera: Aphidiidae)	Guerreri et al. 1993; Du et al. 1996
			Du et al. 1996

[a] Evidence comprises proof that mechanical damage and herbivory elicit different behavioral responses or that a plant from which herbivores have been removed remains attractive while the removed herbivores are not attractive.
[b] This thrips species was erroneously identified and called *Frankliniella pallida* in Dicke and Groeneveld 1986 (W. Mantel, pers. comm.).

Detectability of Herbivore-Induced Plant Volatiles

Herbivory results in cell damage and subsequently in the emission of increased amounts of alcohols, aldehydes, and esters derived from the lipoxygenase pathway and, with time, specific terpenoids may be produced that are not produced in response to mechanical damage (Dicke et al. 1990b; Dicke 1994; Turlings et al. 1993a, 1995; Takabayashi and Dicke 1996). These plant volatiles are highly attractive to carnivorous arthropods (Dicke 1994; Turlings et al. 1995; Takabayashi and Dicke 1996).

Plants represent a much larger biomass than herbivores and thus, if herbivores and plants emit volatiles at a similar rate per unit biomass, plants would be better-detectable odor sources than herbivores. Indeed, carnivores are readily attracted to herbivore-damaged plants from which the herbivores have been removed but not—or to a limited degree—to the herbivores themselves (e.g., Sabelis et al. 1984a; Turlings et al. 1991a; Steinberg et al. 1993; Geervliet et al. 1994). For instance, the parasitoid *Cotesia glomerata* is strongly attracted to plants infested by *Pieris brassicae* caterpillars, even after removal of the caterpillars (fig. 4.1). However, the caterpillars themselves or their feces are only marginally attractive: ca. 50% of the parasitoids do not fly upwind when herbivores or their feces are offered upwind, while more than 90% of the parasitoids fly upwind when (previously) infested plants are offered (fig. 4.1). An infested plant from which the caterpillars have been removed is much more attractive than the removed caterpillars or their feces (fig. 4.1).

Predators and parasitic wasps can discriminate well between volatiles from mechanically damaged plants and herbivore-infested plants. Mechanical damage and herbivore damage evoke a different response in the plant in terms of the composition of the volatile blend induced and/or in terms of volatile quantity. The differential plant response is mediated by herbivore secretions. Application of an herbivore's oral secretion into a mechanical wound results in a plant response that is similar to that in reaction to herbivore damage (Turlings et al. 1990, 1993b; Mattiacci et al. 1994). For caterpillars of *P. brassicae*, this effect could be ascribed to an enzyme in the oral secretion, β-glucosidase; the parasitoid *C. glomerata* is equally attracted to a plant treated with oral secretion and a plant treated with β-glucosidase (Mattiacci et al. 1995).

After herbivores have damaged one leaf of a plant, not only this leaf becomes attractive to carnivores: other leaves from the same plant also emit novel odors (Turlings and Tumlinson 1992) and attract carnivores (Dicke et al. 1990b; Turlings and Tumlinson 1992; Potting et al. 1995). Thus, the plant response is systemic and the odor source is not restricted to the infested plant parts. It is much larger than the herbivores that inflict the damage (fig.

INDUCED INDIRECT DEFENSES

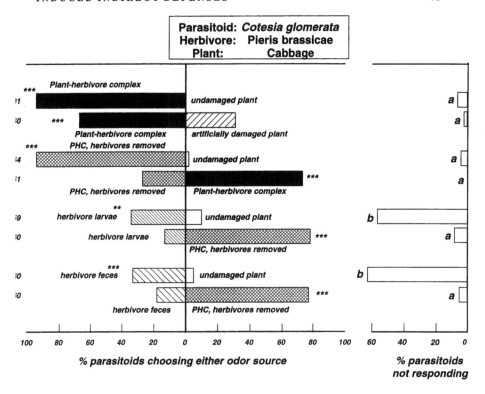

Fig. 4.1. Response of the parasitoid *Cotesia glomerata* to volatiles from different components of the plant-herbivore complex cabbage *Pieris brassicae* in a two-choice flight bioassay. Bars indicating "% parasitoids not responding" show the parasitoids that did not fly toward one of the two offered odor sources. A high "% parasitoids not responding" indicates a low degree of attractiveness of the odor sources offered. Bars of "% parasitoids not responding" of a pair, which are labeled with different letters, are significantly different (contingency table, $\alpha = 0.05$).
** = $0.001 \leq p < 0.01$; *** = $p < 0.001$ (chi-square test; H_0 = parasitoid's distribution over odor sources is 50:50). n = total number of wasps tested in each experiment (responding plus nonresponding individuals). (After data from Steinberg et al. 1993.)

4.2). Induction of the carnivore attractants in the undamaged leaves is achieved by an elicitor that is transported from the damaged to the undamaged leaf (Takabayashi et al. 1991b; Dicke et al. 1993). The odor source may even comprise neighboring plants (fig. 4.2). Exposure of undamaged cotton or bean plants to volatiles from spider mite infested plants makes the undamaged plants attractive to predatory mites (Dicke et al. 1990b, Bruin et al. 1992). It is not clear yet whether this is due to a passive scenario where the exposed plant adsorbs the volatiles from the infested neighbor, or to an active scenario in which production of carnivore attractants is induced in the exposed plant by volatiles from its neighbor. The latter is an important op-

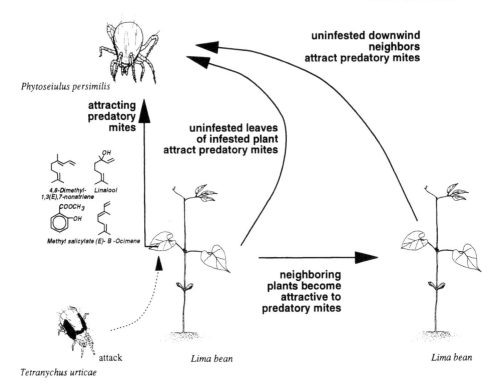

Fig. 4.2. Interactions in a tritrophic system consisting of lima bean plants *Phaseolus lunatus*, two-spotted spider mites *Tetranychus urticae*, and predatory mites *Phytoseiulus persimilis*, mediated by herbivore-induced plant volatiles. (After data from Dicke 1986; Dicke et al. 1990a,b, 1993; Bruin et al. 1992.)

tion, as can be deduced from physiological and biochemical investigations (see Bruin et al. 1995 for review).

Different Types of Plant Response

Plants seem to have one of two different types of carnivore attractants:

1. *Specific volatiles.* Herbivory may lead to the emission of relatively large amounts of plant volatiles that are not emitted—or only in trace amounts—by mechanically damaged plants or undamaged plants (Dicke et al. 1990a; Turlings et al. 1990) (fig. 4.3a). The composition of the blend emitted by herbivore-infested plants may depend on the herbivore species that infests the plant (Takabayashi et al. 1991a; Turlings et al. 1993a). This type of response has been recorded in various plant species, such as lima bean, cu-

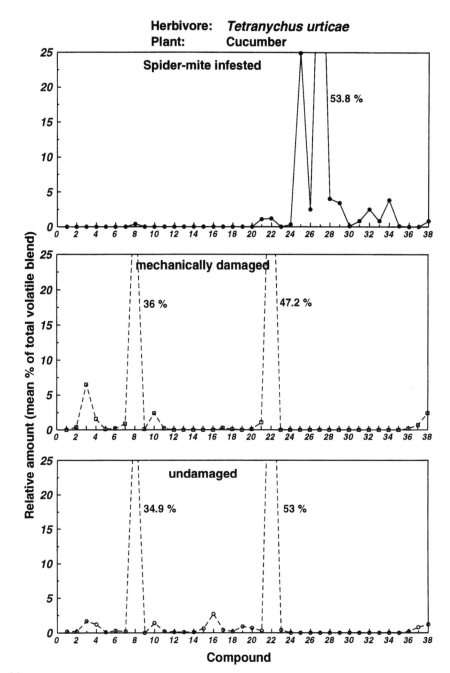

(a)

Herbivore: *Pieris brassicae*
Plant: Cabbage

caterpillar-infested
total peak area: 50,100

mechanically damaged
total peak area: 15,000
46.2 %

undamaged
total peak area: 5,700
39.5 %

relative amount (mean % of total volatile blend)

Compound

(b)

Fig. 4.3.(a) Headspace analyses of leaves from cucumber plants (*Cucumis sativus* cv. Santo F1). *Top panel:* Young leaves infested with spider mites; *middle panel:* Leaves mechanically damaged with a wet cotton wool pad with carborundum powder; *lower panel:* Undamaged leaves. (Data from Takabayashi et al. 1994b.) 1 = pentanal; 2 = 2-pentenal; 3 = 2-hexenal; 4 = hexanal; 5 = heptanal; 6 = decanal; 7 = nonanal; 8 = (Z)-3-hexen-1-ol; 9 = (E)-3-hexen-1-ol; 10 = 1-hexanol; 11 = 2-ethyl-1-hexanol; 12 = 1-butanol; 13 = 1-penten-3-ol; 14 = 1-pentanol; 15 = benzyl alcohol; 16 = 3-octanone; 17 = 1-penten-3-one; 18 = 6-methyl-5-heptene-2-one; 19 = 2-nonanone; 20 = butyl acetate; 21 = hexyl acetate; 22 = (Z)-3-hexen-1-yl acetate; 23 = limonene; 24 = (Z)-β-ocimene; 25 = (E)-β-ocimene; 26 = (3Z)-4,8-dimethyl-1,3,7-nonatriene; 27 = (3E)-4,8-dimethyl-1,3,7-nonatriene; 28 = linalool; 29 = (E,E)-α-farnesene; 30 = (3E,7E)-4,8,12-trimethyl-1,3,7,11-tridecatetraene; 31 = 2-methylpropanenitrile; 32 = 2-methylbutanenitrile; 33 = 3-methylbutanenitrile; 34 = 2-methylbutanal *O*-methyloxime; 35 = 3-methylbutanal *O*-methyloxime; 35 = unknown oxime; 36 = 2-ethylfuran; 37 = caproic acid; 38 = unidentified minor peaks. (b) Headspace analyses of leaves of Brussels sprouts plants (*Brassica oleracea* L. var. *gemmifera*, cv. Titurel). *Top panel:* 100 L1 larvae of the large cabbage white (*Pieris brassicae* L.) had been feeding per leaf for 24 h; *middle panel:* Leaves had been rubbed with a wet cotton wool pad with carborundum powder 20 h before volatile collection; *bottom panel:* Undamaged leaves. (Data obtained from Mattiacci et al. 1994.) 1 = 2-butenal; 2 = (E)-2-pentenal; 3 = (Z)-2-pentenal; 4 = hexanal; 5 = (E)-2-hexenal; 6 = (Z)-2-hexenal; 7 = 2,4-hexadienal; 8 = octanal; 9 = nonanal; 10 = decanal; 11 = dodecanal; 12 = tetradecanal; 13 = 1-cyclopropyl-2-propen-1-one; 14 = 3-pentanone; 15 = 1-penten-3-one; 16 = 3-penten-2-one; 17 = ethanol; 18 = 2-methyl-1-propanol; 19 = 1-butanol; 20 = 3-methyl-1-butanol; 21 = pentanol; 22 = 3-pentanol; 23 = 1-penten-3-ol; 24 = 1-hexanol; 25 = (E)-2-hexen-1-ol; 26 = (Z)-3-hexen-1-ol; 27 = 1-butyl acetate; 28 = 3-methyl-3-buten-1-yl acetate; 29 = iso-pentyl acetate; 30 = 1-pentyl acetate; 31 = 1-hexyl acetate; 32 = (E)-2-hexen-1-yl acetate; 33 = (Z)-3-hexen-1-yl acetate; 34 = 3-hexen-1-yl propanoate; 35 = 3-hexen-1-yl-3-methylbutanoate; 36 = (Z)-3-hexen-1-yl butyrate; 37 = (Z)-3-hexen-1-yl isobutyrate; 38 = (Z)-3-hexen-1-yl isovalerate; 39 = 3-hexen-1-yl caproate; 40 = (Z)-3-hexen-1-yl hexanoate; 41 = isobutyric acid; 42 = caproic acid; 43 = isovaleric acid; 44 = α-pinene; 45 = β-pinene; 46 = α-thujene; 47 = sabinene; 48 = myrcene; 49 = limonene; 50 = 1,8-cineole; 51 = β-elemene; 52 = *E*-sabinenehydrate; 53 = α-farnesene; 54 = 2-ethyl-furan; 55 = dimethyldisulphide; 56 = dimethyltrisulphide; 57 = methylisothiocyanate.

cumber, and corn (for reviews, see Dicke 1994; Takabayashi et al. 1994b; Turlings et al. 1995).

2. *Nonspecific volatiles*. Alternatively, herbivore infestation may lead to the emission of a volatile blend that is qualitatively similar to that emitted by mechanically damaged or undamaged plants. However, the amount released from herbivore-infested plants is larger than from mechanically damaged or undamaged plants (fig. 4.3b). This category is represented by cabbage, potato, and tomato plants (Blaakmeer et al. 1994a; Mattiacci et

al. 1994; Agelopoulos and Keller 1994b; Bolter et al. 1997; Dicke et al. 1998).

Although these types of plant response differ in the mechanism employed, the main effect on carnivores is similar: herbivory leads to carnivore attraction and carnivores can discriminate between mechanically damaged and herbivore-infested plants.

Variation in the Production of Herbivore-Induced Carnivore Attractants

Plant genotypes vary in their response to herbivory. This may be reflected in the amount of volatiles emitted in response to herbivory (Loughrin et al. 1995a) and/or in the relative contribution of each compound to the total blend (Takabayashi et al. 1991a). Differential attraction of carnivores can be the result of such variation in volatile emission (Dicke et al. 1990b). These differences provide the basis on which natural selection can work. Future studies should investigate how and to what degree natural selection can and does modify the production of carnivore attractants. This may be done through manipulative studies with artificial selection.

Herbivore-induced carnivore attractants may also vary within a plant individual. Young leaves of cucumber plants become attractive to predatory mites after being infested by spider mites, whereas old leaves do not: chemical differences that may explain this have been recorded (Takabayashi et al. 1994a). This may be a mechanism of plants to direct carnivores to their growing points, which are important in keeping up in competition for light with neighbors (Edwards et al. 1992). Also, different plant tissues such as flowers or leaves that are infested by the same herbivore species differ in the chemical blend emitted (Turlings et al. 1993a).

The response of a plant can vary with the herbivore species that infests the plant. Carnivorous mites discriminate between plants infested by different herbivorous mites (Sabelis and van de Baan 1983; Sabelis and Dicke 1985; Dicke 1988) and their odor preferences are correlated with prey preference (Sabelis and Dicke 1985; Dicke et al. 1988). Chemical differences in blends among conspecific plants infested by different herbivore species appear to be quantitative rather than qualitative: the blends contain the same chemicals but in different ratios (Takabayashi et al. 1991a; Geervliet et al. 1997).

Even different herbivore instars may result in different plant responses. If corn plants are infested by young caterpillars (*Pseudaletia separata*), the plants produce several novel compounds and they attract parasitic wasps (*Cotesia kariyai*). However, if corn plants are infested by old caterpillars, the plants emit the same chemicals as when mechanically damaged; in this case they produce only a few novel compounds. The same difference was observed when oral secretions of young and old caterpillars were applied on mechanical damage (Takabayashi et al. 1995). Both old and young caterpillars can be parasitized by the wasps. However, the plant benefits greatly when young caterpillars are parasitized, but not when old caterpillars are parasitized. Food consumption by parasitized hosts from 3rd larval instar until egression of the parasitoids is 33% (parasitized in 3rd larval instar) and 95% (parasitized in 6th larval instar) of food consumption by unparasitized hosts from 3rd larval instar until pupation (Tanaka et al. 1992; Takabayashi et al. 1998). Thus, infested corn plants seem to attract parasitoids when it benefits the plant. This is an important observation that supports the hypothesis that induced indirect plant defense is an active plant response. Further support for this hypothesis comes from the observation that the emission of carnivore attractants can fluctuate throughout the day. Cotton plants infested by caterpillars displayed a clear diurnal rhythm, with high emission rates during the day, when parasitoids are active, and low emission rates during the night (Loughrin et al. 1994).

Most pronounced differences in volatile production are recorded when different plant species are infested by the same herbivore species. Many qualitative differences are found in addition to quantitative differences (Dicke et al. 1990b; Takabayashi et al. 1991a; Turlings et al. 1993a; Geervliet et al. 1997). Carnivores use this variation in information during foraging for herbivores and can discriminate among plant species infested by the same herbivore species (Vet and Dicke 1992; Turlings et al. 1993a; Geervliet et al. 1998).

In addition to this variation in induced volatiles that result from biotic interactions, herbivore-induced carnivore attractants are likely to be influenced by abiotic conditions such as light intensity, day length and water stress (Takabayashi et al. 1994b).

Variation in Behavior of Carnivores

Carnivores are confronted with what may seem to be a daunting degree of variation in herbivore-induced carnivore attractants. How do carnivores cope with all this variation? An important solution that carnivores have is temporal specialization through learning (Vet and Dicke 1992; Turlings et al. 1993a; Vet et al. 1995; Vet 1996). Learning of plant stimuli is common among insect parasitoids (Vet et al. 1990; Vet and Groenewold 1990; Vet and Dicke 1992; Turlings et al. 1993a; Vet et al. 1995) and has also been recorded for arthropod predators (Dicke et al. 1990b; Takabayashi et al. 1994b).

Whether or not learning is an adaptive solution is dependent on the degree of variation in information encountered by the carnivore, which is dependent on the degree of specialization, both with respect to the herbivore and with respect to the plant level. The more specialized the carnivore is, the less it is expected to (have to) learn. In specialists, strong innate responses are expected. In carnivores that are generalistic at the plant level, a strong effect of learning on responses to herbivore-induced plant volatiles is expected and observed (Vet and Dicke 1992; Turlings et al. 1993a; Geervliet et al. 1997).

Experiences with hosts/prey or their products may lead to discrimination between volatile blends, even when the differences are very subtle (Turlings et al. 1993a; Geervliet et al. 1998). Consequently, learning can affect foraging efficiency (Papaj and Vet 1990), which is interesting in terms of carnivore optimality on the one hand and its population dynamic consequences on the other (Vet 1996).

Costs of Induced Indirect Defense of Plants

The production of plant chemicals has costs in terms of ATP, related to biosynthesis, maintenance of the biosynthetic pathway, storage, etc. Volatiles have a higher cost than non-volatiles because volatiles are disseminated and have to be renewed constantly. However, a first estimate of biosynthetic costs of herbivore-induced volatile production in lima bean yielded a very low value of 0.001% of leaf production costs per day (Dicke and Sabelis 1989). Although not all costs were included, it seems that costs related to herbivore-induced volatiles are not energetic in nature, but that ecological costs are much more important (see Dicke and Sabelis 1989 for discussion). Once a plant produces an increased amount of volatiles it stands out among its neighbors. Herbivores are well known to employ plant volatiles, to locate their food plant, and thus the increased volatile production may attract herbivores, especially those that are not affected by the attracted carnivores. Indeed, attraction of herbivores to volatiles from infested plants has been recorded for several beetle species (Harari et al. 1994; Loughrin et al. 1995b; Bolter et al. 1997), but in other studies herbivores avoided infested plants (Bernstein 1984; Dicke 1986; Turlings and Tumlinson 1991).

Moreover, herbivore-induced carnivore attractants may also have costs for plants in competitive interactions with their neighbors. Neighboring plants that do not pay the costs of producing the volatiles but gain the benefits of the attracted carnivores improve their fitness at the expense of the odor-producing plant (Sabelis and de Jong 1988; Tuomi et al. 1994; Augner 1994; Godfray 1995). If the volatiles provide too large a benefit for neighbors that do not produce volatiles, nonproducing plants can invade and outcompete volatile-producing plants or polymorphic populations may become established. A wide range of conditions for polymorphic populations is predicted from a model study (Sabelis and de Jong 1988).

Evolution of Induced Indirect Defenses

Research on induced indirect defense has been mostly restricted to mechanistic studies demonstrating how this interaction between plants and carnivorous arthropods operates. Considerable effort has been expended to show unambiguously that the carnivore attractants are plant products and that plants are actively involved in their production (Dicke 1994; Turlings et al. 1995; Takabayashi et al. 1995). This was essential in evaluating which organisms are under natural selection so as to understand how natural selection may mold plant-herbivore-carnivore interactions.

Some initial data are known on variation among genotypes at both the level of infochemical production (Takabayashi et al. 1991a; Loughrin et al. 1995a) and at the level of carnivore response (Prevost and Lewis 1990; see also Papaj 1994). But this is just the beginning. Information is needed on the level of variation as well as on heritabilities. In addition, manipulative studies aiming at artificial selection of the production of and response to herbivore-induced carnivore attractants will have to be performed.

Evolution of Plant and Carnivore Response

An important question is how communication between plants and carnivores evolved. It seems obvious that once the volatiles are there, the carnivores will use them. However, it is less obvious whether the plant has evolved to emit the information.

Induced attraction of carnivores has clear benefits to the plant. In addition, herbivore-induced plant volatiles can have other benefits. For instance, the fatty-acid derived volatiles are toxic to phytopathogenic fungi (Zeringue and McCormick 1989; Croft et al. 1993) or to herbivores (Hildebrand et al. 1993). This has raised the question of whether herbivore-induced plant volatiles are a mechanism of direct defense that has been exploited by carnivores to facilitate herbivore location (Turlings and Tumlinson 1991). Although it is difficult to reconstruct the original function of herbivore-induced plant volatiles, there are good arguments to support the hypothesis that even if the first function of the volatiles was to intoxicate herbivores or pathogens, carnivore responses have molded the plant response, with the result that plant responses can be specific for different herbivore species or herbivore instars that damage the plant. After all, carnivores can have severe effects on herbivore population dynamics (Cappuccino and Price 1995). Carnivores are hampered in finding herbivores by low detectability of cues from herbivores and low reliability of cues from uninfested plants (Vet and Dicke 1992). Therefore, plant characteristics that improve the location of herbivores by carnivores can initialize carnivore-herbivore interactions that lead to local extermination of herbivores (Vet and Dicke 1992; Dicke 1995). Aspects of herbivore-induced carnivore attractants that are considered important in the evolution of induced indirect defense are discussed below.

Information Specificity and Evolution of Plant Response

The most passive scenario of plant involvement is that plants emit the same information in response to herbivory as to mechanical damage. This scenario may be true to some extent for plant species such as cabbage, tomato, and potato (e.g., Blaakmeer et al. 1994a; Mattiacci et al. 1994; Geervliet et al. 1997b; Bolter et al. 1997; Dicke et al. 1998; fig. 4.3B), although carnivores can discriminate between mechanically damaged plants and herbivore-infested plants on the basis of quantitative differences in the blend emitted (Steinberg et al. 1993; Geervliet et al. 1994; fig. 4.1). The passive scenario is definitely not true for many other plant species that produce new volatiles in response to herbivory (e.g., Dicke et al. 1990a,b; Dicke 1994; Turlings et al. 1990, 1995; fig. 4.3A). In one of the latter systems, a clear difference between a plant's response to herbivores and its response to pathogens has been observed. Bean plants (*Phaseolus vulgaris*) respond to pathogen (*Pseudomonas syringae pv phaseolica*) infestation with the emission of nonspecific volatiles, i.e., fatty-acid derivatives (Croft et al. 1993), while in response to spider-mite infestation the same bean cultivar produces specific volatiles, i.e., homoterpenes and methyl salicylate, which are known to attract predatory mites (Dicke and Posthumus, unpubl. data). Thus, herbivory results in the emission by bean plants of a different odor bouquet than mechanical damage or pathogen infestation. The observation that carnivores can discriminate among plants infested by different herbivore species (Sabelis and van de Baan 1983; Sabelis and Dicke 1985; Turlings et al. 1993a; Geervliet et al. 1998) and even different herbivore stages (Takabayashi et al. 1995) indicates that the specificity can be even more subtle. The specificity of the disseminated information is a remarkable difference between induced indirect defense and induced direct defense (Tallamy and Raupp 1991). In this context it is striking that the parasitoid *Cotesia kariyai* is not attracted to volatiles from corn plants infested by old *P. separata* caterpillars that are suitable hosts; corn plants infested by old *P. separata* caterpillars emit a nonspecific mixture of volatiles that resembles the mixture emitted from mechanically damaged corn plants (Takabayashi et al. 1995). Specificity may be a vital characteristic of carnivore attraction because the animals use the emitted information in their foraging decisions and may learn to discriminate between different odor blends (Vet and Dicke 1992; Turlings et al. 1993a; Vet et al. 1995). Various predator species and most parasitoid species are specialists; they benefit from specific information and it pays plants to emit such information if it ensures that effective natural enemies are attracted. The attraction of ineffective natural enemies obviously has no selective advantage and may even be disadvantageous if they outcompete effective natural enemies (Bruin et al. 1995; Dicke and Vet 1998). If the information is nonspecific, the information value is qualitatively similar to general volatiles emitted by uninfested plants; there is only a quantitative difference. Any mutant that provides more specific information than conspecifics that disseminate general information will allow the predators to use this more reliable information and thus imposes selection on conspecific plants to increase the specificity as well. It is important to realize that predation pressure is severe in those systems where the effects of herbivores and carnivores on the populations of their respective food items are strong. This is the case for systems consisting of plants, spider mites, and predatory mites. It is exactly in such a system that a clear case of specificity has been reported: predators are attracted when a suitable prey infests the plant, while they are not attracted when an unsuitable prey infests the plant (Sabelis and van de Baan 1983; Sabelis and Dicke 1985).

Dimension of Odor Source and Evolution of Plant Response

Herbivore-induced carnivore attractants are produced systemically (Dicke et al. 1990b; Takabayashi et al. 1991b; Turlings and Tumlinson 1992). Stem borers that feed in the stem of corn plants induce the emission of parasitoid-attracting volatiles from the foliage (Potting et al. 1995). This plant response cannot be explained in terms of a direct defense, because the caterpillars that feed in the stem will not feed from the leaves. For parasitoids that are faced with the reliability-detectability problem (Vet and Dicke 1992; see above), the systemic emission of herbivore-induced plant volatiles leads to a larger odor source than when the odor would be emitted only from the damaged stem. Thus, by producing induced volatiles systemically, the plant increases the chance that carnivores reach the infested plant. The concentration gradient of the volatiles as well as other sources of information may be used by the carnivores to locate the infested sites on the plant. Thus, the systemic effect can be understood in terms of an interaction with carnivores. This is not true for a directly defensive interaction with the herbivore: the systemically emitted volatiles make the plant potentially better detectable for herbivores (Harari et al. 1994; Bolter et al. 1997).

Evolutionary Aspects of Interaction of Direct and Indirect Defense

Herbivores excel in developing resistance to plant defenses (Kim and McPheron 1993). The stronger the effect of the plant characteristic on herbivore fitness, the stronger the selection. It is remarkable that induced direct defense has only small effects on herbivore fitness, which relates not only to herbivore specialists but also to generalists (for review, see Tallamy and Raupp 1991). For instance, herbivore-inducible proteinase inhibitors only reduce digestion activity in the generalist herbivore *Spodoptera exigua* for a short while, because the caterpillars respond by producing different proteinases that are not inhibited (Jongsma et al. 1995). The small effect on herbivore fitness has been used as an argument against an important role of induced plant defense in herbivore population dynamics (Fowler and Lawton 1985). However, plant effects on herbivore developmental rate may interact with natural enemies by augmenting their effect (Price et al. 1980; Loader and Damman 1991). The interaction of partial host plant resistance and carnivore effectiveness leads to increased rates of herbivore adaptation to plant resistance factors, albeit at a slower rate than to strong host plant resistance (Gould et al. 1991). Thus, an interaction between small effects of induced direct defense with larger effects of induced indirect defense may have evolutionary advantages to plants.

Future Needs in Studies on Evolution of Induced Indirect Plant Defense

Most studies on induced indirect plant defense have dealt with agricultural plant species (table 4.1). Because plant breeding practices have never aimed at increasing induced indirect plant defense, it is likely that wild relatives of agricultural plants have a higher level of induced indirect defense than cultivated genotypes. In the only study so far addressing this issue, larger amounts of herbivore-induced carnivore attractants were indeed recorded for wild cotton plants compared to cultivated genotypes (Loughrin et al. 1995a). To increase our understanding of the evolution of induced indirect defense, studies of wild species are highly needed.

Effect of Induced Indirect Defenses on Population Dynamics

Although research on induced indirect defense has been mostly restricted to mechanistic studies on how this defense is induced and to what extent it varies, this research is profoundly anchored in population dynamics. Spider mites and predatory mites have been model organisms in population-dynamic studies as well as in studies on induced indirect defense, which is

INDUCED INDIRECT DEFENSES 87

not coincidental. The studies by Huffaker and colleagues on interactions between spider mites and predatory mites have influenced ideas on population dynamics to a great extent (Huffaker 1958; Laing and Huffaker 1969). Their studies showed that one of the factors influencing persistence of spider mite–predatory mite interactions is dispersal of predators and prey. More recently, population dynamic studies on spider mites have been an impetus to the study of induced indirect defense. Sabelis (1981) developed a stochastic system-specific population model based on characteristics of individual predators and prey. This model comprised parameters such as prey and predator life-history components, predation parameters, and predator behavior in prey patches that had all been extensively quantified by Sabelis (1981). In addition, he experimentally quantified population dynamics of spider mites and predatory mites in greenhouse crops such as rose and cucumber. A striking discrepancy between model and reality was that spider mite populations in the model exploded at the expense of overexploitation of the food plant, while in the greenhouse the predators exterminated their prey mite populations (Sabelis et al. 1983; Sabelis and van de Meer 1986). Only after modifying the model so as to confine the predators in a prey patch until prey were exterminated in the patch did the model show realistic population dynamics. The mechanism underlying such predator behavior was soon identified as a behavioral response to a volatile cue emanating from spider mite infested leaves (Sabelis and van de Baan 1983; Sabelis et al. 1984b). This cue was later identified as being a mixture of herbivore-induced plant volatiles (Dicke et al. 1990a,b). Thus, predator-prey interactions appear to be critically influenced by predator behavior that is affected by plant-herbivore interactions. It is clear that the question about relative importance of top-down and bottom-up effects is irrelevant in this context. Because herbivore-induced carnivore attractants are a widespread phenomenon (table 4.1), their effects should be given more attention in population-dynamic studies concerning other tritrophic systems. However, in some tritrophic systems, herbivore-induced carnivore attractants have never been found to play a role and population dynamics in these tritrophic systems can be well modeled without this factor but with the inclusion of parasitoid residence times on infested and uninfested leaves (van Roermund 1995).

Population-dynamic interactions in metapopulations are essentially influenced by the migrations of predators and prey (Sabelis and Diekmann 1988; Sabelis and van der Meer 1986; Nachman 1991; Walde 1995). The moment of dispersal during local predator-prey interaction seems to be especially critical. Dispersal of carnivores and herbivores is highly influenced by plant characteristics, as has been shown in this chapter. Thus, the elucidation of plants as an important mediator of critical population-dynamic processes shows that an interaction of top-down and bottom-up effects comprises a vital population-dynamic factor. However, it is important to carry out manipulative studies to support the role of herbivore-induced carnivore attractants in the observed predator-prey interactions at the population level (Anholt 1997).

Conclusions

Plants may respond to the presence and attack of herbivorous arthropods with the induction of *indirect* defense through the emission of volatiles that attract carnivorous enemies of the herbivores. This has been studied for a range of terrestrial tritrophic systems and appears to be a general phenomenon. Induced *direct* defense is usually not a fully watertight solution to plants: despite induced direct defense, overexploitation of plants may still occur in the absence of the herbivore's enemies (Karban and Carey 1984; Sabelis and van der Meer 1986). However, the attraction of carnivorous arthropods to herbivore-infested plants may decisively affect local plant-herbivore population dynamics, leading to the extermination of the herbivore population (Sabelis and van der Meer 1986). Whether this occurs is dependent on the timing of attractant emission as well as the availability of carnivores in the habitat. The first aspect is under direct control of the plant, but the latter is not. Carnivore presence can at best be influenced by plants by offering shelter or alternative food supplies such as extrafloral nectar (Koptur 1992; D. Walter and O'Dowd 1992). To date, little attention has been paid to the interplay of induced direct defense and induced indirect defense (but see, e.g., Vrieling et al. 1991). Yet this interaction may be important in providing sufficient defense to a plant. From an evolutionary point of view, an interaction of small effects of induced direct defense in combination with indirect defense may result in a reduced rate of resistance evolution in the herbivore.

Induced indirect defense in several tritrophic systems is characterized by a high degree of specificity: the plant's response can depend on herbivore species and even herbivore instar (Sabelis and Dicke 1985; Takabayashi et al. 1991a, 1995). This specificity is an important difference with induced direct defense of plants and is likely the result of evolution where plants are under selection to give more specific information than their competitors. The effectiveness of indirect defense is dependent on the timing of the attractant emission as well as the availability of carnivores in the habitat.

5

Consumer-Induced Changes in Phytoplankton: Inducibility, Costs, Benefits, and the Impact on Grazers

ELLEN VAN DONK, MIQUEL LÜRLING, AND WINFRIED LAMPERT

Abstract

While inducible defenses of terrestrial plants against herbivores have long been known, only recently have corresponding responses of phytoplankton to their consumers been reported. Several chlorococcal algal species are able to adjust their phenotype (colony form, spines) to grazing pressure within hours or a few generations. The responses are triggered by information chemicals based on an immediate perception of consumer density. Several zooplankton species induce spines or colony formation in various species of *Scenedesmus* and *Coelastrum*.

Induced changes in algae confer grazing resistance against small zooplankton and can be interpreted as an adaptive antiherbivore strategy. Reduced algal palatability results in life history effects (lower growth and fecundity) on small *Daphnia* species, but not on large ones.

Forming colonies does not seem to lower algal growth rates, but algal colonies may have higher sinking rates. Inducible phenotypic responses may be a consequence of a trade-off between higher sinking rates and grazing resistance of colonial forms. They may have evolved in response to the variable conditions in the lake pelagic zone.

Introduction

Many phytoplankton species are notoriously plastic in morphology, growth form, and biochemical composition and variable traits have been interpreted as defense mechanisms against grazing. Pelagic algae can use two defensive strategies against grazing, either avoid being ingested or, if ingested, pass unharmed through the grazer's gut. Zooplankton feed with differing success on various phytoplankton species, primarily due to parameters like size, shape, cell-wall structure, and production of toxins. Clear relationships exist between the grazer's body size and the maximum size of spherical beads that can be ingested (Burns 1968). Also, the hardness of algae influences their ingestibility; flagellates are more readily ingested than diatoms (DeMott 1995). Gelatinous chlorophytes may be ingested but are poorly digested by zooplankters like *Daphnia* (Porter 1975), resulting in depressed zooplankton growth rates (Vanni and Lampert 1992; Stutzman 1995). Zooplankton rarely feed on cyanobacteria because they are large and can be toxic (Lampert 1987a; Fulton and Pearl 1987). Further, extracellular substances released from cyanobacteria inhibit the grazing activity of daphnids (Ostrofsky et al. 1983; Haney et al. 1994). Mucous excretion by diatoms also inhibits copepod grazing (Malej and Harris 1993). Nutrient-deficient algae may also be grazed with decreased efficiency, either owing to reduced ingestion rates (Sterner and Smith 1993) or reduced assimilation efficiency (Van Donk and Hessen 1993, 1995), increasing their probability of persistence under periods with low growth rates.

Variability in defensive traits in the field can partly be explained by clonal replacement as conditions change (Wood and Leatham 1992). However, there is also evidence for phenotypic plasticity. For example, the dinophyte *Ceratium* shows considerable variation in horn length (Hutchinson 1967), while the cyanobacterium *Microcystis* may vary in toxicity (Benndorf and Henning 1989). The green algal genus *Scenedesmus* is notoriously phenotypically flexible (Trainor 1991). Individual strains of various *Scenedesmus* species can grow as unicells or form colonies (coenobia) of four or eight cells. The cells can also vary with respect to the number and size of the spines. Trainor (1992) suggested that the phenotypic change in *Scenedesmus* is an ordered sequence of ecomorphs that can be defined as cyclomorphosis (sensu Black and Slobodkin 1987) driven by environmental factors. Various abiotic factors (nutrients, pH) and the age of the culture may affect colony size (Egan and Trainor 1989). Also, temperature is particularly effective in controlling *Scenedesmus* phenotypes, unicells are predominant at warm temperatures, while colonies dominate at low temperatures (10 °C) (Trainor 1993).

It is well known that many algal species isolated as clones from the field change morphology or growth form after some generations in laboratory cultures, suggesting that some unknown factor triggers their "typical" appearance in the field. For example, spiny algae like *Staurastrum* lose their bizarre form, large colonies like *Microcystis* grow as single cells, and "grass-blade" flakes of *Aphanizomenon* grow as single filaments. In the field, large flakes of *Aphanizomenon* are frequently found in the presence of large *Daphnia* (Lynch 1980a). A similar phenomenon has been observed in the diatom *Synedra* that occurred as colonies consisting of dozens to hundreds of cells when *Daphnia* was present, but as single cells in its absence (R. Sterner, pers. comm.). However, with these studies it is not possible to de-

DEFENSES IN PHYTOPLANKTON 91

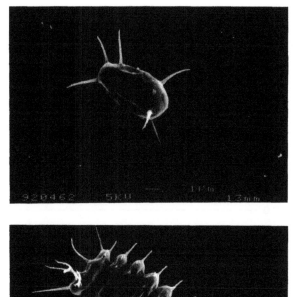

Fig. 5.1. SEM photographs of a typical single cell (*top*) and a *Daphnia*-induced eight-cell colony (coenobium) (*bottom*) of the green alga *Scenedesmus subspicatus*. (From Hessen and Van Donk 1993.)

cide whether the observed effect is the result of selective grazing on small flakes and single cells or an active response to the grazers' presence.

A new mechanism has recently been discovered by Hessen and Van Donk (1993). A kairomone released from grazing *Daphnia* induced the formation of colonies in the green alga *Scenedesmus*. *Scenedesmus subspicatus* formed numerous large, four- to eight-celled coenobia with more rigid and longer spines when exposed to water in which daphnids had been cultured (fig. 5.1). This work stimulated a series of studies testing the relevance of the phenomenon and the mechanisms involved. Although kairomone-induced phenotypic reactions have been reported frequently in zooplankton (see Tollrian and Dodson, chapter 10), very little is known about grazer-induced phenotypic changes in phytoplankton until now. In this chapter we will review consumer-induced defenses in phytoplankton from the recent literature and report new results not yet published.

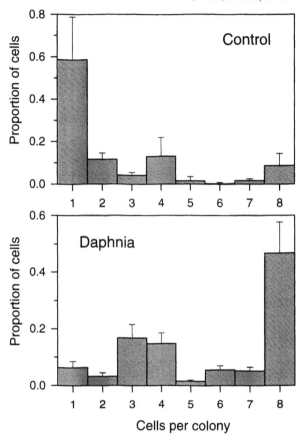

Fig. 5.2. Colony induction in a *Scenedesmus* culture by a kairomone released by *Daphnia magna*. Two ml of filtrate from a *Daphnia* culture (200 individuals l^{-1}) were added to 50 ml of algal culture. Proportions of the total number of cells distributed to aggregates of varying cell numbers after 48 hours of growth. Single cells dominate in the controls (*upper panel*) while eight-cell colonies dominate in the infochemical treatments (*lower panel*). Error bars represent 1 S.D. (n = 3). (From Lampert et al. 1994.)

Induced Phenotypic Changes in Phytoplankton

Induction by Daphnia

Lampert et al. (1994) repeated the experiments of Hessen and Van Donk (1993) with the spineless *Scenedesmus acutus*. They found comparable results for coenobia formation mediated by chemicals released from *Daphnia magna* (fig. 5.2). This *Daphnia* kairomone appeared to be a nonvolatile organic substance of small molecular mass (< 500 MW) that was heat stable, and resistant to pH within the range from 1 to 12. The substance was not affected by treatment with Pronase E, an enzyme that reacts with peptides. The colony-inducing factor is probably not a constituent of the algae themselves, because algal homogenate was not effective (Lampert et al. 1994).

The structure of the colony-inducing kairomone has not yet been identified. It is evidently not cyclic AMP, which has been reported to stimulate cyanobacterial mat formation when added to a dense suspension of *Spirulina platensis* trichomes (Ohmori et al. 1992). *Daphnia* may release significant quantities of cAMP as a result of digestive degradation of algal cells (Francko and Wetzel 1982). However, no flakes or colonies were formed when cAMP was added to *Aphanizomenon flos-aqua*, *Oscillatoria agardhii*, and *Scenedesmus acutus* (Lürling and Van Donk, unpubl. data).

In screening experiments, the effect of *Daphnia* infochemicals on the morphology of fifteen strains of Chlorophyceae, two strains of Bacillariophyceae, and three strains of Cyanophyceae was examined. *Daphnia*-induced colony formation did not appear to be restricted to the genus *Scenedesmus* since *Coelastrum* strains showed a similar response. On the other hand, not all *Scenedesmus* species respond to *Daphnia*, e.g., *S. quadricauda* was not responsive at all (table 5.1). Table 5.1 suggests that colony formation as a response to *Daphnia* is unique to the Chlorophyceae, but the physiological state of different species may obscure effects, as evidenced by the highly variable response of *S. subspicatus*. This species only occasionally responded to *Daphnia* chemicals, probably depending on the growth phase of the alga.

Until now, an increase in spine length of *Scenedesmus* has only been observed in connection with colony formation (Hessen and Van Donk 1993), but it is possible that spine length in unicells is also affected.

Induction by Zooplankton other than **Daphnia**

A reliable bioassay to test induced colony formation in algae by zooplankton has been developed by Lampert et al. (1994). *Scenedesmus acutus* is incubated with filtered test water under standard light and temperature conditions in 50 ml flasks on a shaking table. Particle size spectra of controls and treatments are measured by a particle counter after 48 hours. Mean particle size can be used as a measure of algal response as it is linearly related to the number of cells per colony (Lampert et al. 1994). The assay has been used to investigate the ability of different zooplankton to induce colonies in the test alga, *S. acutus*. Many of the examined animals were capable of inducing colonies at ecologically relevant densities (table 5.2). This broad taxonomic distribution indicates that the ability to induce colonies in *Scenedesmus* is widespread, but the exceptions suggest that it is not universal. It remains

TABLE 5.1
Phytoplankton Species Tested for Inducible Colony Formation

Algal Taxon	Grazer Taxon	Colony Induction
Chlorophyceae		
Scenedesmus acuminatus NIVA-CHL 58	Daphnia pulex	YES
Scenedesmus acutus Max-Planck-Institut	Daphnia galeata	YES
Scenedesmus obliquus NIVA-CHL 6	Daphnia magna	YES
Scenedesmus obliquus SAG 276–1	Daphnia magna	YES
Scenedesmus quadricauda NIVA CHL 7	Daphnia magna	NO
Scenedesmus subspicatus NIVA CHL 55	Daphnia galeata	YES
Coelastrum microporum SAG 217–1a	Daphnia magna	YES
Coelastrum sphaericum SAG 32.81	Daphnia magna	YES
Ankistrodesmus falcatus NIVA CHL 8	Daphnia magna	NO
Ankistrodesmus bibraianus SAG 278–1	Daphnia magna	?
Chlorella vulgaris NIVA CHL 19	Daphnia magna	?
Raphidocelis subcapitata NIVA CHL 1	Daphnia magna	NO
Micractinium pusillum CCAP 248/1	Daphnia magna	NO
Pediastrum duplex SAG 261–3a	Daphnia magna	NO
Planktosphaeria maxima CCAP 65/1	Daphnia magna	NO
Bacillariophyceae		
Synedra tenuis CCAP 1080/2	Daphnia pulex	?
Asterionella formosa CCAP 1005/9	Daphnia pulex	NO
Cyanophyceae		
Microcystis aeruginosa NIVA-CYA 43	Daphnia magna	NO
Oscillatoria agardhii NIVA-CYA 116	Daphnia magna	NO
Aphanizomenon flos-aqua CYA 142	Daphnia magna	NO

Notes: YES = colonies induced by *Daphnia*; NO = no inducible colony formation observed; ? = possible colony formation. Methods are described in Lampert et al. 1994 and Lürling and Van Donk, 1997.

unclear whether the response in *Scenedesmus* is a general zooplankton effect or specific to herbivorous zooplankton. Experiments with carnivorous zooplankters, i.e., *Polyphemus*, *Bytothrephes*, and *Leptodora* must still be performed.

Factors Affecting Colony Induction

The strength of the algal response depends on factors such as size of the zooplankton, available food quantity and quality, animal density, and physiological state of the animals. Colony formation increased nonlinearly with animal densities reaching a plateau at 200 *Daphnia* l^{-1} (fig. 5.3), but the

TABLE 5.2
Zooplankton Species Tested for Ability to Induce Colony Formation in *Scenedesmus*

Zooplankton	Colony Induction in *Scenedesmus*	Density (ind. l^{-1})
Cladocera		
Daphnia magna	YES	20
Daphnia pulex	YES	20
Daphnia galeata	YES	50
Daphnia hyalina	YES	100
Daphnia cucullata	YES	400
Bosmina longirostris	YES	1000
Chydorus sphaericus	NO	500
Simocephalus vetulus	YES	200
Copepoda		
Eudiaptomus gracilis	YES	200
Cyclops agilis	NO	200
Rotifera		
Brachionus calyciflorus	YES	100
Keratella cochlearis	YES	1000
Ostracoda		
Herpetocypris reptans	NO	200
Cypridopsis vidua	NO	200

Note: Densities for positive responses represent minimum effective numbers l^{-1} while densities for negative results represent maximum numbers tested.

effect can be further enhanced when more water containing infochemicals is supplied (Lampert et al. 1994). A clear relationship was found between the length of *D. magna* and the strength of colony formation in *Scenedesmus* (fig. 5.4). This relationship may indicate an important correlation with the grazing activity of the animals. Smaller daphnids like *D. cucullata* do not induce colonies at densities below 400 l⁻¹ probably due to their relatively low biomass rather than the lack of an ability to induce colonies.

Daphnia that feed actively on *Scenedesmus* induce more colonies than either starved daphnids or animals grazing on less edible algae, e.g., the filamentous cyanobacterium *Oscillatoria* (fig. 5.5). The food quality of grazable particles does not seem to be important for the excretion of the colony-inducing substance. Water from cultures of *Daphnia* feeding on unicellular *Microcystis* (fig. 5.5), nutrient-deficient *Scenedesmus*, or yeast induced as many colonies of *Scenedesmus* as water from animals feeding on nonlimited *Scenedesmus*. This means that the inducing factor probably originates from the daphnids digestive system. Eventually bacteria present in the gut of

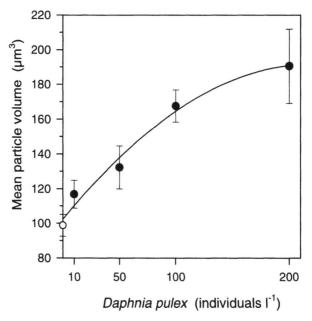

Fig. 5.3. *Daphnia*-density dependent colony formation in *Scenedesmus acutus*. Effect of the addition of 10% (v/v) water (filtered through 0.1 μm) from *Daphnia pulex* cultures varying in density (0–200 animals l^{-1}) to algal cultures. The increase in mean particle volume reflects the formation of colonies. Error bars represent 1 S.D. The second-order regression is $Y = 102.3 + 0.803 X + 0.0018 X^2$ ($r^2 = 0.983$).

daphnids are involved since the ability to produce colony-inducing chemicals can be blocked by exposing the animals to strong antibiotics like carbenicillin or cefotaxime that do not affect algal growth (Lürling and Van Donk, unpubl. data).

In situ Responses

The minimum concentration of the colony-inducing infochemical was equivalent to a density of about 20 large *Daphnia* l^{-1} which is not uncommon in the field. Fresh natural water from eutrophic Lake Zwemlust (The Netherlands) evoked colony-formation in *Scenedesmus*. Especially during spring, a clear relationship between rotifer abundance and colony formation in test algae (*Scenedesmus*) in the laboratory was found. Surprisingly there was no clear relationship between *Daphnia* abundance and the induction strength of the water (Lürling and Van Donk, 1997). Probably the daphnids were starving during this period, while algal biomass was low due to nitrogen limitation (Van Donk and Gulati 1995). Whether the weak relationship was

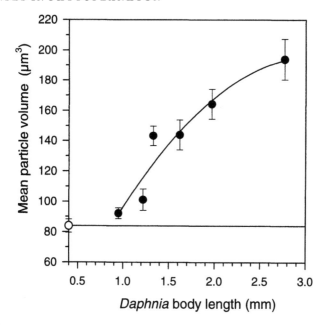

Fig. 5.4. Relationship between body length of *Daphnia magna* and the induction of colony formation in *Scenedesmus acutus*. All algal cultures received 10% filtered water from cultures with 200 animals 1^{-1}. Open circle and horizontal line represent controls without kairomone. The second-order regression is $Y = -20.6 + 139.2 X - 22.6 X^2$ ($r^2 = 0.923$).

caused by low food quantity, high microbial turnover rates of the inducing chemical, other factors, or a combination of factors remains unclear.

Similar results were found with filtered water taken from mesotrophic Schöhsee (northern Germany) enriched with nutrients. From spring to fall, test algae (*Scenedesmus*) produced more colonies in lake water than in the controls, but the addition of water from a *Daphnia* culture further increased the particle size (fig. 5.6).

Benefits of Induced Grazing Resistance in Phytoplankton

As discussed earlier, morphology has consequences for the grazability of algal cells, i.e., for their mortality. Benefits of induced defenses can be estimated by comparing zooplankton grazing rates on noninduced and induced phenotypes. Only a few grazing studies have been performed with algae of different morphology caused by the presence or absence of infochemicals excreted by grazers. Hessen and Van Donk (1993) found lower grazing rates for relatively small *D. magna* when the proportion of colonies of *S. subspicatus* in the food was high. They assumed that the spined colonies were

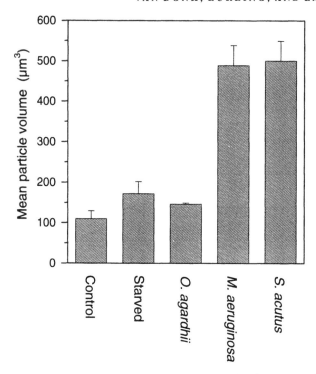

Fig. 5.5. Induction of colonies in *Scenedesmus acutus* by *Daphnia magna* feeding on various food types or starving. 10% (v/v) of 0.1 − μm filtered *Daphnia* culture (300 animals 1^{-1}, incubated for 24 hours) added to algal cultures. Error bars represent 1 S.D. (n = 3).

larger than the maximum size that could be ingested by a 1.75 mm daphnid. However, contrary to Hessen and Van Donk (1993), Lampert et al. (1994) did not detect differences in the uptake of unicells and colonies of *S. acutus*. This strain of *S. acutus* does not have spines, and it may be ingestible for the daphnids tested (above 1 mm body size) even if it forms colonies. The experiments differed also in the concentration of food particles. Lampert et al. (1994) used low food concentrations (below the incipient limiting level), where daphnids filter at their maximum rates, while Hessen and Van Donk (1993) worked at very high food concentrations (above ILL).

Lürling and Van Donk (1996) performed grazing experiments with two size classes of *D. pulex* (adult female approximately 3–4 mm) and the smaller *D. cucullata* (adult female approximately 0.8–1.1 mm) feeding on *S. acutus* (unicells and colonial cells). *D. pulex* showed no differences in clearance rate when fed either unicellular or colonial *Scenedesmus*, while *D. cucullata* had significantly depressed clearance rates when fed colonies compared to unicells. They also repeated the experiment with unicells and

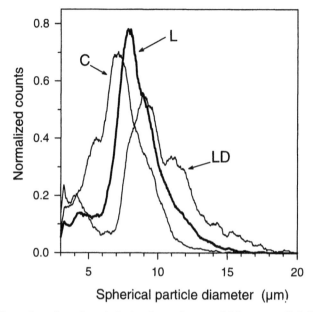

Fig. 5.6. Detection of a colony-inducing factor in natural lake water (Schöhsee, Plön, 4 October 1994). The shift of the *Scenedesmus* size spectrum represents increasing proportions of colonies after 48 hours of algal growth. C = controls; L = lake water enriched with nutrients; LD = lake water with nutrients plus 2 ml filtered water from a *Daphnia magna* culture (200 individuals 1^{-1}).

colonies of *S. obliquus* and found reduced clearance rates for *D. cucullata* and small *D. magna* but not for large *D. magna* (table 5.3). Hence protection against grazing for colonial *Scenedesmus* depends on the grazer's size; colony formation is only effective against small grazers.

Costs of Defense

Contrary to terrestrial and littoral systems, primary producers in pelagic systems are small and short-lived as they are faced with the problem of sinking out of the euphotic zone. Relative to their prey, most pelagic grazers are large and, unlike terrestrial grazers, they eat their entire prey and do not remove parts of the plants. They are often filter feeders that usually ingest many individual phytoplankton at a time. For a typical phytoplankter, being attacked by a grazer thus does not mean a reduction in fitness, but death. However, planktonic primary producers are not defenseless food particles, easily harvested by the consumers. They do, analogous to their terrestrial counterparts, have a set of relatively sophisticated defense mechanisms. As

TABLE 5.3
Feeding of *Daphnia* on Unicellular (U) or Colonial (C) *Scenedesmus* Expressed as the Volume of Water Swept Clear of Algae per Individual per Time (clearance rate, ml ind^{-1} h^{-1})

Zooplankton	Animal Size (mm)	Food Type	Mean Particle Volume (μm^3)	Clearance Rate (ml ind^{-1} h^{-1})
Daphnia pulex	1.0	*S. acutus*	89 U	0.24 ± 0.10
	1.0	"	312 C	0.19 ± 0.08
	2.5	"	89 U	0.97 ± 0.05
	2.5	"	312 C	0.87 ± 0.26
Daphnia cucullata	0.6	*S. acutus*	89 U	0.32 ± 0.06
	0.6	"	312 C	0.03 ± 0.24*
	1.1	"	89 U	0.61 ± 0.11
	1.1	"	312 C	0.19 ± 0.12*
Daphnia magna	1.0	*S. obliquus*	89 U	0.33 ± 0.07
	1.0	"	224 C	0.22 ± 0.11
	1.0	"	476 C	0.05 ± 0.10*
	1.6	"	89 U	0.95 ± 0.11
	1.6	"	224 C	0.95 ± 0.08
	1.6	"	476 C	0.96 ± 0.11
Daphnia cucullata	0.7	*S. obliquus*	89 U	0.24 ± 0.12
	0.7	"	224 C	0.08 ± 0.22
	0.7	"	476 C	0.01 ± 0.24
	1.2	"	89 U	0.41 ± 0.15
	1.2	"	224 C	0.17 ± 0.09*
	1.2	"	476 C	0.07 ± 0.03*

*Clearance rates are significantly lower than rates of animals feeding on unicells.

all defenses have their costs, a mechanism to mobilize them only when necessary would be advantageous.

Grazing resistance is thought to involve some metabolic costs or otherwise reduced competitive abilities. Therefore, one would expect a trade-off between grazing resistance and the tax paid by metabolic costs. The advantage in favor of one or another may cause rapid shifts throughout the seasons or even diurnally as grazing pressure changes. If colony formation is only phenotypically induced when a chemical signals the presence of grazers, one might expect costs associated with colony growth (Dodson 1989b). Growth in colonies may reduce nutrient uptake capabilities of the individual cells, which should result in lower growth rates. However, a reduction in growth rate due to the infochemical has not been found yet (Hessen and Van Donk 1993; Lampert et al. 1994).

According to Stokes' law, larger particles sink faster than small ones

(Reynolds 1984), i.e., colonies should experience higher sinking losses. Hence the costs of colony formation can probably be attributed to higher loss rates rather than reduced growth rates. There is, in fact, preliminary evidence for differential sinking rates in unicells and colonies of *Scenedesmus* (Lampert and Jäger, unpubl.), but these studies are not complete. Due to the higher viscosity of the water at low temperatures, the size effect on the sinking rate may be tolerable at low but not at high temperatures. This is consistent with a tendency toward larger colonies at low temperatures in absence of zooplankton (Trainor 1993). If the trade-off between sinking and grazer resistance is dependent on environmental conditions, one might expect a stronger phenotypic response in warm water, when sinking losses are important.

Effects on Grazers' Life Histories

Life-table experiments reveal body-size-dependent differences in the feedback of colony formation on the grazer. Several species of *Daphnia* were fed *S. acutus* unicells or induced coenobia, respectively. As expected from the feeding experiments, there was no difference in growth for the large *D. pulex* and *D. magna*, but the small *D. cucullata* grew slower with colonies (fig. 5.7). Likewise, the intrinsic rate of population growth did not differ between treatments for the large daphnids, but was reduced for *D. cucullata* that were fed colonial food. Algal growth form did not affect age at maturity in either species, but *D. cucullata* released more live neonates per mature female when fed unicells.

These species-specific effects on the grazer indicate an interesting indirect interaction. *D. pulex* and *D. magna*, but not *D. cucullata*, are able to produce sufficient amounts of infochemical at ecologically relevant densities (table 5.2). On the other hand, *D. cucullata* but not the large daphnids are affected by the colonies. Hence, large *Daphnia* could modify their algal food so that the life history of their small congeners is influenced. This implies that both phytoplankton-zooplankton and zooplankton-zooplankton interactions are altered by the induced colony formation (Lürling and Van Donk 1996).

Although the reasons for the disappearance of large zooplankton species in the presence of fish are clear, the second part of the size efficiency hypothesis, dominance of large zooplankton in the absence of fish (Hall et al. 1976), is still under debate. One explanation is that large zooplankton can utilize a larger spectrum of particle sizes and are, thus, competitively superior. By chemically shifting the algal size spectrum to larger particle sizes, large daphnids would reinforce the competitive advantage. We have called the *Daphnia* exudates (infochemicals) "kairomones," implying that the emitted infochemical evokes a response that is favorable only to the receiver (cf.

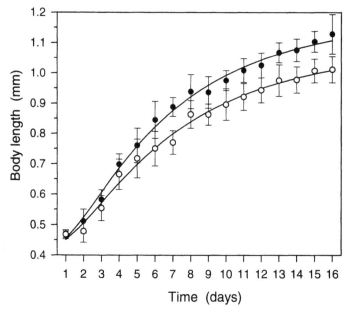

Fig. 5.7. Growth curves of *Daphnia cucullata* fed equal amounts (biovolume $8 \cdot 10^6$ μm^3) of unicellular (filled circles) or colonial (open circles) *Scenedesmus acutus*. Error bars represent 1 S.D.

Tollrian 1993; Reede 1995). In the interaction between large *Daphnia* and *Scenedesmus*, the infochemical can possibly also be called a synomone (Dicke and Sabelis 1988a) if both the emitter and the receiver may gain an advantage from the chemical interaction.

Perspectives for Future Research

Unlike terrestrial plants, phytoplankton cannot respond individually with defense mechanisms when they are attacked by a herbivore. In order to react with induced phenotypic changes, they need a reliable cue that signals the presence of grazers before an encounter. This can only be a chemical cue dissolved in the water.

Research on consumer-induced defenses in phytoplankton is in its infancy, but it already shows great potential for new discoveries and investigations in ecology and population biology. Morphological changes induced by herbivores, like colony and spine formation, have been observed in only a few phytoplankton species until now. It will be necessary to screen a large number of algal species that show defenses and phenotypic variability in order to test whether inducible defenses are restricted to chlorococcal algae. There are many other important anti-grazer properties of algae that may be inducible besides morphology. For example, it may be worth examining the toxicity of Cyanophytes and Prymnesiophytes, the formation of resting spores and gelatinous sheaths, sexual reproduction and formation of zygotes, and vertical migration.

Although the chemical nature of the colony-inducing *Daphnia* factor is not yet known, some progress has been made with its characterization. The infochemical can be extracted from the water and it can be purified on HPLC (E. von Elert, unpubl.). This allows more quantitative experimentation and it may even provide possibilities for experiments at the community level. If the infochemical can be identified, it will be possible to determine its occurrence in the field and to evaluate its ecological relevance.

The *Daphnia-Scenedesmus* interaction is a good model system for the study of inducible defenses in clonal populations. The Malthusian parameter is a good fitness parameter in such populations. As the growth and loss components of the population growth rate can be easily measured in the laboratory, costs and benefits can be quantified directly in terms of relative fitness, which proves to be extremely difficult in terrestrial plants.

6

The Immune System as an Inducible Defense

SIMON D. W. FROST

Abstract

Vertebrate hosts induce powerful immune defenses in response to parasite infection. The vertebrate immune response can be roughly divided into nonspecific immunity, which can recognize relatively few parasites but can respond rapidly, and specific immunity, which can recognize many parasites but is slower to respond. The inverse relationship between the speed and the specificity of the response arises from the way the receptors that are used to recognize parasites are arranged in an immune cell population. Different effector mechanisms are induced at varying rates against different parasites. The way in which nonspecific and specific immunity "decide" on which effectors to use also varies as a consequence of the different arrangement of receptors on nonspecific and specific cells. The combined use of nonspecific and specific immunity provides the "best of both worlds" in terms of mounting rapid, powerful, appropriate responses to a diversity of parasites.

Inducibility of the immune response may be favored evolutionarily by reducing energy costs and immunopathology, but it may also be important in allowing the magnitude of the immune response to change depending on the size of the parasite population. Since the immune system can also discriminate between different parasites, then inducibility also allows the immune system to fine-tune effector mechanisms to each type of parasite. The ability of the immune system to respond to the size and type of parasite on a timescale similar to the turnover of parasites is central to the ability of vertebrates to overcome parasites despite potentially rapid parasite replication and evolution.

The specificity of recognition of parasites by the immune system, which is central to the speed and type of response induced by different parasites, varies tremendously between host species. While the energetic costs involved in expanding individual clones of specific immune cells may be important in limiting specificity, the need to have a functional immune repertoire early in life and/or the need to mount rapid responses when a parasite infects may be more important due to the inverse relationship between speed and specificity.

Introduction

Immune responses share with other inducible defenses the characteristics of being inducible, amplifiable, sometimes possessing a memory component, and having a varying degree of specificity. The vertebrate immune response is a good example of how an inducible defense can track a complex, changing biotic environment. This chapter will focus mainly on the vertebrate immune response, which probably arose only once (in the ancestors of jawed fishes), rather than attempting to review invertebrate internal defense systems, which are likely to be of polyphyletic origin (Sima and Vetvicka 1990; Hoffman et al. 1994).

The first part of this chapter describes how the immune system responds to infection by parasites, focusing on the inducibility and specificity of the response. For further background in immunobiology and an excellent source of references, I refer the reader to Janeway and Travers (1994). There is an inverse relationship between the speed and the specificity of the response that arises as a consequence of the way in which receptors are arranged in an immune cell population. Vertebrate immunity achieves a high level of specificity by possessing many receptor variants arranged in clones of immune cells. This means that before an antigen is encountered, only a few cells will react to the antigen. Launching a specific immune response involves a delay while these rare clones of cells are expanded and selected. Higher vertebrates have reduced this delay by keeping specific cells in high concentrations in specialized lymphoid tissues that offer a microenvironment well suited for bringing together immune cells and antigens, with subsequent clonal expansion and selection. However, there is still a delay in inducing a specific response, as it takes time for antigens to be transported to these tissues and for the immune cells to reach the site of infection. Different, often appropriate, effector mechanisms are induced by different types of parasite. I describe what information the immune system may use to "decide" on an appropriate response. The nonspecific immune response, which recognizes antigenic "patterns" indicative of a diversity of parasites, is very important in instructing specific immune responses, which lack the innate ability to discriminate between parasites and self.

In the second part of the chapter, I ask what selective pressures favor an inducible immune response over a constitutive one. I focus on the relative importance of (1) a strong, variable parasite selection pressure, (2) reliable recognition, (3) an effective response, and (4) a cost to defense (Harvell 1990a,b). The amplifiability of an inducible response allows the magnitude of the immune response to be fine-tuned to the size of the parasite population. Specificity allows an inducible response to employ different effector mechanisms against different parasites. These advantages may be more important in favoring an inducible defense over a constitutive one than the reduction of energetic or immunopathological costs.

In the third part of the chapter, I ask what selective pressures favor a more specific response over a less specific one. Specificity of responses varies greatly between species, with vertebrates but not invertebrates possessing specific immunity, and higher vertebrates having a more specific immune response than lower vertebrates. For some organisms, the need to have a functional immune repertoire early in life and to mount rapid immune responses upon infection may be more important than a more expensive, slower, but highly specific response.

How Is the Vertebrate Immune Response Induced?

The vertebrate immune system can recognize and launch a specific response to a diverse array of parasites whilst generally avoiding a response that may be unnecessary, e.g., to food, or even injurious, e.g., to self-encoded antigens. This inducible specificity not only reduces costs associated with an immune response, but also allows the magnitude of the response to react to the size of the parasite population, and if the immune system is specific enough, to allow the type of response to be fine-tuned to the type of parasite. Here "parasite" is defined as an obligatory symbiosis ("living together") where the parasite benefits at the expense at the host (Cheng 1991). It should be noted that the degree of symbiosis and the net effect on the host are continuous variables, and hence the minimal reliance on and minimal damage to the host above which an organism is considered a parasite are somewhat arbitrary (Toft 1991). Defining what is meant by "self" is problematical: I follow Matzinger (1994) in defining self as any part of the body, and nonself as everything else, noting that not all "self" is seen by the immune system and not all "nonself" elicits an immune response. Antigens are defined as molecules that react with antibodies (although despite what their name suggests, they may not have the ability by themselves to GENerate ANTIbodies).

A typical human immune response can be divided into three phases: an immediate phase (0–4 hours after infection), an early induced phase (4–96 hours), and a late phase (after 96 hours) (Gibbons 1992). Table 6.1 summarizes how recognition and effector components of the immune response depend upon the phase of the immune response and the type of parasite. There are two main points to note. First, during the course of an immune response, the recognition mechanisms change but are similar for different kinds of parasite. Second, the effector mechanisms launched depend upon the type of parasite (see also table 6.2) but are functionally similar during the different phases of the immune response. For example, complement, a mechanism for

TABLE 6.1
The Components of the Three Phases of the Immune Response Involved in Defense Against Three Major Classes of Parasites

Phase of the Immune Response		Immediate (0–4 hours)	Early Induced (4–96 hours)	Late Induced (>96 hours)
Recognition mechanisms		Nonspecific Innate No memory No specific T cells	Nonspecific and specific Inducible No memory No specific T cells	Specific Inducible Memory Specific T cells
Barrier functions	Mechanical	Tight junctions between cells and flow of air or liquid across epithelia	Local inflammation	IgA antibody in luminal spaces IgE antibody on mast cells
	Chemical	Lysozyme in saliva, sweat and tears. Pepsin in gut. Low pH in stomach. Antibacterial peptides in intestine	Local TNF-α	Neutralization[a] Sensitizes mast cells to recruit effector cells and increases transport of antigen to lymphoid tissues
	Microbiological	Normal flora compete for nutrients and attachment	Recruits immune cells to site of infection Activates macrophages and induces nitric oxide production	

TABLE 6.1 (*Continued*)

Phase of the Immune Response	Immediate (0–4 hours)		Early Induced (4–96 hours)		Late Induced (>96 hours)	
Response to extracellular pathogens	Phagocytes	Engulf parasites	Mannose binding protein	Binds to mannose on bacteria, activating complement and opsonising[b] them	IgG antibody and Fc receptor-bearing cells	Neutralises[a] bacterial toxins and viruses
	Alternative complement pathway	Punches holes in parasite cell walls, recruits phagocytes, opsonises[b] parasites, aiding their removal by phagocytes	C-reactive protein	Binds to phosphorycholine on bacteria, activating complement to opsonise[b] them	IgG, IgM antibody and classical component pathway	Complement punches holes in parasite cell walls, recruits phagocytes, opsonises[b] parasites, aiding their removal by phagocytes
			T-cell independent B-cell antibody plus complement	Coat bacteria and promote ingestion by phagocytes		
Response to intracellular bacteria	Macrophages	Activated macrophages lyse bacteria in phagosomes and produce nitric oxide, which is a potent bacteriocide	T-cell independent macrophage activation	Activated macrophages lyse bacteria in phagosomes and produce NO	T-cell activation of macrophages by IFN-γ	Activated macrophages lyse bacteria in phagosomes and produce NO
			IL-1, IL-6, TNF-α	Stimulates liver to produce acute phase proteins (C-reactive and mannose-binding), mobilizes		

	Natural killer cells	IL-12	IFN-α and IFN-β	IL-12 activated NK cells	Killer T cells	IFN-γ
Response to virus-infected cells	Nonspecifically lyses virally infected cells with low expression of certain MHC class I molecules	neutrophils which phagocytose bacteria, increase body temperature and activate the immune system to decrease parasite replication, increase antigen processing and increase specific immune response Stimulates type1, cellular immunity	Inhibition of viral replication in uninfected cells and enhances ability of virus infected cells to present antigen to specific killer T cells	Stimulates type 1, cellular immunity	Specifically lyses virally infected cells	Upregulates expression of MHC class I and II, allowing more powerful specific recognition

Source: Adapted from Janeway and Travers 1994, with permission.
[a]Neutralization is where antibodies bind to intracellular parasites to prevent entry into cells.
[b]Opsonization is where antibodies bind to bacteria to allow better nonspecific recognition.

TABLE 6.2
Different Kinds of Parasites Induce Different Effector Mechanisms That Are Appropriate for Parasite Clearance

	Intracellular		Extracellular	
Site of Infection	Cytoplasmic	Vesicular	Interstitial Spaces, Blood, Lymph	Epithelial Surfaces
Parasites	Viruses *Chlamydia* spp (one species is sexually transmitted and can cause sterility in women) *Rickettsia* spp (e.g., typhus, Rocky Mountain spotted fever) *Listeria monocytogenes* (bacteria) Protozoa	Mycobacteria *Salmonella typhimurium* (food poisoning) *Leishmania* (leishmaniasis) *Listeria* spp (food poisoning) *Trypanosoma* spp (e.g., sleeping sickness (T. brucei brucei) and Chagas disease (T. cruzi)) *Legionella pneumophila* (Legionnaire's disease) *Cryptococcus neoformans* (fungal opportunistic parasite) *Histoplasma* (fungal opportunistic parasite) *Yersinia pestis* (plague)	Viruses Bacteria Protozoa Fungi Worms	*Neisseria gonorrhoeae* (gonorrhoea) Worms *Streptococcus pneumoniae* (otitis media, pneumonia, bacteremia, or meningitis) *Vibrio cholerae* (cholera) *Escherischia coli* (can cause food poisoning) *Candida albicans* (thrush) *Helicobacter pylori* (can cause ulcers)
Effectors	Killer T cells Antibody dependent cell-mediated cytotoxicity Natural killer cells	T-cell dependent macrophage activation	Antibodies Complement Phagocytosis Neutralization	Antibodies, especially IgA Inflammatory cells

Source: Adapted from Janeway and Travers 1994, with permission.

punching holes in bacterial cell walls, is activated rapidly after bacterial infection by nonspecific cells (the "alternative" pathway of complement activation) but later in the infection is activated by specific cells (the "classical" pathway). Complement also binds to bacteria, allowing phagocytes to engulf them more easily—a process called "opsonization." Late in infection, specific antibody binds to bacteria to opsonize them. The way in which the recognition and effector mechanisms vary with parasite identity and course of infection determine which parasites are recognized by various components of the immune system.

The Different Recognition Mechanisms Used during Different Phases of Infection Arise from the Inverse Relationship between Speed of Induction and Specificity

The vertebrate immune response can be roughly divided into nonspecific immunity and specific immunity, which have different dynamics and are induced in different ways.

Nonspecific immunity is launched against many conserved parasite antigens. It is rapidly induced but wanes after the parasite has been cleared with second-set responses to a particular antigen being identical to the first response. Nonspecific cells possess several genetically encoded receptors, each of which can recognize a different parasite antigen, the function (and hence structure) of which has to be conserved. This simple "pattern recognition" allows nonspecific immune cells to reliably recognize many parasites while not reacting to self (Janeway 1992, 1994).

Specific immunity, on the other hand, can potentially recognize a vast number of antigens but takes longer to mount than nonspecific immunity (Janeway and Travers 1994). After the parasite is cleared, effector mechanisms wane, but unlike nonspecific immunity, second-set responses are induced more rapidly, are more specific, and employ effector mechanisms that are better at clearing the particular parasite–the phenomenon of immune memory. Each specific immune cell possesses only one receptor that is used to recognize antigens. The ability of specific immunity to recognize many parasites arises from the diversity of receptors across the population of specific immune cells, with large receptor diversity (10^7–10^9 different specificities in birds and mammals; Du Pasquier 1982) being generated by somatic gene rearrangement. Different specific cells recognize different antigens in different ways. B cells, which are important in patrolling the body cavity, possess receptors that recognize conformations on whole antigens, which can be carbohydrates, proteins, or simple chemical groups. T cells, which are important in monitoring cellular changes in solid tissues, can only recognize peptide antigens in association with cell surface molecules called major histocompatibility complex (MHC) molecules (Matzinger and Zamoyska 1982). There are two classes of MHC: class I tends to be associated with antigens produced within the cell, and class II tends to be associated with antigens taken up from outside the cell. Killer T cells recognize antigen associated with MHC class I, allowing these cells to detect intracellular parasites such as viruses. Antigen-presenting cells take up antigen from outside the cell and present it in association with MHC class II to helper T cells, allowing these cells to detect extracellular parasites.

Natural killer (NK) cells, which are normally included in the nonspecific immune response, can nonspecifically lyse virally infected and cancerous cells. They are induced by virus much sooner than specific antiviral immunity (e.g., killer T cells), and they control virus replication until killer T cells are produced which can eliminate the virus. Unlike other nonspecific cells, NK cells have clonally arranged receptors, similar to specific cells, although of much more limited diversity, leading to the suggestion that they are related to specific T cells (Versteeg 1992). The rapid speed of induction (like nonspecific immunity) to many kinds of virus (like specific immunity) has been hypothesized to arise from the way in which they discriminate infected from uninfected cells (Raulet and Held 1995). NK cells have two types of surface receptor: one that recognizes carbohydrates on self, which triggers NK cells to kill; the second recognizes certain cell surface molecules (MHC class I), which inhibits the killing signal from the first receptor. Some viruses, e.g., herpes viruses, downregulate expression of MHC class I molecules (which are used by specific immunity to recognize infected cells) to avoid specific immunity, but will be killed by NK cells.

The time course of a typical immune response is now described, and the inverse relationship between the speed and specificity of nonspecific and specific recognition of antigens is related to the different way in which nonspecific and specific immunity recognize antigen.

Immediate Responses Are a Result of Nonspecific Recognition

Preexisting nonspecific effector cells already at the site of infection rapidly (0–4 hours after infection) recognize parasites and help to remove the infectious agent. The rapidity of response is due to each nonspecific cell being "hard-wired" to recognize parasites. There is often a large number of nonspecific immune cells at potential sites of infection, such as at the skin and epithelia, allowing rapid clearance of parasites.

Some subsets of so-called specific immune cells are more similar in their speed of response to nonspecific cells than to truly specific B and T cells, and are characterized by low receptor diversity and high frequency at potential sites of infection. One subset of T cells, γδ T cells, is found in surface epithelia, with essentially homogeneous receptors in any one epithelium, and may act by killing infected epithelial cells to prevent further spread of the parasite. One subset of B cells, CD5+ B cells, is found at high frequency in the body cavity (which offers an ideal environment for the growth of extracellular parasites) and appears to recognize common bacterial polysaccharides. These "first lines" of defense may constitute the most evolutionary ancient form of specific immune cells (Stewart 1994).

Early Induced Responses Are a Result of Nonspecific and Specific Recognition

An immediate response may not be sufficient to clear the parasite. This could be due to insufficient nonspecific effectors being present at the site of infection. In such a case, an early induced response is seen (4–96 hours after infection) where nonspecific effectors are recruited to the site of infection by nonspecific effectors already at the site. In addition, nonspecific immunity passes on information to specific immune cells and begins the induction of specific immunity (Fearon and Locksley 1996). This is important as specific cells do not possess the innate ability to determine what is a parasite and what is not. Specific cells require not only antigens to be activated, but also a second signal. Nonspecific cells such as macrophages can provide this second signal to specific cells when they detect parasites. The nonspecific immune response is very important in alerting specific immunity to the presence of the many parasites that display conserved antigenic patterns, allowing specific immunity to discriminate between "infectious" and "noninfectious" (Janeway 1992, 1994).

However, many parasites may initially avoid the "pattern recognition" of specific immunity, although once they grow to a certain population size, it is unlikely that they will avoid signaling their presence to the immune system as they may cause damage to surrounding tissue. It has been proposed that detection of "danger," for example by the detection of heat shock proteins, which are produced by cells of many organisms in response to almost any kind of stress, may trigger specific immunity (Matzinger 1994). Activation of antigen-presenting cells, which pick up and process the antigen and transport it to lymphoid tissues to present it to specific immune cells, may respond to signs of damage, allowing the immune system to discriminate between "safe" and "dangerous."

Late Immune Responses Are Mainly a Result of Specific Recognition

During the early induced phase of an immune response, although specific recognition occurs, specific effector mechanisms are not yet mounted. This is a consequence of two processes. As individual specific cells possess only one variant of their antigen receptor, efficient functioning of the specific immune response depends on expansion of individual clones of specific immune cells in response to antigens and subsequent selection among clones, causing a delay in the induction of specific immune effector mechanisms. Higher vertebrates have evolved specialized lymphoid tissues where immune cells are concentrated and which provide a microenvironment well suited to this expansion and selection, reducing this delay. However, a further delay is involved in transporting antigens to the lymphoid tissues and for effector cells to migrate to the site of infection. Consequently, specific immune responses are seen later than nonspecific responses, after about 96 hours.

During an immune response, the specific immune system becomes better at recognizing the parasite. This is due to two processes: the clonal expansion of B and T cells which possess receptors that bind strongly to parasite antigens, and the somatic hypermutation of B cell receptors, with further selection of B cell clones that bind more strongly to the parasite. There is also a shift in the type of antibody produced during responses against extracellular parasites, a process called "isotype switching." IgM, which by forming polymers binds strongly to repetitive epitopes (e.g., bacterial cell wall polysaccharides), dominates early primary responses, with IgG, IgA, and IgE dominating secondary and subsequent responses. These changes facilitate antigen uptake and presentation, allowing specific responses at lower doses of antigen rather than qualitatively changing the immune response.

Reinfection

If there is reinfection, recognition and effector mechanisms are launched much more rapidly than during the primary response. Depending on the gap between primary infection and reinfection, this so-called "protective immunity" consists of preexisting immune effectors and immunological memory. Preexisting immune effectors offer rapid but often short-term defense against reinfection. Specific memory against localized mucosal infections caused by viruses such as rotavirus, rhinovirus, and respiratory syncytial virus is usually very short. In these cases, protective immunity may be afforded only by preexisting effectors (Ahmed and Gray 1996). After a parasite has been cleared, effector mechanisms wane, presumably due to the cost of maintain-

IMMUNE SYSTEM 115

ing these responses. Possible costs include the energy required to maintain a response, the possible immunopathology an effector may cause, and the diminishing of responses to other pathogens while the immune system is concerned with pathogens that it has already cleared. However, parasite-specific clones of "memory" cells remain expanded, although in a resting state, allowing more rapid induction and a more powerful effector response on reinfection. For example, the primary antibody response consists of low-affinity antibodies from a large number of cells, whereas the secondary antibody response comes from far fewer, high-affinity precursors that have undergone clonal expansion (a four- to tenfold increase) and extensive somatic mutation. Over time, specific immunity becomes "hard-wired" to recognize the parasites that have attacked the host during its lifetime (and hence may attack in the future).

In summary, generally there is an inverse relationship between the speed of recognition and the specificity of recognition. Nonspecific cells and certain subsets of B and T cells with low receptor diversity at potential sites of infection recognize parasites early in infection. Specific cells with high receptor diversity in lymphoid tissues recognize parasites later in infection.

Effector Mechanisms Induced by Parasites Are Often Appropriate for Their Removal and Remain Relatively Unchanged during the Three Phases of Infection

Over the time course of an immune response, specific immune responses are added to the already powerful nonspecific immune responses. Functionally, effector responses are similar throughout the infection but different mechanisms are launched against different parasites (see tables 6.1 and 6.2), e.g., during the early stages of viral infection, NK cells nonspecifically lyse infected cells, whereas later specific killer T cells lyse infected cells. In this section, the ways in which nonspecific and specific immunity "decide" on which effector mechanisms to employ are described.

Nonspecific Immune Cells Are Hard-Wired to Launch Appropriate Responses against Particular Parasites

The "hard-wiring" of each nonspecific immune cell to recognize different kinds of parasites also allows the "hard-wiring" of effector responses to different kinds of parasites. For example, when mannan, a constituent of the cell walls of many bacteria, is detected, macrophages produce interleukin 6 (IL-6), which induces the liver to produce molecules such as mannose binding protein, involved in triggering the activation of complement, a mechanism to punch holes in bacterial cell walls. Macrophages also possess receptors that recognize β-1,3 glucans, a common constituent of fungal cell walls.

Specific Immune Cell Populations Become Hard-Wired to Launch Appropriate Responses against Particular Parasites over the Course of Primary Infection and Any Subsequent Reinfection

During the initial response, naïve CD4+ T cells, which help orchestrate the specific immune response, differentiate into either inflammatory T cells (T$_H$1) or helper T cells (T$_H$2) (Mosmann et al. 1986; Mosmann and Sad 1996). T cell responses against intracellular parasites tend to be dominated by inflammatory cells, which stimulate effectors that attack intracellular parasites such as killer T cells. T cell responses against extracellular parasites tend to be dominated by helper T cells, which help B cells to produce antibodies, an important defense against extracellular parasites. The real situation is far more complex, with multiple overlapping effector mechanisms (see Allen and Maizels 1997 for a recent critique), probably reflecting parasite escape from host defenses and host evolution to clear parasites. However, this simple division of specific responses into "type 1" (against intracellular parasites) and "type 2" (against extracellular parasites) can help illustrate how the specific immune system "decides" on an appropriate response.

The nonspecific immune system can help determine whether a response is type 1 or type 2 by the production of chemical messengers called "cytokines" (Romagnani 1992; Fearon and Locksley 1996). Macrophages and NK cells produce interleukin-12 and gamma interferon, two cytokines that stimulate type 1 immunity, in response to viruses and some intracellular bacteria such as *Listeria*. Mast cells produce interleukin-4, which stimulates type 2 immunity. Another source of IL-4 comes from a subset of "natural" CD4+ T cells that, unlike other CD4+ T cells, possess a nearly invariant T cell receptor (Bendelac et al. 1995), although the nature of antigens recognized by these (nonspecific-like) T cells is unclear. However, even when parasites escape providing antigenic clues as to their identity, we often see specific effector mechanisms "fine-tuned" to particular kinds of parasites.

Antigenic dose, and how it is presented, can shape the specific immune response. Type 2 immune responses are induced by low levels of antigens on antigen-presenting cells, whereas type 1 immune responses are induced by higher levels of antigens. This may be due to the need to respond more rapidly against extracellular parasites than intracellular ones. Extracellular parasites often proliferate rapidly, whereas intracellular parasites initially infect a small number of cells, proliferate slowly, and cause little damage early in infection. However, at very high levels of stimulation, type 2 immunity is induced (Bluestone 1995). The role of this high dose effect in parasite defense is not clear, and it may simply be a mechanism of reducing the immunopathology caused by an overly aggressive type 1 response.

Lee Segel (pers. comm.) has suggested that a "parasite destruction feedback" mechanism could allow the specific immune system to choose the most effective response in the absence of any specific cues from the nonspecific immune system. Dead, dying, or doomed parasites are hypothesized to provide positive feedback to the effectors most responsible for their death and/or negative feedback to the effectors least responsible. This requires that a range of effector mechanisms are "tried" in tiny simulations within lymphoid tissue, with spatial compartmentalization ensuring that the most effective responses receive the most positive (or least negative) feedback on parasite destruction. The production of heat shock proteins, which are produced by many organisms in response to almost any kind of stress, either directly (by the parasite) or indirectly (e.g., by infected cells) could provide such a positive feedback signal. Such a mechanism, if it really exists, would be an extremely powerful way to choose an appropriate response, as the most effective responses available would always be chosen.

In summary, the nonspecific immune system is extremely important in providing information about the type of parasite to the specific immune cells, which lack the "hard-wired" ability to detect certain kinds of parasite. Dose effects do appear to be important in determining the type of specific response, although the role of these effects in host defense is not clear. It is possible that during an infection, the population of specific immune cells becomes hard-wired to employ the most effective effectors via a positive feedback mechanism.

Why Is Vertebrate Immunity Inducible Rather than Constitutive?

When should a defense be inducible rather than constitutive? Harvell (1990a,b) has pointed out several factors that may be important in the selection of inducible responses which, in the context of immune systems and parasites, are as follows: (1) parasites often exert a strong, variable selection pressure; (2) parasites often provide reliable, nonlethal cues; (3) the immune system constitutes an effective defense; and (4) mounting an immune response is often costly. The importance of these factors in maintaining the inducibility of both the nonspecific and specific components of the immune response are discussed, as well as parasite counteradaptations to overcome the inducible nature of the immune response.

(1) Parasites Often Exert a Strong, Variable Selection Pressure

WHAT IS THE EVIDENCE TO SHOW THAT PARASITE SELECTION PRESSURE ON HOSTS HAS BEEN IMPORTANT?

Parasitism, by definition, is associated with morbidity and mortality of the host. There are many theoretical studies suggesting that parasites may be involved in the regulation of host abundance and the maintenance of host-genetic diversity via the effects of infection on fecundity and mortality (Anderson and May 1978; May and Anderson 1978; Hamilton 1982; Gulland 1995; Lively and Apanius 1995; Read 1995). However, empirical evidence of regulation and maintenance of genetic diversity by parasites is sparse, especially in wild animal populations. Evidence that parasites can regulate host abundance in a free-living species comes from data on the interaction between *Trichostrongylus tenuis* in red grouse, *Lagopus lagopus* (Hudson et al. 1985; Hudson and Dobson 1995). Correlational (but not causal) evidence for parasites exerting a strong selection pressure on human populations comes from studies of MHC polymorphism (see Kelsoe and Schulze 1987 for background on how MHC evolves). As T cells can only recognize peptide antigens in association with MHC molecules, the structure of the MHC is very important in determining how well specific immunity recognizes parasites. Several hypotheses have been advanced to account for the extreme polymorphism of MHC (Nei and Hughes 1991), all of which rely at least in part on diversification due to parasite selection pressure (Potts and Wakeland 1990; Hughes and Nei 1988, 1989; Jones et al. 1990; Ohta 1991). For example, in individuals from West Africa where malaria is endemic, the allele HLA-B53, which is associated with recovery from a lethal form of malaria, is very common (Hill et al. 1992). The pattern of MHC alleles across North American Indian populations shows evidence of selection suggested to be caused by parasite selection pressure (Parham and Ohta 1996). Evidence for parasite selection pressure in a nonhuman population comes from a study of Soay sheep on the island of St. Kilda, Scotland, where a high frequency of heterozygotes for adenosine deaminase (deficiency of which is associated with defective T and B cell production) appears to be maintained by a gastrointestinal nematode worm (Gulland et al. 1993).

WHY MIGHT PARASITE SELECTION PRESSURE BE VARIABLE?

During its lifetime, the vertebrate host encounters many different types of parasites with varying levels of exposure. Varying parasite selection pressure may arise for several reasons. For example, host-parasite systems, in a similar fashion to predator-prey or host-parasitoid systems, show a propensity to oscillate. Spatial and stochastic effects can also contribute to the variability of parasite selection pressure. A large difference in the generation time, as is common between microparasites and their hosts, provides a large window of time within which parasite selection pressure can be variable. A high level of host motility may increase the risk of parasite encounters and a high level of complexity of the host may permit a high level of parasite diversity (due to the higher number of "niches" within the host).

(2) Parasites Often Give Reliable, Nonlethal Cues

The nonspecific immune response is induced by common microbial constituents that provide a reliable cue of certain parasites. Many parasites express more than one of these constituents, the function (and hence structure) of which have to be conserved. The specific immune system can be triggered by the nonspecific response, and provide better recognition of these parasites. Parasites may not be able to avoid providing more general signals of "danger" that may be important in inducing an immune response (Matzinger 1994), but this view is not universally accepted.

Parasites often provide cues before the host dies. Infecting doses of parasites are often low (although quantitative data on exact numbers are scant), and hence there is often a window of time before appreciable damage to the host can accumulate. This is especially true for intracellular parasites, which rely upon replication within the host to reach high densities and initially infect only a small number of cells. In addition, many parasites (such as directly transmitted parasites) require the host to live (at least for a short time) in order to transmit to another host, and so may evolve to a level of virulence that does not result in the rapid death of the host (although how parasite virulence evolves is a controversial topic; Frank 1996).

PARASITE EVASION OF RECOGNITION

Some parasites may avoid nonspecific "pattern recognition" if they do not exhibit conserved antigens. Trypanosomes and some bacteria are covered in a coat that hides conserved antigens from nonspecific immunity. Specific immunity can recognize these antigens due to the immense diversity of the clonal receptors, but needs to be alerted to the presence of a parasite. Some parasites have evolved to mimic host antigens and exploit the host's need to remain tolerant of its own tissues. Many DNA viruses (vaccinia, Epstein-Barr virus, cytomegalovirus in humans) and RNA viruses (feline and murine sarcoma viruses) mimic host defense proteins. Host defense proteins, e.g., interleukins and interferons, are much more diverse than other proteins not involved in defense, and molecular mimicry by parasites could have selected for this high diversity (Murphy 1993). Parasites may also evolve to escape recognition by particular MHC molecules—if parasite antigens are not presented by MHC, T cells cannot "see" them. In southeastern China and in

Papua New Guinea, where about 60% of the populations carry the HLA-A11 allele, many isolates of Epstein-Barr virus have mutated to no longer bind to MHC. Virally infected cells often downregulate MHC expression, effectively making them "invisible" to T cells. Natural killer cells may have evolved to complement the activity of T cells, as they nonspecifically lyse cells with low MHC class I expression. Some parasites may avoid providing reliable cues by hiding in immunologically privileged sites or by existing at levels too low to be detected. Herpes simplex and herpes zoster viruses both have a latent stage where they infect nerves, to be reactivated later, usually by some stress or alteration in immune status. *Mycobacterium tuberculosis,* an etiological agent of tuberculosis, may enter a dormant stage that would aid persistence until the immune response weakens.

In summary, the combined use of nonspecific and specific immune responses makes it very difficult for the parasite to entirely escape detection. However, recognition is only one component of the immune response; the rapid deployment of effectors is just as important. The ability of the immune system to "choose" an appropriate immune response, and to mount the response rapidly, together with parasite counteradaptations to the effector component, is now discussed.

(3) The Defense is Effective

For an inducible response to be effective, it should be mounted rapidly after detection of the threat. The combination of nonspecific and specific components provides the best of both worlds in terms of rapid response kinetics and an ability to recognize a large number of antigenically diverse parasites. The response should also be appropriate to clear the parasite. As already discussed, the immune system can translate information on the parasite (such as the site of entry, the size of the parasite population, whether parasites are extracellular or intracellular) into an appropriate response.

EFFECTIVENESS OF THE NONSPECIFIC IMMUNE RESPONSE

The nonspecific immune system can respond rapidly (within 0–4 hours) after infection to a large number of parasites, due to the nonclonal nature of its receptors that can recognize a variety of invariant parasite antigens. The tight coupling of rapid nonspecific recognition to nonspecific effectors allows rapid responses to many parasites, buying time for the induction of a specific response, which is more powerful but takes longer to mount (> 96 hours). The nonspecific immune system was probably a very important prerequisite for the evolution of specific immunity, as specific immunity alone would involve time lags so long as to be ineffectual. The nonspecific immune response is also "hard-wired" to recognize different kinds of parasites, and hence can launch appropriate nonspecific effector mechanisms. The nonspecific immune response also relays this information to the specific response. However, not all antigens can be recognized by nonspecific immunity.

Since many parasites are cleared quickly without triggering an early induced response or other detection, it is difficult to quantify exactly how effective nonspecific immunity is. Some data illustrating the effectiveness of nonspecific immunity come from individuals with genetic defects of nonspecific immunity. There are many genetic defects that lead to phagocyte and complement deficiencies, associated with a susceptibility to bacterial infections (Colten and Rosen 1992, Rotrosen and Gallin 1987). A defect of natural killer cells is associated with a susceptibility to herpes viruses, which can evade specific recognition by downregulation of MHC class I.

EFFECTIVENESS OF THE SPECIFIC IMMUNE RESPONSE

The clonal nature of specific receptors allows many antigens to be exactly recognized, but specific responses take time to mount. To compensate for this, there is strong positive feedback during the immune response that increases the rate of induction. Activated antigen-presenting cells activate T helper cells, which in turn activate antigen-presenting cells, and so on. After the primary response, the initially small clone of antigen-specific cells has expanded, and can respond more strongly and rapidly to the next antigenic challenge; the phenomenon of immune memory. Isotype switching and affinity maturation during the primary response to an antigen also help the secondary response to be faster and stronger. By clonal selection, the immune system can generate powerful, hard-wired responses to parasites over the timescale of the host lifetime, rather than over an evolutionary timescale.

Data showing the effectiveness of the specific response come from individuals with genetic defects of the specific immune response and those suffering from immunodeficiency due to cancer, certain radiotherapy and chemotherapy treatments, or infection with human immunodeficiency virus. Defects in T cell function lead to severe combined immunodeficiency, resulting in susceptibility to many parasites (K. Schwartz et al. 1991; Bosma and Carroll 1993). A defective antibody response, such as a lack of B cells or no isotype switching, is associated with increased susceptibility to extracellular bacteria (Bruton 1952; Preudhomme and Hanson 1990; Volanakis et al. 1992). IgA deficiency is associated with an increased susceptibility to respiratory infections. The destruction of CD4+ T cells in HIV infection results in defects in both cellular and humoral immunity, resulting in an increased susceptibility to many parasites, e.g., cytomegalovirus, *Pneumocystis carinii* (a fungal parasite causing pneumonia) and *Mycobacterium avium intra-*

cellulare (M. Adler 1993). Many of these parasites are "opportunistic," being controlled in healthy individuals by robust T cell responses.

PARASITE COUNTERSTRATEGIES TO IMMUNE EFFECTOR MECHANISMS

The delay in mounting a response allows the parasite a period of relatively unrestricted growth during which it can establish itself. If this delay is too long, then too much irreversible damage may have been done, and constitutive defenses may be favored over inducible ones. Measles and influenza viruses adopt a strategy of replicating rapidly during the primary response, relying upon a high transmission rate between hosts to perpetuate.

The threshold of activation and the magnitude of the immune response are under tight control. If the threshold of activation of an immune response is too low, then the risk of launching an immune response in the absence of a parasite is high. Some parasites may exploit this by growing only very slowly. Though not achieving high levels of parasitemia, such parasites may avoid aggressive immune responses, allowing persistence at low levels (McLean and Kirkwood 1990). *Mycobacterium tuberculosis* may adopt such a strategy (Antia et al. 1996). Once induced, a very aggressive immune response may cause more damage to the host than the parasite, and so at high levels of stimulation the immune response is inhibited. Some parasites have evolved to overstimulate the immune response using so-called superantigens that stimulate 2–20% of all T cells, resulting not in an efficient parasite defense, but in the pathological overproduction of cytokines and the concomitant reduction in adaptive immunity.

A major benefit of inducibility coupled with specificity is the ability to fine-tune effector mechanisms to different kinds of parasite. However, many parasites have evolved the ability to divert immune responses into launching inappropriate effectors. HIV and tuberculosis are effectively cleared by type 1 responses, but type 2 responses are often induced during infection, allowing persistence of the parasites (Mosmann and Sad 1996). It is possible that the switch to type 2 responses is a result of manipulation by the parasite. The complex nature of vertebrate immunity, with multiple, overlapping mechanisms, may reflect selection pressure by parasites to manipulate effector mechanisms.

To summarize, the combination of nonspecific and specific effector mechanisms that are tightly coupled to nonspecific and specific recognition mechanisms allow the vertebrate immune response to respond reliably, rapidly, and appropriately to many (but not all) parasites, overcoming two important constraints for a response to be inducible. The benefits of an inducible defense compared to a constitutive one are now discussed.

(4) There Are Costs to Defense

The immune response may be costly in terms of energy and immunopathology.

Benefits of inducibility are often discussed in relation to cutting costs associated with the defense (e.g. see Lively and Apanius 1995). Every aspect of an immune response requires energy, yet few data are available on energetic cost. Chickens undergoing induced immune responses have retarded growth rates (Klasing et al. 1987). The induction of fever, which can help clear parasites directly (due to the higher mortality of parasites) and indirectly (through faster immune responses at higher temperatures) is energetically costly. A fever of 3 degrees Celsius above baseline in birds and mammals is associated with an increase of 20% in metabolic rate above baseline (Kluger 1979). However, after an immune response and during chronic stress, the immune response is downregulated, a process which itself requires energy, suggesting that energy may not be the only cost associated with an immune response (Sapolsky 1994). An immune response is often a two-edged sword resulting in damage to the host as well as the parasite, and this is an important cause of pathology in infections such as HIV and tuberculosis. A classic example of this is infection of mice with lymphocytic choriomeningitis virus (LCMV), where the immune response causes all the damage–immunosuppressed mice are completely healthy (Zinkernagel et al. 1986). However, it remains unclear whether the reduction of costs of an immune response is sufficient to select for inducibility.

A very important benefit of inducibility is the flexibility it confers on the response. An inducible specific system, such as vertebrate immunity, can fine-tune its response to the size and type of the infecting parasite population. The ability of the immune response to be stronger when stimulated by a larger parasite population allows it to deliver a "short sharp shock" to the parasite population, which may be more effective in clearing parasites than a constant, constitutive defense and may also slow the evolution of escape from host defenses. By responding to parasites as they infect the host, a specific system can choose particular effectors to efficiently clear the parasite. The need to respond sufficiently and appropriately to a diverse parasite population may be a very important advantage of inducibility over constitutiveness. Due to their short generation time, parasites can potentially evolve very rapidly. An inducible defense allows the host to respond to these evolutionary changes in the parasite on a timescale equivalent to that of the turnover of the parasite, and this may be an important reason why parasites may not have necessarily "won the arms race" exemplified by host-parasite coevolution.

In summary, the immune system in its interaction with parasites fits the requirements for the evolution of inducibility: parasites exert a strong, variable selection pressure; the immune system can recognize parasites reliably; the immune system can rapidly mount sufficient, appropriate responses; and there are costs to defense, not only in terms of energy and immunopathology, but also the potential cost of responding insufficiently and/or inappropriately to a parasite threat. Consequently, attempts to quantify the cost of an immune response will have to be measured in the context of parasite infection. Inducibility allows the immune response to be flexible in the face of a diverse, potentially rapidly evolving parasite population. Note that successful antibiotic therapy often bears similarities to a healthy immune response; rapid, correct diagnosis (especially of any preexisting drug resistance) and the rapid administration of several appropriate complementary antibiotics with different effects.

The specificity of the immune response underlies the speed at which the response is induced, and when coupled with inducibility, allows effectors to be fine-tuned to different parasites. In order to fully understand the inducible nature of immunity, we must understand what factors determine the specificity of the immune response. The next section speculates what these factors may be.

What Determines the Specificity of the Immune Response?

The specificity of immune responses varies tremendously among animals. With the exception of the jawless fishes, all vertebrates, but no invertebrate studied so far, possess specific immunity. Even within vertebrates, there is considerable variation in the specificity of the immune response. Lower vertebrate immune systems have a much smaller number of specificities than birds and mammals (fewer than 500,000 compared to between 10^7 and 10^9; Du Pasquier 1982).

Parasite selection pressure and the need to remain tolerant to one's own tissues may make high levels of specificity selectively advantageous. An animal may require many specificities as it may encounter a very varied parasite population over its lifetime. This could arise for many reasons. A higher level of antigenic complexity may offer more niches for different kinds of parasites. A long lifespan and high levels of motility may also increase the diversity of parasites to which the animal is exposed. However, relatively few specificities are required to recognize all parasites. De Boer and Perelson (1993) have suggested that in organisms that are more antigenically complex, more specificities are required to ensure that many parasites are recognized while remaining tolerant to self.

There are several costs of specificity that may limit the number of antigens recognized by the immune system. If lymphocytes are energetically expensive to produce, the expansion and selection of a few clones of cells, with destruction of many cells in the process, may make the generation of high levels of diversity too energetically costly. The number of MHC alleles found in a given population may be limited by the need to tolerate many T cells to each variant of MHC. There is also a cost of the delay in setting up a diverse immune repertoire early in life, and a cost of the delay in launching a specific response. If the animal is exposed to many pathogens early in life, there may not be enough time to generate high levels of diversity before infection occurs, so the immune system must rely more on germ line encoded receptors and any mutants among them. If the delay in inducing a response is long, as is the case for a highly specific response, launching a late response may confer little benefit.

There are several barriers to the production of high levels of diversity. First, a mechanism for generating diversity has to evolve. Janeway (1994) suggested that the complex mechanism of somatic gene rearrangement used to produce specific receptors may have evolved originally to regulate gene expression rather than to generate diversity per se. Even if such a mechanism were present, if cells cannot proliferate rapidly, if the number of lymphocyte generations is small compared to the lifetime of an individual, or if lymphocyte populations are small, somatic recombination will generate little diversity and so the animal would have to rely more on its germ line cells and any mutants among them.

In order to understand the relative importance of the benefits, costs, and constraints of generating a given level of specificity, more quantitative data are needed. Understanding why invertebrates do not possess specific immunity, and why specific immunity in vertebrates is so much higher than in lower vertebrates requires a more detailed knowledge of (a) the time course of infection, (b) the time lags involved in setting up a fully functional repertoire, and (c) the delay in inducing a specific response after infection.

Conclusions

The vertebrate immune response provides us with an excellent example of an inducible defense tracking a complex biotic agent. In keeping with other inducible defenses, there is evidence to show that the selection pressures maintaining the inducibility of the immune response are (1) that parasites exert a strong, variable selection pressure, (2) parasites often provide reliable cues, (3) the immune response is usually effective at clearing parasites, and (4) there are costs in mounting an immune response, although the data needed to quantify these factors are sparse.

Benefits of inducibility are often discussed in terms of saving costs of energy and immunopathology, although it is unclear whether these costs are high enough to select for inducibility. A very important benefit of inducibility is the flexibility it confers. Immune responses are amplifiable, i.e., responses are stronger when parasite loads are higher (generally), and the delivery of a "short sharp shock" may be more effective at parasite clearance than a constitutive defense. The ability to recognize many different parasites allows (through mechanisms that are currently unclear) the immune system to "choose" which effectors will be most effective. The ability to mount sufficient, appropriate responses without error to a changing parasite population may be an important advantage of inducibility over constitutiveness. Though parasites may rapidly evolve mechanisms to escape host immunity due to their short generation time, hosts can respond rapidly to these changes due to the inducible nature of immunity. The ability of the immune system to respond on a timescale similar to the turnover of parasites is central to the ability of vertebrates to overcome parasites despite potentially rapid parasite evolution (B. Levin and Bull 1994).

The true costs of inducibility may be the need to recognize parasites reliably, the need for a threshold of activation, and the time lags involved in launching an appropriate response. The specificity of the response lies at the heart of these constraints. Vertebrate-specific immunity can recognize many parasites but is slow to mount. It is unlikely that specific immunity could have evolved in the absence of nonspecific immunity, which provides rapid defense against many parasites and provides important information to the specific immune system about the type of parasite. The different levels of specificity generated in different host species may reflect the balance between the need to recognize many parasites and the need to respond rapidly to parasites on infection. The phenomenon of immune memory results in a shift in this balance toward faster responses over the lifetime of an individual, with relatively little cost in reduced recognition of other parasites due to the focus on parasites that have attacked in the past and hence are likely to attack in the future. Quantifying costs so intimately linked to the dynamics of an immune response is likely to be extremely difficult.

7

Kairomone-Induced Morphological Defenses in Rotifers

JOHN J. GILBERT

Abstract

Nine species of planktonic rotifers are known to exhibit morphological responses to kairomones produced by various predators (*Asplanchna*, copepods) and interference competitors (cladocerans). These responses involve the development of spines or appendages. The kairomone released by the predator *Asplanchna* is the only one so far investigated and seems to be a protein. It accumulates rapidly in laboratory cultures and occurs in inducing concentrations in natural systems. It directly or indirectly stimulates the oocytes of *Brachionus calyciflorus*, and presumably other responsive rotifers, to develop into long-spined females. Kairomone-induced responses can provide effective post-encounter defense against predators or interference competitors producing the kairomone. The response of *B. calyciflorus* to the *Asplanchna* kairomone increases its ability to coexist with competitively superior congeners that do not respond to the kairomone and are more susceptible to *Asplanchna*. Some species of *Keratella* respond to at least two different kairomones—one from *Asplanchna*, and one or more from copepods and cladocerans. Life-table experiments indicate that morphological responses to the *Asplanchna* kairomone involve no energetic cost in *B. calyciflorus*, and an energetic cost in *Keratella testudo* only at high food concentrations. The design of these experiments, however, may underestimate energetic costs. Other, unknown types of fitness costs may be associated with the morphological responses.

Introduction

Species from several genera of freshwater planktonic rotifers have evolved the ability to develop defensive phenotypes in direct response to chemical messengers, or kairomones (W. Brown et al. 1970), released by some predators and interference competitors. These nongenetic, phenotypic responses can involve a *de novo* induction or elongation of spines, an elongation of

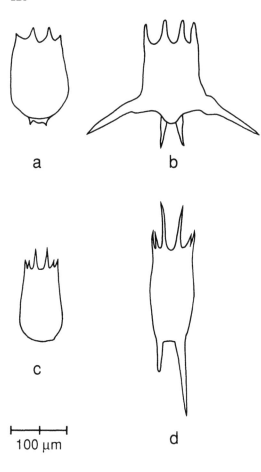

Fig. 7.1. Basic and *Asplanchna*-induced, exuberant phenotypes of *Brachionus calyciflorus* (a, b) and *Keratella slacki* (c, d). (Redrawn from Gilbert 1967 and Gilbert and Stemberger 1984.)

appendages, and an associated increase in body size. With increasing kairomone concentration, spine development increases continuously up to some maximum response. Phenotypes with no environmentally induced spine development have been termed "basic," and those with pronounced environmentally induced spine development have been called "exuberant" (Stemberger and Gilbert 1987a). Kairomone-induced, exuberant phenotypes generally are less likely to be captured, damaged, or ingested following encounters with the predators or competitors that induce them.

The most extensively investigated, kairomone-controlled polymorphism in rotifers was first described by Beauchamp (1952a,b). It involves the induction and elongation of spines in *Brachionus calyciflorus* by a kairomone from the predatory rotifer *Asplanchna* (Beauchamp 1952a,b; Gilbert 1966, 1967, 1980; Gilbert and Waage 1967; Halbach 1969a,b, 1970, 1971a,b; Halbach and Jacobs 1971; Stemberger 1990) (fig. 7.1a,b). An *Asplanchna*

TABLE 7.1
Effects of Medium Conditioned by Potential Predators or Interference Competitors on Spine Development in *Keratella*

Keratella Species	Predator or Competitor and Its Effects	Induced Spines Effective Against	Reference
slacki	*Asplanchna* (I) *Tropocyclops* (N)	*Asplanchna*	1
tropica	2 copepods (I) 5 cladocerans (I) Notonectid (R)	*Acanthocyclops* copepodites (but not adults)	2,3
cochlearis	*Asplanchna* (I) *Tropocyclops* (I) *Mesocyclops* (I) 6 cladocerans (I)	*Asplanchna*, *Daphnia*, and *Tropocyclops* (but not *Mesocyclops*)	4–6
testudo	*Asplanchna* (I) *Daphnia* spp. (I) *Bosmina* (I) *Tropocyclops* (I) *Diaptomus* (I) *Epischura* (I) *Synchaeta* (N) *Chaoborus* (N)	*Daphnia pulex* and *Asplanchna girodi* (but not *A. brightwelli* or *Epischura*)	7
quadrata	*Diacyclops* (I)		8

References: (1) Gilbert and Stemberger 1984; (2) Marinone and Zagarese 1991; (3) Zagarese and Marinone 1992; (4) Stemberger and Gilbert 1984; (5) Gilbert (1988a); (6) Gilbert and MacIsaac (1989); (7) Stemberger and Gilbert (1987b); (8) Stemberger (unpubl. obs.).

Note: Some media increased spine length (I), some had no effect on spine length (N), and one reduced spine length (R).

kairomone also induces spine development in three other species of *Brachionus* (*B. urceolaris* var. *sericus*, *B. bidentata*, *B. patulus*) and setae elongation in *Filinia mystacina* (Pourriot 1964, 1974; Sarma 1987). Finally, recent research has demonstrated spine induction in several species of *Keratella* by kairomones from *Asplanchna* and various crustaceans (table 7.1). Spine development in some of these *Keratella* species is highly asymmetric. For example, in the most exuberant forms of *K. slacki* (fig. 7.1d) and *K. tropica*, the right posterior spine is always very much longer than the left one, which may be lacking entirely (Gilbert and Stemberger 1984; Marinone and Zagarese 1991; Zagarese and Marinone 1992). This asymmetry may provide effective defense with minimal cost (Gilbert and Stemberger 1984).

Polymorphic rotifers can be readily cultured and provide excellent model systems for studying ecological and evolutionary aspects of phenotypic plasticity. Particularly interesting questions involve the production, chemical nature, and mode of action of inducing kairomones, the effectiveness of the phenotypic responses in reducing mortality from predators and interference competitors, the long-term maintenance of different phenotypes in natural populations, and fitness trade-offs in the production of exuberant phenotypes.

The literature on kairomone-controlled polymorphisms in rotifers has not yet been reviewed, although some aspects of the subject have been summarized or considered (Stemberger and Gilbert 1987a; Havel 1987; Harvell 1990a). The present chapter is the first synthesis of much of the available information. It emphasizes kairomone dynamics and also the nature, ecological significance, and cost of developmental responses to kairomones. Particular attention is devoted to problems regarding the design and interpretation of life-table experiments conducted to assess the fitness cost of these responses.

The *Asplanchna* Kairomone

Of the kairomones that affect rotifers, only the one from *Asplanchna* has been investigated. All studies of the *Asplanchna* kairomone involve the phenotypic response of *Brachionus calyciflorus* (Gilbert 1966, 1967; Gilbert and Waage 1967; Halbach 1970). This response involves the induction of long, articulating, posterolateral spines, sometimes *de novo*, and an elongation of three other pairs of spines (Gilbert 1967) (fig. 7.1). Adult *B. calyciflorus* with *Asplanchna*-induced spines also may have a larger body size (Gilbert 1967, 1980; Stemberger 1990). However, increased body size seems to be attributable not to the kairomone, but to some other factor(s) present in *Asplanchna*-conditioned medium. When cultured in conditioned medium from which the kairomone had been removed by passage through, and probably adsorption onto, a membrane filter, adults with the basic phenotype attained the same, relatively large body size as exuberant adults cultured in conditioned medium with the kairomone (Gilbert 1967). In another study, body sizes of the basic and exuberant phenotypes, cultured without and with *Asplanchna*-conditioned medium, were similar (Halbach 1970).

The kairomone is a heat-stable, nondialyzable, pronase-sensitive factor that probably is a protein (Gilbert 1967). Its mechanism of action is not known. Since exposure of recently oviposited, uninucleate eggs to the kairomone does not induce them to develop into more spinous individuals, induction must occur when eggs are within the maternal body cavity (Gilbert 1967). There, the kairomone may penetrate the cell membrane of the egg and directly affect embryonic development. Alternatively, it somehow may alter maternal physiology, perhaps by triggering the release of another factor, which in turn promotes spine development. When *B. calyciflorus* neonates without posterolateral spines are cultured with inducing concentrations of kairomone, their oocytes develop into females with posterolateral spines (Gilbert 1967).

The kairomone does not affect postnatal growth of the exuberant phenotype of *B. calyciflorus*. Both with and without it, the posterolateral spines grow much less rapidly than the body. The rate of growth of these spines relative to that of the body (k) was found to be 0.43 by Gilbert (1967) and 0.54 by Halbach (1970). This allometry may be ecologically significant, as small individuals are especially susceptible to *Asplanchna* predation (Gilbert 1967, 1980) and would benefit from disproportionately long spines.

The concentration of *Asplanchna* kairomone in the environment should be the result of the rates at which it is produced, incorporated by the rotifer prey and perhaps other organisms, and degraded by bacteria. These three processes were addressed by Gilbert (1967), but the most extensive experiments on kairomone production and decay rates were conducted by Halbach (1970). In laboratory cultures, *Asplanchna* rapidly produces inducing levels of kairomone. One adult *A. sieboldi* placed in one ml medium at 20°C for 25 seconds produces enough kairomone to induce just detectable posterolateral spines. The activity of the medium, as assessed by bioassay with *B. calyciflorus*, increases with *Asplanchna* density and exposure time, and with temperature. By determining the decay rate of the kairomone and assuming from empirical observations that the kairomone does not continue to accumulate after 24 hours, Halbach calculated that, at equilibrium, the density of *A. sieboldi* necessary to induce just detectable and also maximally developed spines at 20°C would be about 0.5 and 10 individuals per liter, respectively. These estimates are of course rough, because kairomone production should be greatly affected by the size and age of the *Asplanchna* in a population, and also by the amount of food available to the *Asplanchna*. However, they do indicate that natural densities of *Asplanchna* often are sufficiently high to produce inducing concentrations of the kairomone.

Several field studies have shown that *Asplanchna* does produce enough kairomone to affect the phenotype of co-occurring *B. calyciflorus*. The most direct evidence comes from the biological assay of water from ecosystems with *Asplanchna*. In these assays, water is screened and centrifuged to remove seston, and then employed as medium for a laboratory strain of *B. calyciflorus*. Neonates of the basic phenotype are placed in this medium and cultured in it until they produce several offspring. Then, the phenotypes of these offspring are characterized by a spine index that indicates the length of the posterolateral spines relative to the length of the body. Such assays have demonstrated that natural populations of *Asplanchna* produced inducing concentrations of kairomone in Lake Washington (Gilbert 1967), in a pond in

New Jersey (Gilbert and Waage 1967), in two basins in a botanical garden in Würzburg, Germany, and in a fish pond in Winterhausen, Germany (Halbach 1970). Furthermore, when conducted over a period of time, these bioassays show a positive correlation between the spine index of the laboratory animals and the phenotypes of the *B. calyciflorus* in the natural system (Gilbert and Waage 1967; Halbach 1970).

In addition, much circumstantial evidence supports the view that the *Asplanchna* kairomone influences the phenotype of *B. calyciflorus* in natural systems. When *Asplanchna* is present, *B. calyciflorus* typically is long-spined (Gilbert 1967; J. Green and Lan 1974). Also, in systems where both species co-occur, the spine length of *B. calyciflorus* tends to increase as the population density of *Asplanchna* increases (Gilbert 1967; Gilbert and Waage 1967; Halbach 1970). The most extensive data on this relationship are those of Halbach, who examined twenty-two water bodies every few days for up to several months. In two of these systems, Halbach also showed that the spine lengths of the *B. calyciflorus* were similar to those predicted from his laboratory studies on the effects of *Asplanchna* density and temperature on spine development.

The ability of *B. calyciflorus* to develop spines in response to the *Asplanchna* kairomone is of very clear ecological significance. Compared to the basic phenotype, long-spined phenotypes are much less likely to be captured and ingested by *Asplanchna* and require more time to be ingested (Gilbert 1967, 1980; Halbach 1971a). The spines mechanically interfere with the ability of *Asplanchna* to draw the *Brachionus* into its mouth and then to pass it from its pharynx down its esophagus into its stomach. In *B. calyciflorus*, the long posterolateral spines are especially effective, because they extend laterally when the contacted rotifer withdraws its corona and becomes more turgid (Beauchamp 1965; Gilbert 1967). The degree of protection afforded by long spines depends on the size of the *Brachionus* as well as on the size and species of the *Asplanchna*.

The morphological response of *B. calyciflorus* to *Asplanchna* greatly increases its probability of coexisting with this predator (Beauchamp 1952a, 1965; Halbach 1969a,b, 1971b). In mixed-species laboratory cultures with *A. sieboldi*, *B. calyciflorus* can outlast the *Asplanchna* while *B. rubens*, which has no morphological response to the *Asplanchna* kairomone, is rapidly driven to extinction (Halbach 1969b). Further, the presence of *Asplanchna* may enhance the ability of *B. calyciflorus* to coexist with superior congeneric competitors. When *B. calyciflorus* and *B. rubens* are cultured together, *B. rubens* outcompetes *B. calyciflorus*; however, when they are cultured together in the presence of *Asplanchna*, only *B. calyciflorus* can persist (Halbach 1969b). Temporal changes in the abundance of *Brachionus* species co-occurring with *Asplanchna* in natural communities also indicate that *B. calyciflorus* is much less inhibited by *Asplanchna* than species that have no morphological response to its kairomone: *B. rubens*, *B. urceolaris*, *B. angularis* (Halbach 1969a, 1971).

Kairomone Specificity

The *Asplanchna* kairomone probably is common to all species of the genus, as it is produced by *A. brightwelli*, *A. girodi*, *A. priodonta*, and *A. sieboldi* (Beauchamp 1952b; Gilbert 1967). The ability of other asplanchnid genera (*Asplanchnopus*, *Harringia*) to produce a similar kairomone has not been tested. The predatory rotifer *Eosphora najas* does not induce spine development in *Brachionus calyciflorus* (Beauchamp 1952b), and so must not contain this kairomone.

Kairomones from *Asplanchna* and a variety of crustacean predators and competitors can induce spine development in several species of *Keratella* (table 7.1). Zooplanktivorous insects do not seem to produce such kairomones. Medium conditioned by *Chaoborus* larvae had no affect on *K. testudo* (Stemberger and Gilbert 1987b), and that conditioned by the notonectid *Buenoa* slightly but significantly decreased spine length in *K. tropica* (Zagarese and Marinone 1992).

Examination of the information in table 7.1 permits the deduction that at least some of the predators and competitors that induce spine development in *Keratella* must produce chemically different kairomones. For example, *K. slacki* responds to an *Asplanchna* kairomone but not to one produced by the copepod *Tropocyclops* (Gilbert and Stemberger 1984), while both *K. cochlearis* and *K. testudo* respond to kairomones from both *Asplanchna* and *Tropocyclops* (Stemberger and Gilbert 1984, 1987b). If these kairomones were identical, *K. slacki* also would respond to the *Tropocyclops* kairomone. The failure of *K. slacki* to evolve a response to *Tropocyclops* may be due to lack of sympatry, or to the fact that the basic phenotype is already resistant to this small copepod. *K. cochlearis*, *K. testudo*, and *K. tropica* respond to kairomones from cladocerans as well as copepods. It is not known if the kairomones from these two groups of crustaceans are identical. Thus, *K. cochlearis* and *K. testudo* respond similarly to at least two different kairomones, one from *Asplanchna* and one or more from copepods and cladocerans. *K. tropica* may be similar in this respect but has not been tested with *Asplanchna*.

Significance of Response to Crustacean Kairomones

Development of spines in response to crustacean zooplankton can decrease the ability of a predator or interference competitor to capture, handle, injure, or ingest the rotifer. Calanoid and cyclopoid copepods eat a variety of rotifers (Williamson 1983, 1984, 1987; Williamson and Butler 1986; Stemberger 1985; Roche 1990a,b), and they may be less likely to capture and ingest the exuberant phenotypes of polymorphic species. This has been demonstrated for cyclopoids with both *Keratella cochlearis* and *K. tropica* (Stemberger and Gilbert 1984; Marinone and Zagarese 1991). However, this is not always the case. Spine development did not decrease the susceptibility of *Brachionus calyciflorus* to adult female *Mesocyclops* (Gilbert 1980), or that of *Keratella testudo* to adult female *Epischura* (Stemberger and Gilbert 1987b). Spines may be effective against some predators only when these predators are small or juvenile. Also, the spines of rotifers, such as *Keratella*, with a relatively thick lorica may be more rigid and hence more effective deterrents than those of rotifers, such as *B. calyciflorus*, which have a thinner, more flexible lorica.

Spine development can also protect rotifers from mechanical interference by cladocerans. When drawn into the branchial chamber of *Daphnia pulex*, the exuberant phenotypes of *K. cochlearis* and *K. testudo* are less likely to be killed than the basic ones (Gilbert 1988a; Stemberger and Gilbert 1987b). The mechanism for the reduced risk probably is that the more spinous forms provide a greater irritant in the branchial chamber and are rejected more quickly (Gilbert 1988b). The extent of damage inflicted on rotifers increases with their residence time in the branchial chamber (Gilbert and Stemberger 1985; Burns and Gilbert 1986b), and spine development decreases the residence time of *K. testudo* in *Daphnia pulex* (Stemberger and Gilbert 1987b).

While large-bodied cladocerans are most likely to interfere with rotifers (Burns and Gilbert 1986a,b), small cladocerans probably also can kill or damage very young rotifers, which should be much more susceptible because of their smaller size and less rigid lorica. For example, *Ceriodaphnia dubia* with a body length of 0.7 mm imposed a higher mortality rate on *K. cochlearis* neonates than on adults (Gilbert and MacIsaac 1989). Therefore, spine development may increase the survivorship of rotifers exposed to both small and large cladocerans. This may explain why several species of *Keratella* have evolved developmental responses to a variety of different-sized cladocerans (table 7.1). However, the broad response of these polymorphic *Keratella* to cladocerans simply may reflect the fact that the kairomone produced by large-sized species of cladocerans also is produced by small-sized ones.

Cost of Developmental Responses to Kairomones

It has been suggested that kairomone-induced morphological responses in rotifers should involve some cost and evolve only when the benefits of the response are intermittent and outweigh the costs (Gilbert 1980). When predators or interference competitors become sufficiently abundant to produce inducing concentrations of kairomone, the more well-defended, exuberant phenotypes should be more likely to survive and reproduce. When predators and interference competitors are rare or absent, basic phenotypes should be more fit. Constitutive rather than induced morphological defenses should evolve if effective predators or interference competitors were present most of the time, or if there were no cost associated with the morphological response. It is noteworthy that the occurrence of *Asplanchna* in the ponds studied by Gilbert and Waage (1967) and Halbach (1970) was intermittent. Such a pattern of predator abundance is consistent with selection for an *Asplanchna*-induced defense in *B. calyciflorus*.

The first attempt to evaluate the cost of a kairomone-induced phenotype in zooplankton was with the *Asplanchna*-induced, long-spined phenotype of *Brachionus calyciflorus* (Gilbert 1980). Since then, two additional studies with *Asplanchna*-induced phenotypes have been conducted: a second one with *B. calyciflorus*, and one with *Keratella testudo* (Stemberger 1988, 1990). In each of the three studies, life-table experiments were employed to determine the survivorship, fecundity, and reproductive potential (r_m) of the basic and exuberant phenotypes. In the two studies with *B. calyciflorus*, the clones that were used typically developed short posterolateral spines in the absence of the *Asplanchna* kairomone and much longer ones with the kairomone; thus the "basic" phenotype of *B. calyciflorus* in these studies had short spines.

Some methodological problems are associated with conducting life-table experiments to assess fitness costs of exuberant phenotypes in rotifers and other taxa. These have received little attention. One problem involves a decision of whether to initiate experiments with neonates of the basic or the exuberant phenotype. Another problem involves the necessity of using crude extracts or conditioned medium to obtain a medium with the kairomone. Since no kairomone has been chemically identified, culture medium with the kairomone is prepared by adding extracts of the predator or competitor to the medium, or by allowing the predator or competitor to live in the medium for some time. Such media contain not only the kairomone but also a variety of other organic compounds extracted from, or released by, the predator or competitor. These compounds may adversely affect the target organism, either directly or indirectly, by supporting bacterial growth. Therefore, their presence may erroneously indicate that a developmental response to the kairomone involves a cost. Tollrian (1995, unpubl. obs.), for example, has shown that extracts of, and medium conditioned by, *Chaoborus* larvae induce phenotypic responses in *Daphnia* but also inhibit the *Daphnia*. Purification of these preparations preserved their inductive ability but greatly reduced their inhibitory side effects.

Three different types of designs could be employed in life-table experiments conducted to determine if costs are associated with developmental responses to kairomones. These experimental designs involve different combinations of neonate phenotype and presence or absence of kairomone. In design I, neonates of both the basic and exuberant phenotypes are cultured in medium without the kairomone and other factors released by the predator or competitor, and so the offspring produced by both phenotypes are of the basic phenotype. The advantage of this design is that individuals of both phenotypes are cultured in the same environment from birth until death. The disadvantage of the design is that any costs involved in the production or development of exuberant offspring are excluded. Also, the neonates of the two phenotypes have to be derived from parents grown under different conditions—predator- or competitor-free medium for the basic phenotype, and predator- or competitor-conditioned medium for the exuberant phenotype. In the latter case, conditioning factors other than the kairomone could influence postnatal survivorship, growth, and reproduction.

Design II is similar to design I except that neonates of the exuberant phenotype are cultured in medium conditioned by the predator or competitor. Thus, they will produce offspring of the exuberant phenotype. This design has the advantage of including any production and development costs associated with the exuberant phenotype. However, it accentuates the problem of confounding the effects of the induced phenotype and those that may be caused by predator or competitor conditioning factors other than the kairomone. For example, as mentioned earlier, medium conditioned by *Asplanchna*, but without the kairomone, contains a factor that can increase body size in *B. calyciflorus* (Gilbert 1967). Design II clearly would be ideal if a kairomone were identified and available in pure form. Then the kairomone would be used for the induction and the culture of exuberant individuals, and possible effects of other conditioning factors would be avoided.

Design III is similar to design II except that neonates of the basic phenotype are used in both treatments. This design has the advantage of using neonates with the same culture history, and should assess the cost of producing offspring of the exuberant phenotype. However, like design II, it may confound effects of the kairomone and those of other conditioning factors. More seriously, any costs of the exuberant phenotype relating to survival and reproduction from birth to death are excluded.

In the first experiments with *B. calyciflorus* (Gilbert 1980), designs I and III, respectively, were used to evaluate the costs of expressing and producing the exuberant phenotype. In the present paper, the originally published intrinsic rates of natural increase (r_m) have been corrected for a systematic program error, and were also statistically analyzed using a bootstrap procedure (Gilbert 1994, 1996). The results are presented in tables 7.2 and 7.3, and they show the same patterns reported in 1980. No fitness costs associated with the exuberant phenotype could be detected when food resources

TABLE 7.2
Population Growth Rates (r_m) of *Brachionus calyciflorus* with Short Spines (ss) and *Asplanchna*-Induced Long Spines (ls) Cultured without the *Asplanchna* Kairomone on *Enterobacter* at 23°C and Producing Short-Spined Offspring

Experiment	Food Concentration (μg/ml)	Parental Female Phenotype	r_m Mean	95% CL
1A	100	ss	0.50	0.45–0.55
		ls	0.61	0.51–0.71
1B	100	ss	0.51	0.47–0.55
		ls	0.69	0.65–0.73
2	10	ss	0.32	0.22–0.42
		ls	0.59	0.53–0.65

Notes: Design I life-table experiment (see text). Cohort sizes of 17 or 18. Cultures changed daily. Bootstrapped r_m values from data in Gilbert 1980.

were abundant and supported high population growth rates. In fact, in the design I experiments (table 7.2), at each of two high food concentrations, long-spined rotifers producing short-spined offspring had much higher population growth rates than short-spined rotifers producing short-spined offspring. In the design III experiments (table 7.3), short-spined rotifers producing long-spined offspring in the experimental treatment and those producing short-spined offspring in the control treatment had essentially identical population growth rates. Thus, at the food concentrations used, there was no evidence of any costs involved with the production and development of the exuberant phenotype.

TABLE 7.3
Population Growth Rates (r_m) of Short-Spined *Brachionus calyciflorus* Cultured without and with the *Asplanchna* Kairomone on *Enterobacter aerogenes* at 23°C and Producing Offspring with Short Spines (ss) and Long Spines (ls), Respectively

Experiment	Food Concentration (μg/ml)	Offspring Phenotype	r_m Mean	95% CL
3	100	ss	0.50	0.46–0.54
		ls	0.52	0.47–0.57
4	20	ss	0.57	0.54–0.60
		ls	0.57	0.55–0.59

Notes: Design III life-table experiment (see text). Cohort sizes of 16 or 18. Cultures changed daily. Bootstrapped r_m values from data in Gilbert 1980.

Similar, but more extensive results with *B. calyciflorus* were obtained by Stemberger (1990). Using design I experiments, he showed that long-spined parents produced more offspring and had higher population growth rates than short-spined ones at each of five food concentrations, with one concentration near the threshold food concentration (where $r_m = 0$) and one below it. Therefore, the exuberant phenotype appears to be more fit. Its population growth rate was higher at all food concentrations. Therefore, its threshold food concentration also was lower, permitting it to reproduce at lower food concentrations.

The higher population growth rate of the exuberant form at high food concentrations is consistent with a known relationship between rotifer body size and maximum population growth rate (r_{max}). The exuberant forms in these experiments had larger body sizes and weights (Gilbert 1980, Stemberger 1990), and a comparison of eight species of planktonic rotifers showed a strong, positive relationship between body mass and r_{max} (Stemberger and Gilbert 1985, 1987b). The lower threshold food concentration of the exuberant form is not consistent with the larger size of this form. The across-species comparison just cited also demonstrated that threshold food concentration increases with body mass (Stemberger and Gilbert 1985, 1987c). Thus, the larger, exuberant form was expected to have a higher threshold food concentration.

The higher population growth rate of the exuberant phenotype at very low food concentrations can be explained by an independent effect of low food concentration on spine development and sinking rate (Stemberger 1990). At low food concentrations *B. calyciflorus* produces short posterolateral spines in the absence of *Asplanchna* (Schneider 1937; Erman 1962; Halbach 1970; Stemberger 1990). Since this rotifer has a thin lorica, its soft and flexible spines contribute little additional mass. Thus, at low food concentrations spine development should decrease sinking rate and provide an energetic advantage (Stemberger 1990). Also, Erman (1962) suggested that long spines may increase clearance rate. While this does not occur at high food concentrations (Halbach 1971a; Bogdan and Gilbert 1982), it may at low ones (Stemberger 1990). The effects of the *Asplanchna* kairomone and low food concentration are additive, and so the posterolateral spines of kairomone-induced phenotypes are especially long at low food concentrations (Halbach 1970; Stemberger 1990). This greater spine development may further decrease the sinking rate and account for the greater ability of the exuberant phenotype to survive and reproduce at low food concentrations.

Kairomone-induced, long-spined *B. calyciflorus* produce larger eggs than similarly sized, short-spined *B. calyciflorus* at high food concentrations (Gilbert 1980). For females with a body length of 161 μm, mean (1 SE) egg volumes of the long- and short-spined phenotypes were 2.79 (0.07) and 1.97 (0.1) × 10^6 μm³, respectively. The greater egg volume of the long-spined phenotype probably reflects a greater energy investment that can be sustained as long as food concentrations are high. Production of larger eggs in the presence of *Asplanchna* may allow the development of larger and more spinous young that can reproduce earlier as well as escape predation. The relationship between body size and egg volume for the two phenotypes at low food concentrations has not been investigated.

The *Asplanchna* kairomone also induces a long-spined phenotype in *Keratella testudo* (Stemberger and Gilbert 1987b). The cost of this response was investigated by Stemberger (1988). Using design I life-table experiments at six food concentrations, he found that the population growth rate (r_m) of the exuberant phenotype was lower than that of the basic phenotype at the four highest concentrations, but similar to it at the two lowest concentrations near the threshold food concentration. This pattern can be explained by changes in the effects of the induced spines at high and low food concentrations. The lorica, and hence the spines, of *K. testudo* are more dense and rigid than those of *B. calyciflorus*. Thus, at high food concentrations, well-fed animals of the exuberant phenotype have a higher specific gravity due to the greater amount of lorica associated with their extra spines. Therefore, they have to expend more energy to counteract sinking.

However, at low food concentrations the exuberant and basic phenotypes of *K. testudo* have much lower and more similar sinking rates. Under these conditions, animals sink in a horizontal position, rather than posterior end downwards at high food concentrations, and the greater spine development of the exuberant phenotype greatly increases its coefficient of form resistance relative to that of the basic phenotype. Thus, at low food concentrations, the greater drag from the posterior spines of the exuberant phenotype may offset the greater specific gravity of this phenotype and explain why population growth rates of the basic and exuberant phenotypes are similar. As the exuberant phenotype of this species is induced by cladoceran competitors, as well as by *Asplanchna* and copepods, it often may be associated with high densities of competitors and hence periods of low food availability. Thus, this phenotype frequently may be induced under conditions where it is not at an energetic disadvantage.

It is clear from these studies with *B. calyciflorus* and *K. testudo* that an assessment of energetic costs associated with defensive, phenotypic responses to predator or competitor kairomones requires examination of the reproductive potential of basic and exuberant phenotypes over a range of food concentrations. Quite different views of fitness costs could result from experiments conducted at high and low food concentrations. For example, as in *K. testudo*, an exuberant phenotype could have a population growth rate lower than that of a basic phenotype at a high food concentration, but a similar one at a low food concentration. Thus, the exuberant phenotype may be less fit at high food concentrations, but have a similar threshold food concentration and be equally able to survive and reproduce under conditions of severe food limitation.

In interpreting the results of life-table experiments to assess the cost of exuberant phenotypes, it is necessary to keep in mind methodological problems involved in the experimental design. For example, all of the experiments with *K. testudo*, and most of the ones with *B. calyciflorus*, employed a procedure (design I) where neonates of both the basic and exuberant phenotypes were cultured without the *Asplanchna* kairomone. As discussed earlier, this design has its advantages but would not include any costs associated with producing offspring of the exuberant phenotype. Thus, these experiments may underestimate the costs of maintaining this phenotype.

Also, it is important to consider the possibility that the cost of exuberant phenotypes in rotifers may be less related to energetics than to other, as yet unknown, fitness parameters. While exuberant phenotypes may show no reduction in reproductive potential, they may be less able to survive or reproduce in natural environments and communities. For example, they may be more susceptible than basic phenotypes to certain predators. Also, there could be very unexpected fitness costs. Exuberant mictic females may be less likely to be fertilized by males, and hence to produce resting eggs from which individuals may initiate new populations in future years. Perhaps long spines could decrease the ability of a male to circle a female during its characteristic precopulatory behavior (Gilbert 1992).

Potential for Behavioral Responses to Kairomones

Behavioral responses to predator kairomones are known in ciliates (Kusch 1993c; Kuhlmann 1994) and various metazooplankton (Larsson and Dodson 1993; Dodson et al. 1994). To date, there has been no attempt to determine if such responses occur in rotifers. However, it seems likely that rotifers could sense chemicals released into the environment by predators and interference competitors, and then either swim away from strata where they are most abundant or exhibit more pronounced escape responses to actual encounters with them. While planktonic rotifers have limited powers of locomotion, with normal swimming speeds ranging from about 0.3 up to about 1.5 mm/s (Stemberger and Gilbert 1987c), they still may be able to adjust their position in the water column to minimize encounters with predators and interference competitors. Also, a number of rotifers have escape responses in which rapid swimming is initiated after contact or near contact with predators and interference competitors. Such responses occur in *Filinia, Hexarthra, Keratella*, and *Polyarthra*, but have been described in detail and analyzed only in *Polyarthra* (Gilbert 1985, 1987; Kirk and Gilbert 1988) and *Keratella* (Gilbert and Kirk 1988). These escape responses may be intensified in the presence of kairomones from predators or interference competitors. While behavioral responses to such kairomones may occur in rotifers with and without associated morphological responses, they may be more likely in the latter. Evolution of an effective morphological response may diminish selective pressure for a behavioral one, and vice versa.

8

Predator-Induced Defenses in Ciliated Protozoa

HANS-WERNER KUHLMANN, JÜRGEN KUSCH,
AND KLAUS HECKMANN

Abstract

Inducible defenses are known for more than ten ciliate species belonging to the genera *Euplotes*, *Onychodromus*, *Lambornella*, *Colpidium*, and *Aspidisca*. Studies on freshwater *Euplotes* species revealed that the presence of uni- and multicellular predators induce the cells to undergo morphogenetic changes that make engulfment by the predators more difficult. *Euplotes* develops, within a few hours, extended lateral "wings" as well as dorsal and ventral projections. The changes do not require cell division and are accompanied by a reorganization of the cytoskeleton. Predation risks of transformed cells were found to be 2–20 times lower than of nontransformed ones. On the other hand, the demographic costs that the prey organisms have to pay for being better protected are reflected in a reduced rate in reproduction.

Some predator-released substances induce not only changes in morphology, but also a change in behavior: the induced cells move backwards upon contact with a predator, then turn around and withdraw. Cells capable of developing this behavioral response were able to coexist with their predator for a much longer period than cells not able to develop this response.

Two predator-released kairomones that induce defenses in *Euplotes* have been isolated. Both factors are polypeptides that differ from each other. They induce morphological changes in *Euplotes* at concentrations as low as 10^{-11} and 10^{-12} M, respectively. Recent data on an avoidance-inducing signal also indicate that it is a polypeptide.

Introduction

The phenomenon that ciliated protozoa may defend themselves by changing their cell shape in response to their predators (overview in Heckmann 1995) was first observed in 1984, when freshwater *Euplotes* cells with an unusual morphology were detected in water samples that contained *Lembadion bullinum*, a ciliate that feeds on *Euplotes* (Kuhlmann and Heckmann 1985). The predator was not able to capture the *Euplotes* cells because of their unusual circular form. When the *Euplotes* cells were put into fresh culture medium they changed back to their typical ovoid form. The ovoid-shaped cells were not protected from becoming swallowed if they were placed in a vessel with predators. Our observation, and the study of the influence of other predators on *Euplotes*, led to the discovery that *E. octocarinatus* and some of its close relatives are triggered by predator-released substances to undergo defensive changes in shape. The changes are not only of ecological importance; they also have interesting developmental implications. The system provides the opportunity to identify the substances released by the predator and recognized by the prey, which is apparently much more difficult in other prey-predator systems.

In the last twelve years predator-induced defense mechanisms have been described for more than ten ciliate species (fig. 8.1). It is unclear at the moment whether this defense strategy is also used by other protozoa. Since for many ciliates two or more morphological forms have been described, we assume that predator-induced changes in cell shape is a widespread phenomenon.

Predator-Induced Changes in Freshwater *Euplotes* Species

Studies carried out with *Euplotes* revealed that ciliates of this genus have at least eight predators (five ciliates, one rhizopod, one turbellarian, and one oligochaete) that induce the cells to undergo defensive morphogenetic changes (fig. 8.1). The changes are accompanied by a reorganization of the cytoskeleton of *Euplotes* and do not require cell division. Transformed *Euplotes* cells maintain their newly developed cell shape for many cell generations if the concentration of the inducing signal substance remains at a high level. Some predators of *Euplotes* induce not only changes in morphology, but also a change in behavior.

Morphological Transformation

If co-cultivated with one of its predators, *Euplotes* cells develop extended "lateral wings" that give them an almost circular appearance. In addition, an extended ventral projection, and a particularly protuberant dorsal ridge are formed (fig. 8.2). The induced changes look very similar in the six *Euplotes* species that showed a response (Kuhlmann and Heckmann 1985). The changes in cell shape can be recognized by light microscopy after 4–5 hours of exposure to the hymenostome ciliate *Lembadion bullinum*, but also in the presence of the rhizopod *Amoeba proteus*, the turbellarian *Stenostomum*

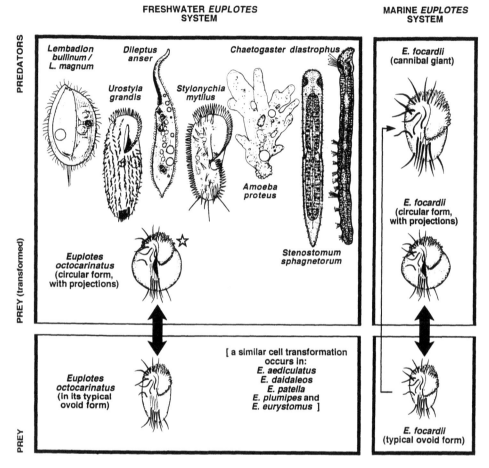

Fig. 8.1. Predator-induced defenses in ciliated protozoa (overview). Presently, five freshwater systems and one marine system can be distinguished. Prey organisms in their typical forms are illustrated in the lower row; prey organisms after transformation are shown above. Predators that induce the transformation of their potential prey if sharing with them the same habitat (symbolized by the rectangles) are listed in the upper row. Note that the predatory ciliate *Lembadion magnum*

sphagnetorum, and other predators of *Euplotes*. The outgrowth of cell protuberances in the prey may continue for 24–36 hours; its speed and extent depend on the concentration and physiological conditions of the predators (Kusch 1993a,c). In an experiment in which two chambers (one of them filled with *Euplotes* and the other with *Lembadion* cells) were allowed to exchange fluid through a polyester net, a factor released by the predator (L-factor) that caused the morphological changes in *Euplotes* was separable from *Lembadion* and capable of being exchanged between the two chambers. Preliminary characterization of the L-factor revealed that it is a protein

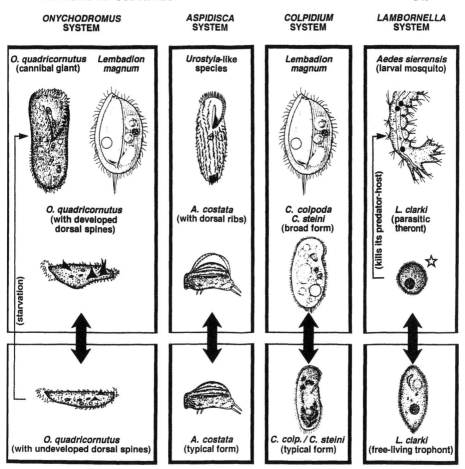

duces a morphological transformation in several ciliate genera. Prey organisms which, in addition to the morphological transformation, exhibit an induced behavioral change in response to certain predators, are marked with asterisks. Magnifications are not the same for all predators and prey (for references, see text).

(Kuhlmann and Heckmann 1985). Retransformation of circular *Euplotes* cells can be induced by putting them into fresh culture medium or by removing all predators from a mixed-species culture; it is completed within two or three days. Usually, *Euplotes* divides into two or four daughter cells during this period.

The mean height of greatly enlarged cells of *E. octocarinatus* was 41 ± 1 μm (standard error of the mean) instead of 22 ± 1 μm in controls, corresponding to almost a doubling of the height in the induced cells. Similarly, the widths differed, with a mean of 100 ± 1 μm in treated cells compared to

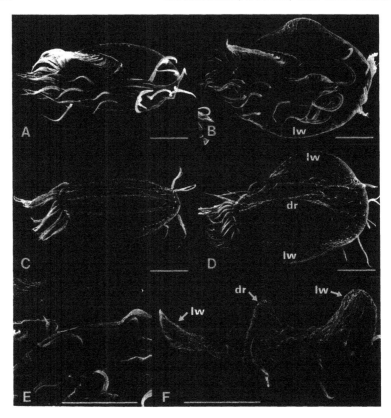

Fig. 8.2. Scanning electronmicrographs of typical and transformed morphs of *Euplotes octocarinatus*. (A) Ventral view of a typical ("ovoid") cell; the triangularly shaped mouth opening as well as cilia attached to each other and forming membranelles are located in the anterior left cell half, while cirri (= bundles of cilia) are predominantly positioned on the right ventral cell surface. (B) Ventral view of a transformed cell ("winged morph") of *Euplotes*, 36 hours after co-cultivation with *S. sphagnetorum*; lateral wings (lw) as well as a ventral projection (vp) are visible. (C) In the dorsal view of an ovoid cell, ciliary membranelles can be seen at the anterior cell pole and cirri at the posterior pole of the cell. (D) Dorsal view of a winged morph with lateral wings (lw) and a protuberant dorsal ridge (dr). (E) Detail of the morphology of the ventral projection (vp) of a transformed cell. (F) Lateral wings (lw) and the dorsal ridge (dr) of a transformed cell as seen from its posterior pole toward the anterior one. Bars = 20 μm. (Kuhlmann and Heckmann 1994.)

61 ± 1 μm in the controls. The mean lengths were 107 ± 1 μm and 86 ± 1 μm, respectively (Kuhlmann and Heckmann 1985). Other *Euplotes* species also enlarged considerably under the influence of predator-released factors. The mouth openings of the four ciliates which were found to prey on *Euplotes* have an average length of less than 100 μm (Kuhlmann and Heck-

mann 1994). In slightly starved cells of *L. bullinum* the average mouth size is about 85 × 35 μm (Kuhlmann 1993). As the mouth is hardly flexible, transformed cells of *E. octocarinatus* become too bulky to be swallowed by *Lembadion*.

Exposure of *Euplotes* cells to the kairomone released by *L. bullinum*, or by other predators, provides an easy-to-handle model system for the study of induced changes in cell morphology. These changes are presumably initiated by the binding of the factor to specific receptors, which induce the change of cell morphology. The first visible step in the chain of transformation events that can be recognized is a transient cessation of cell division in a growing population of *Euplotes* cells. Probably the tubulin pool and other macromolecules needed for cell division are with priority utilized for the assembly of cytoskeletal microtubules required for the expanding cell surface (Jerka-Dziadosz et al. 1987). In transformed cells, a considerable increase in the number of microtubular triads on the dorsal and ventral surfaces and the appearance of extra single microtubules between the dorsal triads was observed. However, certain interconnected groupings of microtubules located on the dorsal surface were greatly diminished after transformation. Intracytoplasmic microtubules were also more abundant in the enlarged cells than in the untreated ones.

Behavioral Changes

Enlargement of a prey organism limits its chance of becoming swallowed by a size-selective predator such as *Lembadion* or *Stenostomum*. Size selective predators induced a change in shape in *Euplotes*. Food consumption by *Amoeba proteus* is less constrained by particle size, since amoebae do not have a defined mouth but take up their food by phagocytosis. An induced change in behavior is here the main effect that protects *Euplotes* cells. In the first hours of co-cultivation, *Euplotes octocarinatus* usually moved, at contact with *Amoeba proteus*, across this predator that then often caught the ciliate. Later, when cells of *E. octocarinatus* came into contact with amoebae, they showed an avoidance reaction by immediately moving backwards, then turning around and withdrawing by moving forward (Kusch 1993c). This "avoidance" of *Amoeba proteus* by *E. octocarinatus* showed basically the same pattern as had been observed earlier by Kuhlmann (1989, 1994) for the so-called escape response of *E. octocarinatus* that occurred when the cells bumped into *Stenostomum sphagnetorum*.

The avoidance behavior of *Euplotes* was quantitatively measured by counting its frequency in relation to the total number of prey-predator contacts. The frequency of avoidance behavior increased in *E. octocarinatus* from 15% to 98% depending on the predator density and the time of coexistence. Medium that was conditioned by 10 amoebae/ml induced significant behavioral changes in *E. octocarinatus* within four hours. Ciliates that avoided *Amoeba* did not avoid other particles, e.g., beads of glass or agarose. Therefore, avoidance behavior is not a constitutive characteristic of *Euplotes* species. Instead, a predator-derived signal induces an "avoidance readiness" in the ciliates. Those ciliate cells that survived the time of induction and development of defensive behavior, and thereafter came into contact with the amoeban predators, recognized these, presumably by a substance on the predator's surface, and immediately withdrew (Kusch 1993c).

Whether behavioral changes are linked to morphological changes was investigated by observing cells of *E. octocarinatus* that had been induced to enlarged size by *Stenostomum*. Those ciliates did not avoid amoebae immediately after placing them together with the predator, but avoidance behavior increased with increasing time of coexistence. *E. aediculatus* did not change its behavior toward amoebae, neither after coexistence with *Amoeba*, nor when it had a *Stenostomum*-induced defensive morphology (Kusch 1993c). On the other hand, cells transformed by the L-factor did show an avoidance behavior when united with *Stenostomum* (Kuhlmann, unpubl.). Thus the behavioral response of *Euplotes* may not be automatically linked to morphological changes. At least sometimes it appears to be specific toward the defense-inducing predator.

Phenotypic Plasticity in Other Ciliates

Morphological variations are known for a large number of ciliate species. To determine whether polymorphism is formed as a response to predatory cues requires laboratory experiments. However, morphological changes in cell shape that occur in laboratory cultures can be due to injuries; physical, chemical, or other parameters, e.g., to the composition and quality of the culture medium; or to the kind and amount of food that is offered. Sometimes, morphologically different types of cells can be observed among clonemates, which are cultivated side by side. In such cases, unidentified signals or intraspecific predation may have induced gigantism, or macrostome forms in some cells (Kuhlmann 1993). Formation of giants and macrostome formation are generally interpreted as survival strategies by which at least a few individuals of a population become able to withstand a period of severe starvation by cannibalism. As both phenomena are not caused by predators, they are disregarded in this chapter, with the exceptions of cases in which the transformed giant cells themselves induce a transformation in smaller clonemates, thereby reducing their susceptibility to predation.

Inducible Defense in a Marine Euplotes Species

In *Euplotes focardii*, an Antarctic ciliate, a marked polymorphism ensues in cultures with a relatively high cell density (Valbonesi and Luporini 1990; Valbonesi et al., 1995). Typical, ovoid cells progressively transform into giants, which are nearly spherical in shape and increase from 72×54 μm to 180×170 μm. Such transformed cells may not only ingest more algal food, but may also prey on clonemates of smaller dimensions. Many of the smaller cells start to develop quite similar lateral wings as well as dorsal and ventral projections, just as *Euplotes octocarinatus* does in the presence of *Lembadion* (Valbonesi, pers. comm.). The "winged form" of *E. focardii* is apparently better protected from becoming swallowed by giants than the typical form. Cell-transforming signals, either soluble or cell anchored, have not been identified.

Predator-Induced Defense in Onychodromus

The hypotrichous ciliate *Onychodromus quadricornutus* is remarkable in its potential for voluminous size (up to 900 μm in length), its possession of a unique set of dorsal spines, and its capability to express two kinds of morphs induced by intraspecific predation (Wicklow 1988). Cell length frequencies in replicates of a well-fed clone show normal distributions; starvation followed by intraspecific predation, however, induces cells within a clone to transform into two size classes: small lanceolate cells and cannibalistic giants. The smaller cells, especially recently divided cells, are susceptible to cannibalism by larger clonemates. Cannibals enlarge and ultimately grow into giants.

Induction experiments indicate that a substance released by cannibalistic giants stimulates defensive spine growth in clonemates within 24 hours. Hypertrophied spines (> 40 instead of < 20 μm in length) appear throughout the population, even on the giant morphs. Like the changes in induced *Euplotes*, spine growth does not require cell division. Giants can also induce spine growth in cells that do not belong to the same clone. Furthermore, *Onychodromus* cells exposed to the predacious ciliate *Lembadion magnum* also develop hypertrophied spines.

Two replicate selection experiments showed that cells with undeveloped spines were preyed on to a much greater extent than cells with prominent spines (Wicklow 1988). When cannibal giants of *Onychodromus* made physical contact with a potential prey cell, there was an immediate avoidance reaction by the prey cell and an attack response by the predator in the form of increased forward speed. Cells are captured as they are overtaken and forced into the buccal cavity during the increased forward movement of giants. Prominent spines protect cells from predation by increasing their effective dorsoventral dimensions, which makes capture by giants more difficult.

Induction of dorsal spine growth in *O. quadricornutus* was the first reported case of a defensive developmental polymorphism induced by cues from a conspecific predator. That cells of the same clone simultaneously prey upon clonemates and defend against clonemates appears paradoxical. When defensive spines are induced within an *O. quadricornutus* clone, the genotype of the predator (cue releaser) is identical to the genotype of the prey (cue receiver). This paradox is resolved by evidence that *Onychodromus* cells are capable of responding with inducible defenses to a cue released by conspecific cannibals that are non-clonemates as well as to a cue released by an interspecific predator, *Lembadion* (Wicklow 1988).

Inducible Morphological Changes in Aspidisca

Aspidisca costata is a small hypotrichous ciliate, which often appears in maturing activated sludge. Besides three morphological varieties, a "mast," a "normal," and a "starved" one, a "special variety" of *A. costata* has occasionally been observed at advancing nitrification processes (Hamm 1964). The "special variety" with two protruding dorsal ribs could not be induced in laboratory cultures of *Aspidisca* without addition of predatory ciliates (*Urostyla*-like species). In mixed-species cultures of prey and predators, however, *Aspidisca* developed the dorsal ribs within 24 hours (Kuhlmann, unpubl.). Whether cells of the "special variety" of *Aspidisca* are better protected against their predators than those showing the normal form is not known.

Predator-Induced Defenses in Colpidium

Induced changes in morphology have also been described in the genus *Colpidium* (Fyda and Wiąckowski 1992). The holotrichous ciliates *C. colpoda* and *C. kleini* change their morphology after being exposed to one of their predators, *Lembadion magnum*, for 24 hours. The induced morph in both species is significantly broader than the normal morph, rendering ingestion much more difficult for *Lembadion*. As in *Euplotes*, transformed cells return to their normal cell shape and size after only a few cell divisions when separated from their predator. Attempts to provoke the reaction with cell-free medium from a *L. magnum* culture have been unsuccessful.

Predator-Induced Trophic Shift of Lambornella

An extreme case of an induced polymorphism was discovered in the tetrahymenid ciliate *Lambornella clarki*, a protozoan living in water-filled tree holes of western North America (Washburn et al. 1988). Larvae of the tree hole mosquito, *Aedes sierrensis*, release a waterborne factor that is recognized by their protozoan prey and transforms them into obligate parasites that attack the insect predator. Free-living pyriform morphs (trophonts) of *L. clarki* are induced to undergo a synchronous response in which cells divide and transform into spherical, astomatous parasitic cells (theronts) that encyst on larval predators. Parasitic ciliates penetrate the cuticle and enter the hemocoel, where they multiply and ultimately kill their predator-host. Soon after, numerous trophonts are released, some of which differentiate into theronts that attack surviving predators.

Using the ecological strategy of shifting trophic levels, trophonts avoid predation and parasitize their would-be predators. This shift allows for changes in the microbial community by reducing populations of the dominant predator. It is not uncommon for *L. clarki* to eliminate mosquitoes from tree holes. When this occurs, predation pressure on free-swimming protozoans including *L. clarki* is relaxed and microbial populations increase. Thus, this predator-prey-host-parasite relationship may be important in determining the structure of tree hole microbial communities. Furthermore, it provides a novel concept in the search for new biological control agents of larval and adult mosquitoes as important transmitters of pathogens (Washburn et al. 1988).

Kairomones Initiating the Defenses of Freshwater *Euplotes* Species

In freshwater *Euplotes* species, predator-derived chemical cues (kairomones) induce defenses. The importance of released versus predator-bound signals for the induction of defensive changes in *Euplotes* was first investigated. The extent of morphological changes correlated to the logarithm of the predator density; even one predator per ml induced significant changes in *Euplotes*. Defensive changes were also correlated with the concentration of kairomones in filtrates of the predator's culture medium. But exposure to the conditioned medium always induced less morphological change than was induced when the corresponding number of predators was present in the culture vessel (Kuhlmann and Heckmann 1985). Therefore, in one experiment, conditioned medium that contained released kairomones was continuously provided to *Euplotes* with the help of a flow-through chamber. In this case direct contact between prey and predators was avoided, but a steady concentration of kairomones influenced the *Euplotes* cells. Under these conditions the prey size reached values that were as high as those reached in the presence of the predators (Kusch 1993a). These results showed that released signal substances suffice to induce the morphological changes in *Euplotes*.

Bacterial or thermic degradation, or incorporation by the prey ciliates probably influence the amount of kairomones present in a sample. In the absence of predators, the time-dependent process of transformation may stop early, when the signal concentration decreases. In the presence of predators, the amount of kairomones is influenced by the predators via the continuous release of new signal molecules (Kusch 1993b). A special predator-bound signal substance does not seem to be important for defense induction in *Euplotes*. This does not exclude the existence of predator-bound signal molecules. The kairomone possibly functions as a surface protein in the predator. Induction of morphological changes via released signal molecules seems to have greater ecological significance than induction via predator-bound signals, because released signals induce defense over a distance, before direct contact between prey and predator (Kusch 1993a).

Isolation and Characterization of Predator-Released Kairomones

Characterization of predator-released substances that induce defensive reactions in their prey are of high priority. It might provide insight into developmental, ecological, and evolutionary questions, and into the mechanisms of interspecific recognition. To determine whether kairomones of different predators are identical or not, the signal substances released from *Lembadion bullinum* (L-factor) and from *Stenostomum sphagnetorum* (S-factor) were purified and analyzed (Kusch and Heckmann 1992; Kusch 1993b). Digestion of the factors by immobilized proteolytic enzymes indicated a protein nature of the kairomones. Their activities were not diminished after incubations with several other enzymes, including nucleases, lipases, and glucosidases. Ultrafiltration of both factors showed that their molecular weight is above 10,000 Da. Gel electrophoresis revealed a mass of 30,000 Da for the L-factor; 31,500 Da were measured by mass spectrometry. Electrophoresis of the S-factor resulted in a molecular mass of 17,500 Da. The absorption spectra of the purified kairomones were that of proteins, with maxima only at 210 nm and at 280 nm. Therefore, both molecules must be peptides with different molecular masses.

The purified L-factor had a specific activity of 5×10^4 U/μg protein. The minimum concentration that has a detectable morphogenetic effect therefore is 10^{-12} Mol/l. The purified S-factor was determined to have a specific activity of approximately 2×10^4 U/μg protein. The minimum concentration with detectable morphogenetic effect is, therefore, 10^{-11} M. The concentration of L-factor in the cultures of *Lembadion* was determined to be approximately 10^{-10}–10^{-11} M. The concentration of S-factor in the predator-cultures was approximately 10^{-11} M. Experiments with the immobilized lectin concanavalin A indicated that the factors do not contain glucose or mannose. However, if concanavalin A was allowed to bind to the cells before one of the factors was added, the cell transformation of *Euplotes* was inhibited. This suggests an involvement of carbohydrate structures located on the cell surface of *Euplotes* in the process of chemical signaling. The purified proteins are heat sensitive. Incubation at 95°C for 10 min destroyed all morphogenetic activity of the L-factor; after 30 min at this temperature, the S-factor was also inactivated (Kusch and Heckmann 1992; Kusch 1993b, 1994a).

These data showed that the kairomones from different predators differ. This raises the question of whether the recognition of predators has evolved in *Euplotes* independently for the various predators. An analysis of the amino acid sequences of the factors is necessary to answer this question definitely, because one could also assume that all the signals share a certain sequence. A short amino acid sequence of the L-factor was determined recently (Kusch and Heckmann, unpubl.). This allowed us to deduce the amino acid sequence of the L-factor from its gene. Analysis of the amino acid sequence, and biochemical as well as immunocytochemical detection, of the L-factor indicates that the L-factor is located on the surface of *Lembadion*. Presumably the factor becomes released by cleavage of the hydrophilic main part of the molecule from a short membrane-bound hydrophobic amino acid sequence (Peters-Regehr et al. 1997).

Culture medium conditioned by *A. proteus* induces morphological and behavioral changes in *E. octocarinatus*. This indicates that a kairomone (A-factor) is released also by this predator (Kusch 1993c). The A-factor was analyzed by enzymatic degradation of its biological activity. The results show that the avoidance inducing signal substance is a peptide, like the L- and S-factors. A peptide nature could not be proven for the low morphogenetic activity, because this was diminished also in the controls of proteolytic digestion. Ultrafiltration experiments showed that the A-factor has a molecular weight between 5,000 and 10,000 Da. A molecule of this mass range also induced a morphological response in *E. octocarinatus*. But whether a morphological response and a behavioral response of *E. octocarinatus* are induced by the same signal substance from *A. proteus*, or by two different kairomones of similar molecular weight, is unclear. The data available for the A-factor confirm that the kairomones from various predators responsible for the species-specific reactions described earlier differ from one another (Kusch 1993c).

Effects of Extracellular Nucleotides

Kuhlmann and Schmidt (1994) attempted to extract a morphogenetically active surface kairomone directly from *Lembadion bullinum* cells. This led to the surprising discovery that purine-containing compounds of low molecular weight can also induce cell transformation and prompted the authors to investigate the morphogenetic activity of commercially available purine and pyrimidine nucleotides. There was a striking similarity between predator-induced morphogenetic changes and changes induced by some of these substances, of which ATP/dATP and adenosine-tetraphosphate were most active. The concentrations of substances needed to induce these changes in *Euplotes octocarinatus* were, however, unphysiologically high (50 μm). Nevertheless, the discovery that the application of extracellular nucleotides induces the transformation of *Euplotes* could be of interest to discern the chain of events occurring from the binding of the kairomone leading to a major morphological reconstruction of the cell.

Distribution and Specificity of Induced Defenses

The responses of *Euplotes* cells toward different predator-released signals vary. The number of predators that are recognized and the extent and the kind of response to one predator species vary in different *Euplotes* species. The spectrum of responses might provide insight into the mechanism of signal recognition and its evolution.

E. daidaleos and *E. octocarinatus* change their cell shape only to organisms that are able to feed on *Euplotes* cells. *Euplotes* does not respond to organisms that do not feed on this ciliate (e.g., *Paramecium*). Also *Daphnia longispina* and *Bursaria truncatella*, predators which are able to swallow the circular form of *Euplotes daidaleos*, showed no capacity to induce morphological changes in *Euplotes* (Kusch 1994b, 1995).

Three potential predators, *Lembadion bullinum* (Ciliophora), *Chaetogaster diastrophus* (Oligochaeta), and *Stenostomum sphagnetorum* (Turbellaria), caused defensive changes to various extents in *E. daidaleos* (Kusch 1995). Induced defense was correlated to different predation risks by the predators *Lembadion bullinum* and *Stenostomum sphagnetorum*. So, at equal densities, *S. sphagnetorum* always induced a larger cell width in *Euplotes* than did *L. bullinum*. The extent of induced defense may reflect either a different sensitivity of *Euplotes* to different predators, or a different amount of signal molecules that is released by different predators.

Some *Euplotes* species (*E. woodruffi, E. crenosus*) do not react to *Lembadion* and other predators tested, whereas others (*E. octocarinatus, E. ae-*

diculatus, E. patella, E. daidaleos, E. plumipes, E. eurystomus) do react (Kuhlmann and Heckmann 1985). In the responding *Euplotes* species one or both types of defenses occur. Besides morphological plasticity in response to size-selective predators like *L. bullinum, S. sphagnetorum,* or *C. diastrophus* (Kuhlmann and Heckmann 1985; Kusch 1993a, 1995), behavioral changes are induced by the predators that are not size selective, e.g., *Amoeba proteus*. Of the three *Euplotes* species investigated, morphological changes are induced by *A. proteus* only in *E. octocarinatus*. The changes are less extensive than those induced by *L. bullinum, S. sphagnetorum,* and *C. diastrophus* (Kusch 1993a,c, 1995).

The differences in the spectrum of responses of *Euplotes* species suggest that the cells must possess more than one receptor for binding the signal substances of the various predators. This is confirmed by the observed differences between the signal substances. The reason for the lack of inducible defense in some *Euplotes* species is unknown. Probably they occupy microhabitats where the defense-inducing predators tested by us do not occur.

Adaptive Significance

A significantly higher probability of rejection by predators was demonstrated for the defensive "winged" form of *Euplotes* compared to the ovoid form (Kuhlmann 1990; Kuhlmann and Heckmann 1994). Transformed cells of *E. octocarinatus* were 2–20 times more likely to be rejected by their predators than untransformed ones.

Predators induce morphological changes in ciliates within a few hours (Kuhlmann and Heckmann 1985; Kusch 1993a; Washburn et al. 1988; Wicklow 1988). This short time required to develop a defensive morphology should enable *Euplotes* and other prey to markedly reduce the risk of predation in natural habitats. But predator-induced changes are graded responses (Harvell 1990b). The adaptive significance of predator-induced phenotypic changes therefore depends on the extent of change that is induced (Kusch 1993a, 1995). Patchy distributions of the predators are common in their natural habitats, and *Lembadion magnum* was found at maximum natural densities of 5 individuals per ml, while average densities reached two individuals per ml. These local densities were high enough to induce morphological changes in *Euplotes* (Kusch 1993a).

The size of *Euplotes* at which the predators are not able to swallow this prey, and the density at which the predators induce this size in the prey ciliate, may give information on the adaptive significance of inducible defense in *Euplotes*. Furthermore, it is of interest to know whether all predators induce morphological changes in *Euplotes* to the same or to different extents, and whether the predation risks of *Euplotes* are the same for various

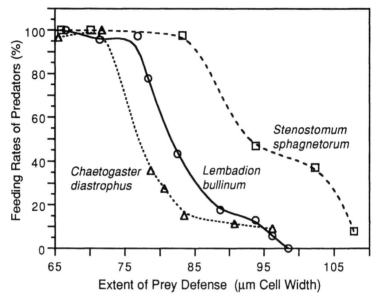

Fig. 8.3. The predation risk of *Euplotes daidaleos* depends on the extent of predator-induced morphological changes (measured by the cell width) and on the predator species. 100% feeding rate is equivalent to 23 ingested *Euplotes* prey per 10 *L. bullinum*, or to 17 prey ciliates per 1 *S. sphagnetorum*, or to 7 prey ciliates per 1 *C. diastrophus* during one hour. (Kusch 1995.)

predators. Even a moderate morphological change in prey influenced the feeding rate of predators considerably (fig. 8.3; Kusch 1994b, 1995). The extent of defense in *Euplotes* that was induced by 10 predators/ml during 24 h (89 ± 8 μm with *S. sphagnetorum*, 82 ± 6 μm with *L. bullinum*, and 85 ± 6 μm with *C. diastrophus*) decreased the predation risk from those predators to 67% in the presence of *S. sphagnetorum*, to 50% with *L. bullinum*, and to 15% with *C. diastrophus*, compared to the typical form of *Euplotes*. A *Euplotes* cell width of 93 μm, 82 μm, or 77 μm reduced the predation risk by 50% in the presence of *S. sphagnetorum*, *L. bullinum*, and *C. diastrophus*, respectively. Perhaps the effects of different coexisting predator species on the morphology of *Euplotes* are cumulative. In a natural population the defensive form of *Euplotes daidaleos* was found with average cell widths of 88 ± 8 μm, sufficient for a 25–87% reduction of the predation risk by the predator species investigated so far. The results indicate that predator-induced defense in natural *Euplotes* populations is beneficial to this prey, and that it is adapted to the predation abilities of *Euplotes* predators (Kusch 1994b, 1995).

Differences in the extent of induced morphological defense by various predators may be compensated by behavioral differences, which are more

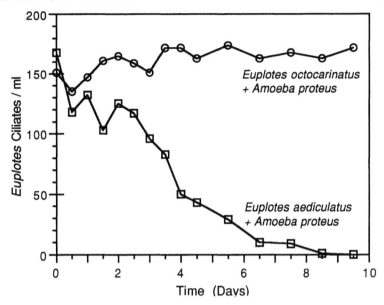

Fig. 8.4. Effects of predation by *Amoeba proteus* on the ciliates *Euplotes octocarinatus* and *E. aediculatus*. *E. octocarinatus*, which responds to the presence of *Amoeba* by developing an avoidance behavior, was able to coexist, whereas *E. aediculatus*, which does not have this capacity, became extinct. Densities of the prey ciliates were counted during co-cultivation with 10 *A. proteus*/ml. (Kusch 1993c.)

effective for defense of less size-selective predators (Kusch 1993c). Experiments with *Amoeba proteus* demonstrated the importance of behavioral changes in reducing the predation risk of *Euplotes*. *Amoeba proteus* is able to prey on *Euplotes* cells that do not have any induced defense. In the presence of ten *Amoeba*/ml, all *E. aediculatus* (160 ciliates/ml) were killed by predators within ten days (fig. 8.4). However, *E. octocarinatus* survived and reproduced in the presence of *A. proteus*. Whereas both *Euplotes* species changed their morphology in the presence of *Lembadion* and *Stenostomum*, only *E. octocarinatus* responded to the presence of *Amoeba proteus*. But the effect of the amoeban predator was low, compared with the other predators mentioned before. The size of *Euplotes octocarinatus* did not exceed that of *E. aediculatus*. Therefore, morphological changes cannot be the reason for differences in the predation risks. Furthermore, *Amoeba* is not a size-selective predator as are *Lembadion* and *Stenostomum*. Even *Euplotes* cells that had a *Stenostomum*-induced enlarged morphology were consumed by the amoebae (Kusch 1993c). Obviously, behavioral differences between *Euplotes* species are the reason for the differences in their predation risk by *Amoeba proteus*. Whereas morphological defense may be crucial for the risk of predation by size-selective predators, behavioral changes considerably lower the vulnerability to the less size-selective predator *Amoeba proteus*. Such behavioral responses and possibly abiotic factors may drastically influence the distribution of those organisms in ponds and may lead to rapid changes of predator and *Euplotes* densities in distinct areas of their habitat (Kusch 1995).

A few mixed-species cultures of one protozoan prey species and one proto- or metazoan predator species was followed during a longer period (Kuhlmann and Heckmann 1994). The time until either the prey or the predator population died out was recorded. Prey and predator populations coexisted if the cultures were occasionally fed with a small flagellate. If no food was provided, about one-half of the mixed-species cultures existed for more than 3 months.

DNA fingerprints of *Euplotes* (Kusch and Heckmann 1996) may be used as molecular markers to test the adaptive significance of predator-induced defense in natural habitats. The possibility to identify individual *Euplotes* clones by DNA fingerprinting allows the introduction of clones with differing defensive abilities into a habitat, and subsequently to measure changes in genotype frequencies within such a population.

Costs for *Euplotes*

Morphological defense in *Euplotes octocarinatus* is associated with a division delay and therefore has a demographic cost (Kusch and Kuhlmann 1994). The generation time of the ciliate increased by 18% during the development of a defensive morphology and remained increased by 16% in the following reproduction cycles of the winged form (98 ± 6 μm cell width). The population growth rate therefore decreased by a maximum of 15%. Yet the benefit of predator-induced morphological defense in *E. octocarinatus* was a decrease of the mortality rate by a maximum of 90–100%, in dependence on the predator species and density, and in the nutritional conditions of the prey organism (Kuhlmann and Heckmann 1994; Kusch 1995). Therefore, in the presence of predators, the development of a defensive morphology is favored in *Euplotes*.

Cost is associated with synthesis of proteins necessary for morphological changes. The protein content of the winged form was found to be higher by 55% than that of the typical form. Protein synthesis inhibitors interfere with morphological changes; this also shows that protein synthesis is necessary. Synthesis of proteins that are necessary for morphological changes take place during the growth phase (G_1) of the ciliates' reproductive cycle. This was prolonged by 38%. The length of the G_1-phase increases with increasing generation time from 25 ± 1 h to 34 ± 2 h, whereas the length of the S-phase does not change (13 ± 1 h to 12 ± 3 h). These results indicate that, in addition to reorganization processes, prolonged cell growth causes prolonged generation times in *Euplotes* (Kusch and Kuhlmann 1994). This is confirmed by the observation that food supply is necessary for defense development (Wiąckowski and Szkarłat 1996).

The feeding rate of winged forms of *Euplotes octocarinatus* is not diminished and therefore does not cause any cost. A stronger swimming effort may be required in *Euplotes* due to hydrodynamical reasons. Enlarged shape and prolonged cell projections suggest that swimming may require more energy, but unicellular organisms operate at low Reynolds numbers. The magnitude of this cost is difficult to evaluate and may be minor compared to other metabolic costs (Kusch and Kuhlmann 1994).

Costs may explain why all individuals of a *Euplotes* population increase their size in relation to the predator density. Size frequency distributions of the populations have only one maximum, so the predator density does not affect the number of cells with a highly transformed morphology. Instead, all individuals of a *Euplotes* population transform their morphology to an extent that is related to the concentration of the inducing signal molecule. Via this mechanism of response, moderate defensive changes may take place early in prey when predator abundance is low and direct encounters are still unlikely. This may lead to only low costs to the prey. If the signal concentration increases by increased predator abundance, a stronger morphological change, sufficient for predator defense, may occur. If the signal concentration drops by the disappearance of predators, the weakly transformed prey organisms would not suffer high costs. Ciliates that would transform to a maximum extent at a low signal concentration presumably often would suffer high cost and would have a low rate of reproduction (Kusch 1993a).

9

Ecology and Evolution of Predator-Induced Behavior of Zooplankton: Depth Selection Behavior and Diel Vertical Migration

LUC DE MEESTER, PIOTR DAWIDOWICZ,
ERIK VAN GOOL, AND CARSTEN J. LOOSE

Abstract

Diel vertical migration of zooplankton is a prime example of a predator avoidance behavior, and there is considerable literature on both ultimate and proximate aspects. We treat diel vertical migration as a special case of depth selection behavior, in which depth preference changes over a diel cycle. We review literature on predator-induced changes in depth preference and diel vertical migration of zooplankton, with special emphasis on the flexibility of this behavior with respect to different types of predators, differences in predation pressure, and the presence of modifying factors such as food satiation. Attention is drawn to the association of day-depth preference (diel vertical migration) with other traits that affect vulnerability to predators, such as body size. Within- as well as among-population genetic differences in predator-induced changes in day depth and diel vertical migration have been reported in *Daphnia*. The maintenance of genetic variation has been attributed to frequency-dependent selection, to the association of habitat selection traits with traits influencing relative fitness in the different habitats, and to the continuous input of new genotypes due to the hatching of resting eggs. We propose two additional factors that may maintain genetic polymorphism for predator-induced behavior in zooplankton, both resulting from the complexity of the possible responses to the presence of predator kairomones. First, the reliability of the proximate environment as a source of information on the most adaptive defense decreases with increasing number of different possible defenses. This is especially so at the among-population level and when responses to different predators are counterproductive. Second, maintenance of genetic polymorphism at the within-population level might also result from the existence of multiple optima, with different strategies (involving combinations of different induced traits) having equal fitness under prevailing environmental conditions. Data stressing the importance of both mechanisms are presented.

Depth Selection Behavior and Diel Vertical Migration of Zooplankton

Diel vertical migration (DVM) is widespread in freshwater and marine zooplankton (reviews by Cushing 1951; Haney 1993). Migration amplitudes vary from less than one meter to several tens of meters in lakes or even hundreds of meters in the open ocean. DVM is a very striking behavior: though zooplankton is considered to be characterized by low locomotory abilities, it is no exception for an individual of, for instance, 1.5 mm body length to migrate over an amplitude of 20 m, covering daily a distance of more than 26,500 times its own body length. Normal or nocturnal migration is when the zooplankton population resides deeper in the water column during the day than during the night, with an evening ascent around dusk and a morning descent around dawn (Cushing 1951). Reverse migrations, with the density peak of the population residing deeper during the night than during the day, have also been reported in the literature, but are less common than nocturnal migrations (Bayly 1986; Ohman 1990). Many zooplankton populations do not migrate vertically in a diel pattern, but have nevertheless a very distinct vertical distribution (Pijanowska and Dawidowicz 1987; Stewart and George 1987). To avoid confusion, we define depth selection behavior (DSB) as the behavior by which the zooplankton maintains a particular (daytime, nighttime) vertical distribution in relation to the vertical stratification of the water column (light, temperature, food, predation pressure). The term *diel vertical migration* is restricted to a special case of depth selection behavior, in which the preferred depth changes in a diel pattern. These daily changes in habitat are mainly accomplished by swimming in response to light changes during dawn and dusk. Both depth selection behavior and diel vertical migration are thus habitat selection behaviors, and the patterns of DSB and DVM may be influenced by similar selective forces. As DVM is a special case of DSB, there is some added complexity to DVM, involving both proximate (secondary phototaxis) and ultimate (adaptive value of timing and speed of migration) aspects.

In this chapter, we first give an outline of the current views on ecology and evolution of depth selection and diel vertical migration behavior of zooplankton. We focus on proximate and ultimate aspects of the behavior, the costs and benefits involved, and the evidence for genetic polymorphism in depth selection behavior. We then review the recent literature on predator-induced changes in depth selection and diel vertical migration behavior, focusing on two main questions: (1) Why are inducible defenses so ubiquitous in zooplankton? and (2) What factors contribute to the maintenance of genetic polymorphism for inducible defenses?

Proximate Factors of DSB and DVM

Experimental work involving *Daphnia* (Ringelberg 1964; Van Gool and Ringelberg 1995), *Artemia* (Forward and Hettler 1992), and marine invertebrates (e.g., crab larvae; Forward 1988) has provided detailed evidence that a fundamental cue for DVM is the relative change in light intensity. An increase in light intensity at a rate surpassing the threshold can lead to a negatively phototactic reaction resulting in a downward movement of the animals. Similarly, decreases in light intensity can induce a positively phototactic response that results in an upward movement. The minimal rate of light change needed to evoke a swimming response (response threshold) depends upon the adaptation light intensity (Ringelberg et al. 1967; Forward 1988). This stimulus response mechanism can explain the timing of DVM satisfactorily, with threshold values found in the laboratory corresponding to the light intensity changes when migration starts in the natural habitat at dawn and dusk (Ringelberg et al. 1991; Ringelberg 1995). Although swimming in response to light intensity change is the primary physiological mechanism underlying DVM, other factors modify this response. These modifying factors include food (satiation level) as well as predator-specific chemicals (induced responses; Ringelberg 1991a; Van Gool and Ringelberg 1995). If, due to the interaction with modifying factors, the response threshold of light intensity change is not surpassed, no DVM takes place, and a DSB pattern with no diel changes in depth preference occurs. Depth maintenance during the day and during the night is also influenced by other factors such as primary phototaxis (the response to a constant light gradient), temperature gradients (e.g., associated with a thermocline), and food gradients (George 1983). So far, however, these factors have been much less well studied from a physiological point of view.

Ultimate Factors of DSB and DVM

The adaptive significance of DSB and DVM has been much debated in the past, but it is now generally accepted that DVM is a predator avoidance strategy (Lampert 1989, 1993b). The predator avoidance hypothesis states that a zooplankter migrates to deeper water during the day to hide from visual predators and comes to surface water during the night to feed and to benefit from higher temperatures (resulting in a higher intrinsic rate of increase). Reverse migrations are a strategy to avoid predation by invertebrates, which are themselves often forced to migrate in a nocturnal pattern in the presence of visual predators (Ohman et al. 1983; Ohman 1990). When food availability or predation risk are not dependent upon depth, zooplankton may exhibit a stable vertical distribution (Pijanowska and Dawidowicz 1987). Three lines of evidence support the predator avoidance hypothesis: (1) a correlation is observed between day depth (migration amplitude) and predation pressure, (2) a strong relationship exists between zooplankton visibility (size, color) and day depth (migration amplitude), and (3) changes in day depth and DVM pattern can be induced by the presence of predators.

Several studies have reported a positive correlation between predation pressure exerted by fish and day depth or DVM amplitude, both in time series within a habitat (interannual changes: Bollens et al. 1992; seasonal changes: Stich and Lampert 1981; Ringelberg et al. 1991) as well as in surveys of different habitats (Gliwicz 1986; Dodson 1990).

There is also a positive correlation between day depth and size of the zooplankters, both at the among- and the within-species level. Studies documenting the vertical distribution of coexisting *Daphnia* species indicate that larger-sized species on average reside deeper in the water column during the day than smaller-sized species (Stich and Lampert 1981; Pijanowska and Dawidowicz 1987), and a similar relationship has been reported in the marine environment (Hays et al. 1994). Small size may explain why rotifers often do not migrate or migrate with a small amplitude (Miracle 1977). Several authors have documented ontogenetic shifts in day depth and DVM behavior, with adults of a given species residing in deeper water during the day and migrating over a larger amplitude than small-sized larvae or juveniles (*Daphnia*: Dumont et al. 1973; copepods: Huntley and Brooks 1982). In addition to body size per se, it has been observed that egg-bearing animals often migrate over a larger amplitude than animals without eggs (Vuorinen et al. 1983; Bollens and Frost 1991). Even when only adults of a given zooplankton species are considered, there is a positive correlation between body size and day depth (field data: Guisande et al. 1991; experimental work involving clones that differ in day depth: De Meester 1994a, 1995; Reede and Ringelberg 1995). De Meester et al. (1995), using clones from the Schöhsee (the lake studied by Guisande et al. 1991), showed that this size-by-depth relationship holds within as well as between clones. For relatively large zooplankton, the daytime vertical distribution will be largely determined by the changes in visual predation pressure with depth, whereas the nighttime distribution will often be determined by food and temperature gradients. For small zooplankton, the reverse may be true, with the nighttime distribution being determined by predation pressure (by invertebrates), and the daytime distribution by food availability. It should be noted that the details of the size-by-depth relationship will be dependent upon characteristics of both the zooplankton prey (i.e., escape capabilities, transparency, etc.) and the habitat (i.e., type of predators present, overall predation pressure, transparency of the water, etc.).

Furthermore, it has been shown that changes in day depth and DVM can be directly induced by the presence of predators. In most cases, the zooplankter recognizes the presence of predator-specific kairomones and responds by a change in behavior (reviewed by Larsson and Dodson 1993). In response to kairomones associated with visual predators (fish), zooplankton become more negatively phototactic, resulting in a nocturnal vertical migration (Ringelberg 1991a; De Meester 1993a; Loose 1993), whereas kairomones from invertebrate predators (e.g., *Chaoborus*) induce a more positively phototactic behavior, resulting in a reverse migration (Neill 1990; see table 9.1). The marine calanoid copepod *Acartia hudsonica* responds to mechanical rather than chemical cues (Bollens et al. 1994).

Although predator avoidance is the main adaptive benefit for the zooplankter residing in deeper water during the day, additional factors may be involved in the development of a particular DSB pattern in a population. It is, for instance, conceivable that the deleterious effects of high light intensities in the surface water layers (e.g., high UV radiation) can select for a deeper day depth under some circumstances (e.g., at high altitudes and latitudes; photoprotection hypothesis: Hairston 1980; see also De Meester and Beenaerts 1993). In shallow ponds and lakes, predators not yet cited above, such as surface-bound insects (Arts et al. 1981) and ducks (*Anas clypeata* L., Pirot and Pont 1987) can be important.

In experimental studies on phototactic behavior of zooplankton, two approaches have been taken: whereas some investigators have studied the behavior of zooplankters under constant light intensity, quantifying primary phototaxis (De Meester 1991, 1993a; Loose et al. 1993), others have studied the response of zooplankton to changes in light intensity (secondary phototaxis; Ringelberg 1964; Van Gool and Ringelberg 1995). Primary phototaxis is relevant to daytime depth distribution (De Meester 1993b), and secondary phototaxis relates directly to the vertical movement of the zooplankton around dawn and dusk (Ringelberg 1964, 1995). Both the daytime vertical distribution and the timing and rate of vertical movement during twilight can determine the ultimate benefit of the predator avoidance behavior. Thus, it is to be expected that primary and secondary phototaxis are both subjected to natural selection. Although both types of phototactic behavior are most probably intimately related in a physiological way, it would be unwise to treat one behavior as being the mere consequence of the other.

TABLE 9.1
Examples of Predator-Induced Changes in Phototactic or Depth Selection Behavior in Zooplankton

Prey Species	Predator (kairomones)	Response	Method	Trait	Reference
Artemia sp. *nauplii*	Fish	−DSB	L	2	Forward and Hettler 1992; Forward 1993
Daphnia ambigua[a]	Fish	+DSB	L	1	Dodson 1988
Daphnia carinata	Fish	−DSB	F	1	Haney 1993
Daphnia galeata mendotae	Fish	−DSB	L	1	Dodson 1988b
Daphnia galeata x hyalina	Fish	−DSB	L	2	Ringelberg 1991a, b;[b] Van Gool and Ringelberg 1995
	Fish	−DSB	P	1	Loose 1993
	Fish	−DSB/NR[c]	P	1	De Meester et al. 1995
Daphnia longispina	Fish	−DSB	L	1	Watt and Young 1994
Daphnia magna	Fish	−DSB/NR[c]	L	1	De Meester 1993a, 1996a
	Fish	−DSB	L	1	Loose et al. 1993
	Fish	−DSB/NR	L	1	Pijanowska et al. 1993
	Fish	−DSB	L	1	Watt and Young 1994
Daphnia obtusa	Fish	NR	L	1	Dodson 1988b
Daphnia parvula[a]	Fish	+DSB	L	1	Dodson 1988b
Daphnia pulex	Fish	−DSB	L	1	Dodson 1988b
Daphnia pulicaria	Fish	−DSB	L	1	Dodson 1988b
	Fish	−DSB	F	1	Leibold 1990b
Daphnia retrocurva	Fish	+DSB	L	1	Dodson 1988b
Acartia hudsonica	Fish	NR	F	1	Bollens and Frost 1989
Diaptomus kenai	Fish	NR	F	1	Neill 1992
Chaoborus flavicans	Fish	−DSB	F/L	1	Tjossem 1990
	Fish	−DSB	L	1	Dawidowicz et al. 1990; Dawidowicz 1993
Chaoborus punctipennis	Fish	−DSB	F	1	Leibold 1990

TABLE 9.1 (*Continued*)

Prey Species	Predator (kairomones)	Response	Method	Trait	Reference
Daphnia ambigua[a]	Chaoborus	NR	L	1	Dodson 1988b
Daphnia galeata mendotae	Chaoborus	+DSB	L	1	Dodson 1988b
Daphnia obtusa	Chaoborus	NR	L	1	Dodson 1988b
Daphnia parvula[a]	Chaoborus	NR	L	1	Dodson 1988b
Daphnia pulex	Chaoborus	+DSB	L	1	Dodson 1988b
	Chaoborus	+DSB	L	1	Ramcharan et al. 1992
Daphnia pulicaria	Chaoborus	+DSB	L	1	Dodson 1988b
Daphnia retrocurva	Chaoborus	NR	L	1	Dodson 1988b
Diaptomus kenai	Chaoborus	+DSB/NR[c]	F	1	Neill 1990, 1992
Daphnia ambigua[a]	Notonecta	−DSB	L	1	Dodson 1988b
Daphnia galeata mendotae	Notonecta	−DSB	L	1	Dodson 1988b
Daphnia longispina	Notonecta	NR	L	1	Watt and Young 1994
Daphnia magna	Notonecta	NR	L	1	Watt and Young 1994
Daphnia obtusa	Notonecta	+DSB	L	1	Dodson 1988b
Daphnia parvula[a]	Notonecta	NR	L	1	Dodson 1988b
Daphnia pulex	Notonecta	−DSB	L	1	Dodson 1988b
Daphnia pulicaria	Notonecta	−DSB	L	1	Dodson 1988b
Daphnia retrocurva	Notonecta (Mechanical Cues)	+DSB	L	1	Dodson 1988b
Acartia hudsonica	Fish mimics	−DSB	F	1	Bollens et al. 1994

Notes: No differentiation is made according to fish species, because none of the studies that have tested kairomones of different fish species found differences (see Larsson and Dodson 1993). *Response:* +DSB, shift to more positively phototactic behavior or to shallower day depth (often corresponding to reverse migration); −DSB, shift to more negatively phototactic behavior or to deeper day depth (often corresponding to nocturnal migration); NR, no response. *Method:* F, field enclosures; L, laboratory experiments in small-scale experimental set-up; P, laboratory mesocosms (Plankton Towers; Max-Planck-Institut für Limnologie, Plön). *Trait measured:* 1, vertical distribution or depth position (including primary phototaxis); 2, secondary phototaxis.

[a]The average body size of the animals of this *Daphnia* species, used in the experiments by Dodson (1988b), measured < 1.2 mm;

[b]The *Daphnia* taxon inhabiting Lake Maarsseveen was originally referred to as *D. hyalina*; allozyme electrophoresis has, however, indicated that this population is largely dominated by *D. galeata x hyalina* hybrids (Spaak and Hoekstra 1993).

[c]Several studies have revealed interclonal or interpopulational differences in response.

Costs and Benefits of Depth Selection Behavior

Diel vertical migration entails important metabolic costs to the zooplankter: instead of remaining in the near-surface waters characterized by a relatively high temperature and high food quality and quantity, the animals migrate to deeper, nutrient-poor, and cold waters. The direct cost of swimming vertically over distances similar to DVM amplitudes encountered in natural habitats has been shown to be negligible when measured in terms of reproductive output and fitness (Dawidowicz and Loose 1992a). However, metabolic costs due to temperature differences are pronounced (Stich and Lampert 1984) and may account for over 50% reduction in individual growth rate (Dawidowicz and Loose 1992b) as well as birthrate in migrating populations (Loose and Dawidowicz 1994). It is the interplay between the costs of staying at a particular depth and the benefits resulting from the reduction in predation pressure that will result in a particular DSB or DVM pattern (Ohman 1990).

It is important to emphasize that there are different strategies to cope with predation pressure by visual predators (Zaret 1975). At least four general mechanisms that are not mutually exclusive can be distinguished: a zooplankter can become smaller (i.e., reproduce at a size that is less conspicuous to predators); more transparent (e.g., by a reduction in eye pigmentation; Confer et al. 1980); exhibit a specific DSB or DVM; or swarm. The first three mechanisms reduce the conspicuousness of an individual, whereas the fourth reduces the chances of the individual being preyed upon (dilution effect). In addition to the dilution effect, swarming may also have a confusing effect on visually hunting predators (Milinski 1977). DSB and DVM may reduce conspicuousness to visual predators or the encounter probability with invertebrate predators. Most of these defense mechanisms cited (body size, DSB, swarming) are inducible by predator kairomones (review by Larsson and Dodson 1993; Pijanowska 1994), and all of them have associated costs. With respect to body size, it has been shown that large-bodied zooplankton have a lower food concentration threshold at which assimilation equals respiration than small-bodied zooplankton, which results in large-bodied zooplankton being superior competitors (Gliwicz 1990). A higher transparency makes animals more vulnerable to photodamage (Hairston 1979) and may reduce starvation resistance, since carotenoid pigments may also function as energy storage (Ringelberg 1980). Diel vertical migration has pronounced metabolic costs (fewer eggs and longer development times; Loose and Dawidowicz 1994). Swarming may also reduce the intrinsic rate of increase of zooplankton through food depletion within the swarms. Since these responses are alternative or complementary mechanisms to reduce predation, and all have associated costs, it is to be expected that associations between these traits will develop. Positive (Hairston 1980) as well as negative (Hays et al. 1994) associations have been reported between pigmentation and DVM, and several studies have reported a positive association between body size and DVM or day depth (Guisande et al. 1991; De Meester 1994a, 1995; Hays et al. 1994; De Meester et al. 1995). Depending on the conspicuousness of the animals, either migration or nonmigration will be most advantageous. This is illustrated by the data on *Daphnia hyalina* and *D. galeata* DVM patterns in Lake Constance (Stich and Lampert 1981, 1984), and by the data of De Meester et al. (1995) on different genotypes of the *D. hyalina x galeata* population in the Schöhsee.

Genetic Polymorphism for DSB and DVM

The substantial variability of diel vertical migration patterns in zooplankton species (Bayly 1986; Haney 1993) can in part be explained by phenotypically plastic responses cued by predator kairomones (Larsson and Dodson 1993) and changes in food concentration (Huntley and Brooks 1982; Johnsen and Jakobsen 1987). However, evidence for genotype-dependent depth preferences and DVM patterns is accumulating, at least in *Daphnia*. Weider (1984) observed repeatable differences in vertical distribution and migration among different *D. pulex* multilocus genotypes as defined by allozyme markers. Similar interclonal differences in vertical distribution have been shown for both lake-dwelling *Daphnia* (Müller and Seitz 1993; King and Miracle 1995) and a *D. magna* population of a shallow pond (De Meester et al. 1994). In addition to these field surveys, interclonal differences in phototactic behavior have been reported for *D. magna* (e.g., De Meester 1991), with intrapopulational broad sense heritability estimates ranging from 0.39 to 0.67 (De Meester 1990, 1996a). In their study on *D. galeata x hyalina* hybrids of Lake Maarsseveen, Ringelberg and Flik (1994) observed that the relationship between the relative change in light intensity and the displacement velocity in *Daphnia* changed over a four-year period, suggesting that there was selection on the phototactic basis of these migrations.

Predator-Induced Changes in Depth Selection Behavior of Zooplankton

Numerous studies have documented induced defenses in zooplankton, including shifts in life history and morphological and behavioral traits (reviewed by Larsson and Dodson 1993; Gilbert, chapter 7; Tollrian and Dod-

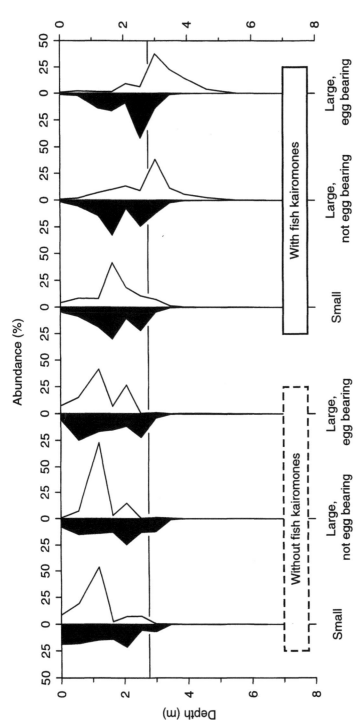

Fig. 9.1. Daytime (white) and nighttime (black) vertical distribution of a population of a *Daphnia galeata × hyalina* clone in the Plankton Towers of the Max-Planck-Institut für Limnologie in Plön, Germany, under high food conditions (POC > 0.26 mgC l^{-1}). The first three diagrams are profiles in the absence of fish kairomones, and the last three are the corresponding distributions in their presence. Body length of small animals: < 1.1 mm; of large animals: > 1.1 mm. The horizontal line indicates the depth of the thermocline at 2.77 m. (For methods, see Lampert and Loose 1992 and Loose 1993. Data from Loose 1992.)

son, chapter 10). With respect to phototactic and depth selection behavior, studies have documented induced responses in the dipteran larva *Chaoborus*, the cladoceran *Daphnia*, the anostracan *Artemia*, and the copepods *Diaptomus* and *Acartia* (table 9.1). Most studies have observed a shift to deeper day-depth (nocturnal migration) in the presence of visual predators such as fish (see fig. 9.1) and *Notonecta*, whereas the presence of *Chaoborus* larvae induces a reverse migration. The preponderance of data on cyclically parthenogenetic organisms with respect to induced defense is due to practical reasons, and should not be interpreted as indicating that inducible defenses are restricted to this particular genetic system. Data on *Chaoborus* (Dawidowicz et al. 1990; Leibold 1990; Tjossem 1990; Dawidowicz 1993) and various copepods (*Acartia*: Bollens et al. 1994; *Diaptomus*: Neill 1990, 1992) demonstrate the importance of predator-induced DSB in obligately sexual organisms.

Why are inducible defenses so ubiquitous in zooplankton? Several factors favor the evolution of predator-induced over constitutive defense strategies (Dodson 1989b; Adler and Harvell 1990). The first is strong but variable predation pressure. Predation by fish and invertebrate predators has a major impact on the community structure of ponds and lakes, and predation pressure is highly variable in time and space (Lampert 1987b). In temperate lakes, predation pressure by fish is low during winter and high during summer, through a combined effect of increased activity associated with higher temperatures and an increased abundance due to recruitment (Gliwicz and Pijanowska 1989). An extreme case is Lake Maarsseveen, where there is a migration of large shoals of perch (*Perca fluviatilis*) to the open water zone during May–June, causing the onset of a pronounced migration of the *Daphnia hyalina x galeata* hybrid population (Ringelberg et al. 1991). Predation pressure by *Chaoborus* similarly varies seasonally, as only third and fourth instar larvae prey upon *Daphnia* (Swift 1992).

A second condition favoring inducible defenses is the presence of a perceptible and reliable cue that can be used as an estimate of predation risk. Specific chemicals produced by the predators (kairomones) certainly are a direct and reliable indicator of predator presence. The high specificity of predator kairomones allows a differential response to different predators (DSB: table 9.1; life history: Stibor 1992). In addition to the mere presence of the kairomone, its concentration may provide information on the density of the predators, and hence on the overall predation risk. Several studies have shown that the amplitude of an induced migration is indeed positively related to the concentration of the kairomone (Loose 1993; Van Gool and Ringelberg, unpubl.).

The effectiveness of an induced defense is strongly dependent on the time lag between detection of predator kairomones and the full development of the defense. Inducible defense is favored when this latency period is short.

Data by Ringelberg and Van Gool (1995) and De Meester and Cousyn (1997) show that predator-induced changes in phototactic behavior occur almost immediately, and that a full development of the defense (more negatively phototactic behavior) is established in less than 12 hours.

Finally, inducible rather than constitutive defenses are likely to develop when there is a cost associated with the defense. We can distinguish between two kinds of defense costs that have to be offset by the reduction in mortality by predation: metabolic costs, resulting in a lower fitness as defined by the intrinsic rate of increase r; and ecological costs, which arise from the ecological settings of the habitat, *in casu* biotic interactions. Staying at a deeper depth during the day entails a high fitness cost (Dawidowicz and Loose 1992b; Loose and Dawidowicz 1994), which may result in the competitive displacement of a constitutively defended species (genotype) by an inducible species (genotype) (Lively 1986b; Adler and Harvell 1990; Harvell 1990a). With respect to ecological costs, it is important to emphasize that a defense which reduces vulnerability to one type of predator may increase vulnerability to another type of predator. The most adaptive DSB behavior with respect to, for instance, visually hunting fish is indeed totally different from the most adaptive DSB behavior with respect to *Chaoborus* larvae. As a consequence, inducible antipredator strategies will result in an increased fitness when the relative importance of different types of predators changes over time.

Factors Modifying the Response to Predator Kairomones

Laboratory experiments have revealed that several factors can influence the extent to which a zooplankter responds to the presence of specific predator kairomones. The presence of multiple predators, for instance, may modify the response to the presence of a given predator kairomone (see Tollrian and Dodson, chapter 10). One of the most important modifying factors in DSB is food satiation (Gabriel and Thomas 1988). Several studies have shown that the amplitude of zooplankton migration decreases when food is limiting (Huntley and Brooks 1982; Johnsen and Jakobsen 1987). Van Gool and Ringelberg (1995) showed that the responsiveness of starved animals to changes in light intensity is much less than that of well-fed animals both in the absence and presence of predators. As a matter of fact, the effect of food availability and predator kairomone were additive (fig. 9.2). When well fed, all animals exposed to fish kairomones responded to a standard increase in light intensity (0.13 min^{-1}) with a downward swimming, whereas starved animals or animals that were not exposed to fish kairomones had a much lower response (60% instead of 100%). Only 6% of the daphnids that were neither fed nor exposed to fish kairomones showed a significant response to

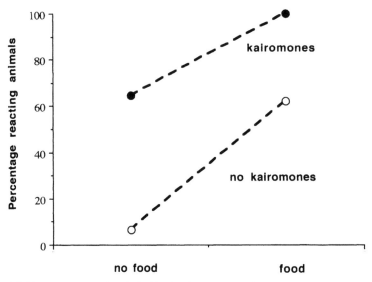

Fig. 9.2. Percentage reacting individuals of one clone of the hybrid *Daphnia galeata x hyalina* from Lake Maarsseveen (the Netherlands) at different combinations of food availability (*Scenedesmus acuminatus* present or absent) and predator (*Perca fluviatilis*) kairomones (present or absent). Reactivity was measured through light-induced swimming in response to a relative light change of 0.13 min^{-1}, which is about the maximum relative light change prevailing at sunrise at Lake Maarsseveen. The effect of food and kairomone on the percentage reacting animals was significant by two-way ANOVA ($p < 0.001$), but there was no significant interaction effect. (Reproduced from Van Gool and Ringelberg 1995, with permission.)

the same change in light intensity. In *Artemia* nauplii that are exposed to fish kairomones, there is no effect of starvation on the response to an increase in light intensity, but starvation does enhance the response to a decrease in light intensity (Forward and Hettler 1992). This difference in responses reflects the fact that the descent to day depth around dawn functions in predator avoidance, whereas the ascent at sunset functions in feeding enhancement.

Genetic Polymorphism versus Phenotypic Plasticity

De Meester (1993a, 1996a) has shown that the change in phototactic behavior of *Daphnia magna* in the presence of fish kairomones is genotype dependent. Such genetic differences in inducible defenses are probably widespread, as several studies have reported interclonal differences in the response to predator kairomones for other traits (behavior other than DSB: De Meester and Pijanowska 1996; life history traits: Weider and Pijanowska 1993; Reede and Ringelberg 1995; morphology: Parejko and Dodson 1991;

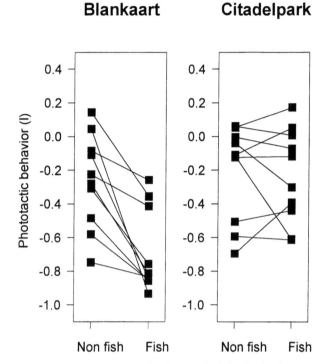

Fig. 9.3. Reaction norms for changes in phototactic behavior in the presence of fish kairomones of ten clones from each of two populations. Clones are random samples from the exephippial generation of the respective populations. Data points are clonal averages based on 4–7 experiments (see De Meester 1996a for methods). Lake Blankaart is characterized by a high predation pressure from fish; there are no fish in the pond at Citadelpark. Most clones derived from Lake Blankaart become negatively phototactic in the presence of fish kairomones, whereas most clones from Citadelpark pond do not show any change in behavior. The behavior of the populations in the absence of fish kairomones is similar. (Data from De Meester 1996a.)

Spitze 1992). Intrapopulational genetic variation in inducible defenses yields the raw material for interpopulational genetic differences to develop. De Meester (1996a) analyzed phototactic behavior in the absence and presence of fish kairomones of a random set of ten *Daphnia magna* clones from each of three populations. Pronounced among-population genetic differences were observed that were in agreement with a priori predictions based on information on the differences in fish predation pressure in the habitats the clones originated from (fig. 9.3), and it was concluded that local adaptation was involved. Parejko and Dodson (1991) similarly reported a pattern of among-population genetic differentiation in *Chaoborus*-induced development of neckteeth in *Daphnia pulex* that was in concordance with the hypothesis of local adaptation. Few studies have been designed to test for interpopulation genetic differentiation and local adaptation with respect to predator-induced defenses, but these phenomena may be widespread in zooplankton.

The Maintenance of Genetic Polymorphism for Induced Defense Mechanisms

Theoretical models have indicated that under a wide range of conditions, generalist genotypes showing inducible defense are superior to specialist genotypes that are constitutively (non)defended (Lively 1986b; Lynch and Gabriel 1987; but see Wilson and Yoshimura 1994). How then can we explain the genetic polymorphism that has been observed in studies such as Parejko and Dodson (1991), Spitze (1992), and De Meester (1993a, 1996a)? Obvious possibilities explaining the coexistence of nondefended, inducible, and constitutively defended genotypes are that building and maintenance costs of the receptor-effector system of inducible response mechanisms are high, or that the proximate cues triggering induced responses are unreliable ($<$ 50% correct choices are made; Lively 1986b). Genetic polymorphism may also be enhanced by a continuous enrichment from the dormant gene pool, with long-term viability of resting eggs reintroducing genotypes that have disappeared from the population through directional selection (Hairston and DeStasio 1988). This nonequilibrium argument cannot in itself explain coexistence in the long term, however. Habitat selection as well as frequency- and density-dependent selection are probably important factors in the maintenance of genetic polymorphism for DSB in zooplankton (Wilson 1989; De Meester 1994a,b). For instance, negatively phototactic *Daphnia* genotypes that remain at greater depth at all times (De Meester 1991, 1993b) may benefit from frequency-dependent selection because their relative fitness in the marginal near-bottom habitat is higher than that of other genotypes (De Meester 1994b).

In addition to the above-mentioned factors, we stress that the complexity of possible responses to the presence of predators may be a major factor in the maintenance of genetic polymorphism for induced defense mechanisms. *Daphnia magna*, for instance, has been shown to respond to the presence of fish by changes in several life history traits (including size and age at maturity, clutch size and egg size; e.g., Weider and Pijanowska 1993) and several behavioral traits (DSB: table 9.1; swarming: Pijanowska 1994; alertness: De Meester and Pijanowska 1996). Studies in which induced changes in several traits have been monitored for the same (set of) genotype(s) have provided evidence that the inducibility is trait dependent, with different genotypes showing a response in different traits (e.g., Spitze 1992; Weider and Pijanowska 1993; De Meester and Pijanowska 1996; see also Tollrian 1995b for further evidence in favor of this "uncoupling hypothesis"). This complexity of possible responses has been overlooked in models that have analyzed the relative fitness of generalist and specialist genotypes, and has two important consequences.

First, the multitude of choices that can be made in response to the presence of predator kairomones results in a lowering of the overall reliability of the available proximate cues with respect to the optimal combination of responses, decreasing the adaptiveness of inducible defense (Lively 1986b). We argue that the information provided by the proximate environment is often insufficiently detailed such that there is no one-to-one relationship between the stimuli received by the zooplankton and the most adaptive response. This factor is expected to be especially important with respect to the development of interpopulational genetic differentiation. Although the ecological settings of a particular habitat are often sufficiently simple and predictable that the environmental stimuli provide a good indication of the most adaptive combination of induced defense responses, this is less so when different habitats (populations) are compared. We therefore expect that in most habitats, so-called generalist genotypes exhibiting the appropriate phenotypic plasticity will evolve, but that different types of generalist genotypes will evolve in different habitats. What are called "generalist" genotypes at the scale of the local habitat are then in fact highly specialized genotypes at a regional scale, as they exhibit phenotypically plastic responses that are tuned to local conditions (see De Meester 1996b).

A second consequence of the complexity of possible responses is that several combinations of induced responses may yield the same relative fitness. Such multiple fitness optima may be a powerful force maintaining genetic polymorphism at the within-population level. In a selection experiment with fish in 11 m deep plankton towers, De Meester et al. (1995) showed that *Daphnia hyalina x galeata* clones that were isolated from the same habitat had nearly identical relative fitnesses in the presence of strong fish predation pressure, through different strategies that involved different combinations of body size and DVM. In other words, the effects of variation in each of the traits were balanced by variation in the other trait.

Conclusions

DSB and DVM are a model system to study habitat selection behavior in zooplankton, with a considerable amount of data on both proximate and ultimate aspects. The adaptive significance of DSB can be understood from the perspective of a trade-off between optimization of reproductive output and reduction of mortality from predation. The trade-off depends on the abundance and types of predators, the characteristics of the vertical gradients in temperature and food quality and quantity, and the presence of competitors. The trade-off is further dependent on the size, sex, reproductive state (all determining conspicuousness), the physiological condition (determining the effect of, for instance, food stress), and escape responses (largely species or instar-specific) of the zooplankton individual. Light (primary and secondary phototaxis) provides the underlying proximate cue for DSB and DVM. We expect responses to light stimuli to be modified by predator kairomones (type, concentration), food characteristics (affected by satiation, food quality, and gradients in food concentration), temperature (temperature gradients), and information on competitors (crowding chemicals, allomones, mechanical stimuli). So far, there is only good evidence for predator kairomones and food concentration as modifying factors. The response to light stimuli is also expected to be dependent upon species, ontogenetic stage, size, sex, reproductive state, and physiological condition of the zooplankton individual. Finally, whether, how, and to what extent the above-mentioned factors modify the phototactic behavior is genotype-dependent, probably because of the complexity of possible responses and the existence of multiple fitness optima. In considering these multiple fitness optima, one should not only focus on the day- and nighttime vertical distribution of the zooplankton, but also on the timing and speed of the descent and ascent phases of DVM. Genetic differences in DSB may further originate indirectly, for instance through genetic differences in body size. Although some aspects of the above scheme are hypothetical and need further study, we believe that this synthesis, which provides a very dynamic view on the micro-evolution of DSB in zooplankton, is consistent with the presently available, massive data set on this important habitat selection behavior.

10

Inducible Defenses in Cladocera: Constraints, Costs, and Multipredator Environments

RALPH TOLLRIAN AND STANLEY I. DODSON

Abstract

Cladocerans provide a model system for studying the evolution of inducible defenses. Several species change morphology, life history, and behavior in response to chemical cues released by predators. Chaoborids, notonectids, and fish are reported to induce specific defenses. In this review we pinpoint five factors that favor the evolution of inducible defenses in Cladocera: (1) Each of the predators can have a strong, but seasonally variable impact on zooplankton communities. During part of the year the predators have no, or a very low impact. (2) Many of the defenses are induced by chemical cues released by the predators. These chemical cues are specific and allow for formation of defenses in response to the actual risk of predation. Chemical cues are good and reliable signals in aquatic environments. Mechanical or visual signals are for inducing escape reactions. (3) Cladocerans can form effective defenses against many predators within their developmental and environmental constraints. Prey body size in relation to the predator's size selectivity is a key factor for the evolution of different types of defense mechanisms in different species. (4) For many of the defenses, costs are reported, which can be saved when the defense is not formed. Often trade-offs are not necessarily direct allocation costs for the formation of the defensive trait, but rather result from interactions of the defense with the environment (environmental costs). (5) Cladocerans often coexist with several predator species, each with a different prey selectivity. Thus, modifying a trait in response to one predator can subsequently increase vulnerability to other predators. Fitness disadvantages of forming or maintaining the wrong defense can be high. Evolution in environments with sometimes changing predation regimes favors inducible and possibly reversible defenses.

Cladocerans evolved a set of different types of inducible defenses against several predators. The defenses against each predator can be mixed in response to co-occurring predators. The phenotypic plasticity allows single clones to survive under heterogeneous environments.

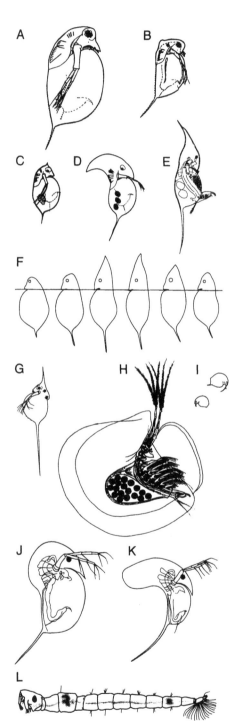

Fig. 10.1. Morphological defenses of several cladocerans: (A) *Daphnia pulex* adult female (Brooks 1957); (B) *Daphnia pulex* adult male (Brooks 1957); (C) *Daphnia ambigua* adult female (Brooks 1957); (D) *Daphnia retrocurva* adult female (Brooks 1957); (E) *Daphnia cucullata* adult female (Lilljeborg 1900); (F) *Daphnia cucullata* adult females from different times of the year (Woltereck 1909); (G) *Daphnia lumholtzi* (Havel et al. 1995); (H) *Holopedium gibberum* (Lilljeborg 1900); (I) *Bosmina longirostris* adult females (Kerfoot 1987); (J, K) *Daphnia cephalata* (called a form of *D. carinata* by Bayly and Williams 1973) adult females showing a range of morphological variation (Hebert 1978); (L) Larva of *Chaoborus* sp.

Introduction

Cladocerans ("water fleas") are a group of several hundred species of small crustaceans (Crustacea, Branchiopoda) classified into eight loosely related orders (Dodson and Frey 1991). Studies of chemical induction have been done almost exclusively with the pelagic orders (*vide* fig. 10.1).

Pelagic cladocerans are a major link between limnetic primary production and higher trophic levels. They are able to suppress algal growth in natural systems (McQueen et al. 1986) and are important prey items (Post et al. 1992), rendering them important tools in the practice of biomanipulation (J. Shapiro 1980). In addition, cladocerans are perfectly suited for general ecological experiments. They are clonal, have short generation times, and can be kept in large quantities in small laboratory space. Several species express morphological, life history, and behavioral defenses in response to chemical cues released by predators. Predaceous chaoborids (phantom midge larvae), notonectids (backswimmer adults and nymphs), and several species of fish have been reported to induce specific defenses (for older reviews see Havel 1987; Dodson 1989b).

In this chapter we will review inducible defenses in Cladocera. We will discuss factors favoring the evolution of phenotypic plasticity in defensive traits for this group. Currently four main factors are under discussion. Inducible defenses occur when (1) the probability of contacts by biological agents is unpredictable, (2) defenses are effective, (3) the cues associated with contact are reliable and not fatal, and (4) fitness costs for the defense offset some of the benefits of the defense (Havel 1987; Harvell 1990a, Adler and Harvell 1990). In our review of cladocerans we will (1) emphasize that the probability of attack has to be rather variable than unpredictable; (2) discuss constraints that limit the effectiveness of defenses and are possibly a major factor preventing the evolution of inducible defenses in many systems; (3) show that cladocerans evolved sensitivity to a wide array of predator kairomones; (4) determine that besides the classical allocation costs, other types of costs might be more relevant.

Cladocerans typically express several defenses simultaneously in response to a single predator. Behavioral, morphological, and life history defenses differ in timing of induction and in reversibility and can probably be finetuned to form a good combination of defenses. Thus, phenotypic plasticity can allow single generalist genotypes to exist as different specialized phenotypes in different environments.

Predators

Predation is an important agent of natural selection in freshwater systems. Hrbácek (1962) showed that zooplankton communities in fishless lakes are dominated by large cladoceran species, while small cladocerans dominate lakes containing fish. Based on the same observation, Brooks and Dodson (1965) developed the "Size Efficiency Hypothesis" (SEH), which postulates that large species are competitively superior to small species but are suppressed or eliminated by visually hunting planktivorous fish. However, experiments often failed to prove the higher efficiency of large species in competition experiments (reviewed in DeMott 1989) which led Dodson (1974) to expand the SEH and to incorporate effects of size-selective (-dependent) invertebrate predators into the theory. Many invertebrate predators are themselves large zooplankters and therefore are suppressed by visually hunting fish. Under these conditions small cladoceran species become dominant. In the absence of fish, invertebrate predators suppress small prey species, leading to a dominance of large cladoceran species.

Thus, cladocerans have to cope with a variety of size-selective/dependent predators (Dodson 1974; Zaret 1980). Large visually hunting predators, such as fish and salamanders, are usually limited in catching prey at the "detection" step in the predator-prey process: prey are selected according to visibility. Smaller predators, which may detect prey visually (larval fish) or mechanically (*Chaoborus*), are often limited by their ability to capture, handle, or ingest prey.

Although many different predators probably induce defenses in Cladocera, there are three well-studied groups of predators: phantom midge larvae, fish, and backswimmers.

Aquatic larvae of the phantom midges (Chaoboridae, Diptera) are important predators in freshwater limnetic environments, sometimes accounting for substantial *Daphnia* mortality (Dodson 1972; Kajak and Ranke-Rybicka 1970; Spitze 1985). In temperate environments chaoborids have one or two generations a year. They pass through four aquatic larval stages, pupate, then hatch to a short aerial life. Larvae overwinter in the sediments in the fourth larval stage. *Chaoborus* larvae migrate vertically in lakes (to avoid fish predation) but not in fishless ponds (Stenson 1980). Chaoborids are size-selective ambush predators, catching and eating prey in the body size range of 0.5 to 2.5 mm (e.g., Parejko 1991; Swift 1992; Tollrian 1995a). The pattern of selectivity is not due to active selection, but due to the opposing effects of encounter probability, which increases with prey size, and strike efficiency, which decreases with prey size (Pastorok 1981). *Chaoborus* larvae use mechanoreceptors to locate prey within their reactive distance (Riessen et al. 1984) and capture prey with specialized mouthparts and antennae (Duhr 1955).

Most planktivorous (plankton-eating) freshwater fish detect prey visually (O'Brien 1987). Larval fish are probably the most important zooplankton predators. They have a high metabolism and reach highest numbers in early summer, when the water is warm and therefore feeding activity is strongest (e.g., Post et al. 1992). By autumn the year class is drastically reduced, usually by predation. Survival for more than one year is seldom greater than 1% of the number of initial free-swimming larvae (reviewed by Gliwicz and Pijanowska 1989).

Backswimmers (Hemiptera, Notonectidae) are important predators in shallow ponds and in the littoral zone of lakes (Murdoch and Sih 1978; McArdle and Lawton 1979). They float at the surface and use their vision and mechanoreceptors to locate prey. They are fast swimmers and grasp their prey with their legs. Like fish, backswimmers locate their prey mainly visually and prefer large prey (Scott and Murdoch 1983), although the largest cladocerans are too big to be captured (Grant and Bayly 1981).

Thus, a similarity between Chaoborids, fish, and notonectids is that they can have a strong impact on cladoceran populations, and their selective impact fluctuates seasonally.

Defenses

Induced Traits

Cladocerans can be induced to express a variety of morphological, life history, and behavioral defenses (table 10.1). Typically, pre-encounter defenses make the cladocerans less conspicuous or less likely to be encountered, and post-encounter defenses make them more difficult to handle or ingest.

MORPHOLOGICAL DEFENSES

Several *Daphnia* species respond morphologically to phantom midge larvae, notonectids, or fish (table 10.2, fig. 10.1).

The best-studied system is the relationship between *Daphnia pulex* and *Chaoborus*, which became a model system for the study of inducible defenses. *D. pulex* is a large (up to 3 mm long) pond-dwelling cladoceran. Juveniles develop small protuberances in the neck region which are termed "neckteeth" (fig. 10.1). These neckteeth have been described from field studies since the last century (e.g., Hartwig 1897), but more recently Krueger and Dodson (1981) found that neckteeth can be induced by the presence of chaoborids or water from a culture of *Chaoborus* larvae. The development consists of a syndrome including discrete small "teeth" and a continuous change of the neck region (Tollrian 1993). Other species forming neckteeth include *D. hyalina* (J. Lüning, pers. comm.; Tollrian unpubl.) and *D. rosea* (A. Sell, pers. comm.; Tollrian unpubl.). Necktooth production varies among clones (Havel 1985a; Parejko and Dodson 1991; Spitze 1992). In some clones, neckteeth are always produced; in others, they appear only when induced, and different clones have different thresholds of

TABLE 10.1
Summary of Kairomone Inducible Antipredation Defenses in Daphnia

Predation	Pre-Encounter Defenses			Post-Encounter Defenses		
Method	Life History	Morphological	Behavioral	Life History	Morphological	Behavioral
Visual	Reduced size and age at first reproduction	Reduced eye diameter; increased transparency				
Tactile				Increased size and age at first reproduction	Neckteeth	"Deadman response"
Both	Production of resting eggs		Diel vertical migration; horizontal migration; shore avoidance; swarming; reduced activity		Tail spines; enlarged helmets; crests	Fast escape swimming; alertness

Notes: Pre-encounter defenses tend to be against visual predation or both visual and tactile types of predation. Defenses oriented toward reducing tactile predation are post-encounter.

TABLE 10.2
Induced Morphological Defenses

Prey	Predator	Induced Morphological Trait	Reference
Bosmina longirostris	*Epischura* (Copepoda)	Mucro and antennule length	Kerfoot 1987
ambigua	*Chaoborus* spp.	Neonate helmet length	Dodson 1989a
ambigua	*Chaoborus* spp.	Adult helmet length	Hebert & Grewe 1985
ambigua	*Lepomis* (fish)	Adult helmet shorter, tail spine	Dodson 1989a
ambigua	*Notonecta* spp.	Tail spine	Dodson 1989a
carinata	*Anisops* (Notonectidae)	Crest	Grant & Bayly 1981
catawba	*Chaoborus* spp.	Neckteeth	Vanni 1987
cucullata	*Chaoborus* spp.	Helmet length	Tollrian 1990
galeata	*Chaoborus* spp.	Helmet length	Hanazato 1991
galeata mendotae	*Chaoborus* spp.	Helmet length	Dodson 1988b
galeata mendotae	*Lepomis* (fish)	Helmet length shorter	Dodson 1988b
galeata mendotae	*Notonecta* spp.	Helmet length, tail spine	Dodson 1988b
hyalina	*Chaoborus* spp.	Neckteeth	Tollrian, in prep.
lumholtzi	*Leucaspius* (fish)	Helmet length	Tollrian 1994
obtusa	*Lepomis* (fish)	Tail spine	Dodson 1989a
obtusa	*Notonecta* spp.	Tail spine	Dodson 1989a
parvula	*Chaoborus* spp.	Adult tail spine	Dodson 1989a
parvula	*Lepomis* (fish)	Tail spine	Dodson 1989a
parvula	*Notonecta* spp.	Tail spine	Dodson 1989a
pulex	*Chaoborus* spp.	Tail spine, neckteeth	Krueger & Dodson 1981
pulex	*Chaoborus* spp.	Neckteeth, fornices, body width, tail spine	Spitze & Sadler 1996

TABLE 10.2 (*Continued*)

Prey	Predator	Induced Morphological Trait	Reference
D. pulex	Chaoborus spp.	Body depth, neckteeth	Tollrian 1995b
D. pulex	Lepomis (fish)	Adult tail spine	Dodson 1989a
D. pulex	Notonecta spp.	Tail spine	Dodson 1989a
D. pulicaria	Chaoborus spp.	Tail spine	Dodson 1989a
D. pulicaria	Lepomis (fish)	Adult tail spine	Dodson 1989a
D. pulicaria	Notonecta spp.	Tail spine	Dodson 1989a
D. retrocurva	Chaoborus spp.	Helmet length	Dodson 1988b
D. retrocurva	Notonecta spp.	Helmet length, tail spine	Dodson 1988b
D. rosea	Chaoborus spp.	Neckteeth	Tollrian, in prep.
D. schodleri	Buenoa (notonectid)	Crest	Schwartz 1991
Holopedium gibberum	Chaoborus spp.	Capsule size	Stenson 1987

Note: If not otherwise stated, the induced morphological trait increases in size.

concentration of *Chaoborus*-conditioned water. The degree of necktooth formation is dependent on kairomone concentration (Parejko and Dodson 1990; Tollrian 1993).

Other species of *Daphnia* form inducible helmets or spikes at the tip of the head in response to *Chaoborus* kairomones (chemical substances released by the predators). Helmets in *D. cucullata* became the textbook example for cyclomorphosis (Wesenberg-Lund 1900). Woltereck (1909) developed his concept of reaction norms on this trait. Inducible helmets range from short spikes in *D. ambigua* (Hebert and Grewe 1985) to long helmets in *D. retrocurva* (Dodson 1988a) or *D. cucullata* (Tollrian 1990) (fig. 10.1).

Notonectids also induce morphological defenses. *D. carinata* and *D. shoedleri* exposed to kairomones from notonectids forms huge cephalic crests (Grant and Bayly 1981; S. Schwartz 1991). These crests reduce vulnerability to notonectid predators.

Fish kairomones induce helmet growth in *Daphnia lumholtzi* (Tollrian 1994). Helmets in *D. lumholtzi* are the largest helmets formed in cladocerans (fig 10.1G). This species currently invades North America and inducible defenses could be one factor favoring successful colonization.

Gelatinous mucus sheets that occur exclusively in *Holopedium gibberum* (fig. 10.1) are reported to increase in diameter when *Chaoborus* kairomones are present (Stenson 1987).

Other inducible morphological changes include tail spine length (Spitze and Sadler, 1996; Tollrian 1994), body width (Tollrian 1995b), and body size (e.g., Dodson 1989a, handled in "life history").

To avoid ambiguous terminology, we suggest calling the toothlike projections in the neck region "neckteeth," defined as a single individual carrying a single necktooth, which can consist of several tiny spikes on top of a transformed neck region (e.g., *D. pulex, D. rosea*). "Spine" should only be used for tail spines and "crest" should refer to a huge bladelike transformation of the head margin (e.g., in *D. carinata*).

Why are there no species forming helmets and neckteeth simultaneously? Helmets and neckteeth probably evolved from the same basic trait—a plastic cell line that is inducible to produce growth factors and mitogenes, which increase mitosis rate and cell growth in neighbor cells via cell-to-cell communication (Jacobs 1980; Beaton and Hebert 1997). Although the same cell line probably leads to either helmets or neckteeth, no species forms neckteeth and helmets simultaneously. Either these traits exclude each other or a double protection is not adaptive. The developmental mechanisms between helmet-forming species and neckteeth-forming species are very similar. Inducibility starts in the embryo shortly before hatching, and inducible stages exist later in juvenile instars. Most probably sensitivity to the kairomones starts when chemoreceptors start to function during ontogenesis (Tollrian, unpubl.). However, direct induction of the cells in the neck region by the kairomone has also been suggested (Parejko 1992). Induction takes place two instars before formation of the trait (Tollrian, in prep.). While the small neckteeth can be formed rapidly between molts, the growth of a protective helmet requires several molts. Thus, neckteeth protect the juveniles of larger *Daphnia* species that are at birth already in a vulnerable size range of *Chaoborus*, though they are already too large for small invertebrate predators such as copepods. Helmets protect smaller *Daphnia* species, as juveniles against small invertebrate predators and as adults against *Chaoborus*. Thus, body size is the principal key to understanding the selection for either helmets or neckteeth as defense (Tollrian, in prep.). A double protection should not be selected for.

LIFE HISTORY SHIFTS

Life history theory predicts adaptive changes in life history traits in response to predation (e.g., Law 1979; Michod 1979; Charlesworth 1980; Lynch 1980b; Kozlowski and Wiegert 1987; Stearns 1992). Taylor and Gabriel (1992) derived a model for optimal resource partitioning under various predator regimes. The model predicts that under invertebrate predation selecting

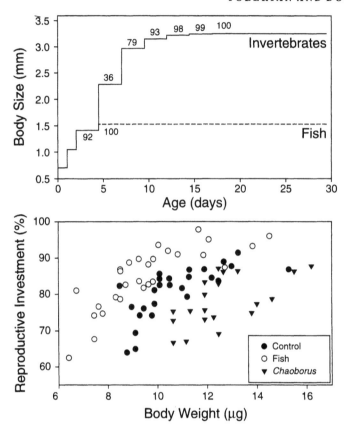

Fig. 10.2. Life history adaptations of *Daphnia* to fish and invertebrate predation. (a) Theoretical optimal allocation of resources to growth or reproduction for predation on small prey size (invertebrate predator) and large prey size (fish) (model from Taylor and Gabriel 1992). Numbers indicate the share of total resources invested into reproduction. (b) Investment into reproduction of *D. hyalina* raised under control, fish, and *Chaoborus* conditions (data from Stibor and Lüning 1994).

small prey sizes (e.g., *Chaoborus* predation), maturity will be delayed to attain a larger body size (fig. 10.2a). Allocation of a higher portion of the energy for growth results in a lower population growth rate (r), in the absence of predation.

Studies with *D. galeata* (Machácek 1991), *D. hyalina* (Stibor 1992; Stibor and Lüning 1994), *D. magna* (Weider and Pijanowska 1993), or *D. pulex* (Dodson and Havel 1988; Dodson 1989a) and *Chaoborus*, fish, and notonectid predators confirm these models (fig. 10.2b). Resource allocation was either shifted toward maternal growth under *Chaoborus* predation (later reproduction, larger size at first reproduction, lower relative reproductive investment) or toward reproduction under fish predation (earlier reproduction, smaller size at maturity, higher relative investment into reproduction). In the latter case, more and smaller eggs were produced. The total energy investments were similar.

However, in other species the relationship between investments into growth and reproduction are not that balanced. Tollrian (1995b) reported that the higher investment into growth in *D. pulex* under *Chaoborus* predation led to a larger body size, which led to a better energy budget for larger animals, as predicted by the SEH (Brooks and Dodson 1965). Stibor (1995) showed that smaller daphnids have relatively lower lipid contents (energy reserves). Thus, if the investment into growth leads to an additional energy bonus, the fitness cannot be expected to be equal between both life history strategies. Size at first reproduction (SFR) is a key parameter in the life history of many species (Lynch 1980b; Stearns 1992). For *D. pulex* exposed to *Chaoborus*, SFR tends to be positively associated with the clutch size (reviewed in Tollrian 1995b), thus producing a higher mean fitness (measured by the intrinsic population growth rate r) for *Chaoborus*-induced phenotypes, even if there is no mortality due to *Chaoborus*. *Chaoborus*-induced daphnids become larger and can gain more energy. Conversely, *Daphnia* exposed to fish tend to be smaller (e.g., Dodson 1989a). Nevertheless some *Daphnia* species/clones induced by fish have a higher potential r value, because more, but smaller, offspring are produced (Dodson 1989a). Higher fitness for induced morphotypes in the absence of direct predation has been reported for several prey species (Gilbert 1980; O'Brien et al. 1980; Dodson and Havel 1988; Stemberger 1990; A. Black 1993; Tollrian 1995b).

Lampert (1993a) showed for a single clone of *Daphnia magna* that a rapid decrease in SFR can be achieved within two generations by elimination of large animals. Smaller adults produced smaller neonates, which produced smaller adults in the next generation. Conversely, under *Chaoborus* predation an elimination of small adults will increase the size at maturity in the following generations. This shift can occur even faster as a maternal effect induced by kairomones (Stibor 1992; Tollrian 1995b). Offspring from mothers grown in *Chaoborus*-conditioned medium were larger and produced larger neonates, while the reverse was true for fish-conditioned medium.

Hairston (1987) proposed predator avoidance as one ultimate reason for diapause. And recently Slusarczyk (1995) reported that fish kairomones even act as proximate factors to induce resting egg production in *D. magna*. Fish can completely eliminate the large-species *D. magna*, except for resting eggs. Resting eggs provide insurance for long-term population persistence and reestablishment during low-predation periods. Although the dark resting eggs make *Daphnia* more visible to fish, resting eggs are dispersed by surviving passage through the fish gut (Mellors 1975).

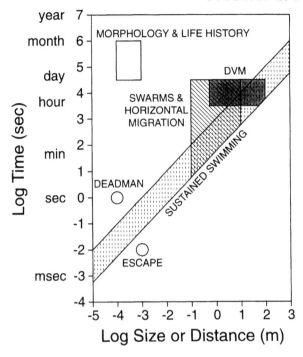

Fig. 10.3. Spatial and temporal scales associated with inducible morphological, life history, and behavioral defenses of cladocerans.

BEHAVIORAL DEFENSES

The best-known induced behavioral defense in Cladocera is diel vertical migration (DVM, reviewed in chapter 9), which is expressed by entire populations at scales of meters and hours (fig. 10.3). At the smaller scales of millimeters and seconds, individual zooplankton swimming behavior is a critical component of zooplankton ecology (reviewed in Larsson and Dodson 1993). Cladocerans filter their environment for food during swimming. They have to swim, otherwise they would sink or reside in a food-depleted microenvironment. On the other hand, a higher swimming speed proportionally increases the encounter rate for ambush predators (Gerritsen and Strickler 1977). Swimming speeds recorded for *Daphnia* range from about 3 to about 20 mm/sec. Large animals tend to swim faster than small ones (Dodson and Ramcharan 1991).

Some variability in cladoceran swimming behavior is induced by predators. For example, several *Daphnia* species changed their vertical distribution in response to kairomones (Dodson 1989a). Furthermore, alertness to predator attacks can be induced. Clones of the *D. pulex* group, when exposed to fish kairomone, become more sensitive to mechanical stimuli (Pijanowska, pers. comm.; Brewer et al., in prep.).

Swarms of pelagic cladocerans have densities several orders of magnitude greater than the lake average. Less extreme swarms probably occur in the open water (Malone and McQueene 1983; Folt et al. 1993). In some cases, prey swarms are induced by predator kairomones (e.g., Kvam and Kleiven 1995).

Behavioral defenses in cladocerans are influenced by internal status and motivation. Under limiting food conditions, behavioral defenses that reduce food uptake are minimized (e.g., Flick and Ringelberg 1993). Motivation to respond behaviorally is induced by predator smell. However, behavioral defenses can be triggered by several external stimuli (light intensity—DVM, reviewed in Ringelberg 1987; mechanical stimuli—alertness, Brewer et al., in prep.; kairomones—spatial avoidance, e.g., Dodson 1988b).

TEMPORAL AND SPATIAL SCALES OF DEFENSES

The three kinds of defenses have different temporal and spatial scales (fig. 10.3). Gabriel and Lynch (1992) calculated that the evolution of reaction norms (assuming a narrow tolerance curve that can be shifted) would be favored over broadly adapted genotypes (a wide tolerance curve that is inflexible) if the between-generation variation in the environment is large compared to the within-generation variation. This is true for irreversible phenotypically plastic responses like a change in resource allocation from fecundity to growth. A large body size remains large for an animal's life span (usually a few weeks). The fixation of the life history strategies late in juvenile stages allows *D. pulex* to respond during ontogenesis to changing environments (Tollrian 1995b). These responses require the better part of the life span to be expressed. If the within-generation variation in the environment is high, reversible responses might be favored. Morphological defenses like helmets or neckteeth can be formed or reduced within 2–3 molts and thus can be considered reversible phenotypic plastic responses expressed for a few days only. Behavioral reactions are reversible, short-term responses. Escape responses typically occupy a few milliseconds. Alertness can be induced within 30 minutes (Brewer, pers. comm.); horizontal or vertical migrations typically require a few hours.

Measured in spatial scales, cladocerans are small animals, between about 0.3 and 5 mm long (fig. 10.1). In some cases, induction of long spines or helmets effectively increase body size two or three times, as a defense against invertebrate predators.

Escape responses, which typically extend only over a few millimeters, include fast swimming (Szlauer 1968; Brewer, pers. comm.) and no swimming (in the case of the *Bosmina* "dead man" sinking response reported by

Kerfoot 1978). Avoidance mechanisms (e.g., migrations) may be as small (e.g., into or out of a plant bed), but can be dramatic (in some lakes, cladocerans show migrations of 10–100 meters).

The evolution of defenses covering different temporal and spatial scales possibly allows tailoring of effective defenses against different densities of predators as well as against different predator species.

Protection

Each of the defenses act in a different way and offers its own unique protection. A predator's feeding cycle can be divided into prey detection, pursuit, capture, handling, and ingestion. Inducible defenses can act during several stages of a predator's feeding cycle. If a predator is searching for food or trying to localize a food source, predator avoidance mechanisms can act to prevent detection. These avoidance mechanisms are mainly behavioral. They include spatial segregation (the result of migration), life history changes (e.g., lower visibility of smaller bodies due to a switch in reproductive effort from growth to reproduction), and morphological defenses (contrast reduction, e.g., reduction of the area of pigments in the eye; see Zaret 1972). Additionally, morphological changes could alter hydrodynamics and possibly could reduce the reactive distance of tactile predators.

Once the prey item is detected, escape mechanisms can defend the prey during "pursuit" (Link 1996). Escape mechanisms before or during an attack but before a contact are exclusively behavioral (e.g., fast start, "dead man response," or swarm formation as predator irritation). A cladoceran captured by a large fish is immediately ingested. However, larval fish and most invertebrate predators need to handle their prey before ingestion. As the predator positions the prey for ingestion, the prey struggles and often escapes (e.g., Havel and Dodson 1984). Morphological defenses, such as long spines or helmets, increase handling time and increase the chance of escape.

PROTECTION BY MORPHOLOGICAL DEFENSES

Neckteeth offer good protection against *Chaoborus* (fig. 10.4). Krueger and Dodson (1981) and Havel and Dodson (1984) reported reduced vulnerability of juvenile *D. pulex*. Results from Tollrian (1995a) indicated that the protection associated with neckteeth is neither caused by simultaneous shifts in body size nor by co-occurring behavioral defenses.

Helmets in *D. galeata mendotae* protect against *Chaoborus* (Mort 1986). Moreover, they additionally protect small species against copepods and predaceous cladocerans (Havel 1985b). However, in small *D. retrocurva* no

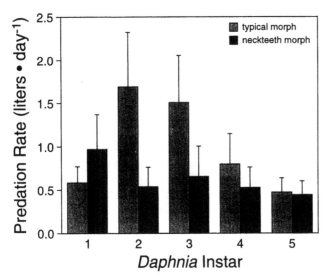

Fig. 10.4. Instar-dependent vulnerability of typical and induced *D. pulex* to *Chaoborus* predation. (Data from Tollrian 1995a.)

protective effect of helmets could be found against large chaoborids (Havel 1985b).

These results illustrate some constraints on the evolution of inducible defenses. Small species simply cannot form morphological defenses against predators that easily can swallow them whole, or at least they are not likely to evolve such traits. The larger an induced trait relative to the body, the higher the chance that it will interfere with other vital requirements (e.g., swimming, gathering food, reproduction).

The protective mechanisms for these traits are still unknown. Havel and Dodson (1984) reported for neckteeth a higher escape rate after the daphnids had been caught. They suggested that neckteeth prevent ingestion. It is also possible that neckteeth interfere with the predator's ability to get a firm grasp on its prey (Tollrian, unpubl.). This would support Dodson's (1974) "anti-lock and key hypothesis."

For helmets, several mechanisms might work. For example, helmets offer more area for muscle attachment, which might allow for stronger escape swimming (Hebert 1978). Other possible (and untested) mechanisms are (1) helmets interfere with the predator's ability to manipulate prey; (2) helmets lead to wrong prey distance estimations by the predator; (3) helmet or neckteeth are mere developmental correlates with thickened exoskeleton; and (3) the "stealth hypothesis"—that helmets influence hydrodynamics in such a way that the prey can pass the predator in a reduced distance without being detected by small-scale turbulences. Personal observations indicate that once captured, helmets in *D. cucullata* do not protect against *Chaoborus* (Tollrian, unpubl.). Viewed from the front, helmets have a bladelike shape. They are easily folded by *Chaoborus* mandibles.

Crests in *D. carinata* protect against notonectids (Grant and Bayly 1981). Crests are associated with improved ability to evade notonectids, but the crests may also interfere with handling as the predator orients the prey for feeding.

Fish-induced helmets in *D. lumholtzi* will prevent ingestion by fish larvae up to a certain size. However, helmets are a defense also against larger predators. Juvenile bluegill sunfish rejected *D. lumholtzi* with giant helmets more often than similar-sized *D. magna* (Swaffar and O'Brien 1996). Furthermore, the coexistence of this relatively large cladoceran species with fish indicates that natively coexisting fish seem to avoid this species (reviewed in Tollrian 1994).

It is likely that most defenses cannot be attributed to a single trait. Tollrian (1993) suggested that neckteeth represent the tip of the iceberg of protection. He reported other co-occurring changes in the neck region. An increase in body depth (Tollrian 1995b) is another possibly protective change in morphology. Swift (1992) showed that prey depth is a better predictor for prey size range in *Chaoborus* than body length. *Chaoborus* usually swallows prey that cannot be deformed only if its diameter is narrower than the larvae's head capsule diameter. The increase in body depth might additionally make prey handling for the larvae more difficult (Tollrian 1995b). Spitze and Sadler (1996) conducted a multivariate analysis of the protective effects of various traits and found that neckteeth, body length, tailspine length, and fornice width were independently positive correlated with a protective effect.

PROTECTION BY LIFE HISTORY SHIFTS

Cladocerans can flexibly allocate their resources to growth or reproduction in response to predation regimes. Furthermore, prey compensate for size-selective predation by producing offspring of a size less likely to be eaten. For example, prey produce larger offspring when exposed to predators that eat small cladocerans.

Changes in body size and in SFR as a result of selection by direct predation have been reported from field studies in a variety of species (Reznick et al. 1990; Vanni 1987; Hutchings 1993; Sih and Moore 1993). Thus it can be concluded that the induced changes are advantageous for specific predation regimes. If the predators are visually hunting fish, a reduced and therefore less visible body size in cladocerans is beneficial throughout the total vulnerable size range. The same should be true for visually and tactile hunting notonectids. Interestingly, cladocerans have two opposing life history defenses to notonectid predators. Smaller species (e.g., *D. pulex*, *D. hyalina*) mature at a smaller size when raised with backswimmer kairomones (Dodson and Havel 1988; Stibor and Lüning 1994). In contrast, larger *D. carinata* and *D. magna* mature at a larger size (Barry 1994; Weider and Pijanowska 1993). Thus selection shaped a predator avoidance strategy for small prey species, and an escape strategy for larger prey species.

Pastorok (1981) showed that prey vulnerability to *Chaoborus* predation is at a maximum at intermediate cladoceran sizes. Tollrian (1995a) suggested that before the peak in vulnerability is reached, a larger body size incurs the disadvantage of a higher encounter rate, whereas later the advantage of the reduced strike efficiency is overwhelming. Therefore, it can be predicted that very small prey species should not respond to *Chaoborus* kairomones by increasing their SFR, while very large prey species should rather respond by producing larger offspring, which can reach a safer size sooner.

PROTECTION BY BEHAVIORAL DEFENSES

Individual behavior affects the outcome of predator-prey interactions, especially in the pelagic environment, where prey movement is important both as a cue to predators and as a determinant of encounter rate.

Visually hunting predators (e.g., fish and salamander larvae, but also invertebrates such as odonate nymphs, notonectids, and *Dytiscus*) are limited by conspicuousness of prey. Size, contrast against the "background," and motion are factors influencing prey visibility (O'Brien 1987). Conspicuous swimming behavior may attract predators (Brewer and Coughlin 1995).

Encounter probability increases with increasing swimming speed of the faster organism for either prey or predator (Gerritsen and Strickler 1977). A reduced swimming speed could reduce prey conspicuousness and predator encounter. Variability in swimming speed among clones (Dodson et al., in prep.) may reflect an adaptive compromise among a large variety of selective forces, including the cost of swimming, predator-prey interactions, and feeding rate.

Swarming behavior can reduce the individual predation risk by a "dilution effect" and additionally by confusing the predators. It affects both visually hunting and tactile predators. Aggregation in swarms could reduce predation by randomly distributed predators (e.g., model by Young et al. 1994). The assumption is that, in the case of swarming prey, only predators in the vicinity of the swarms would catch prey. These predators would be saturated, and the total number of prey eaten per night would decrease.

Induced alertness could help prey to escape attacks (Brewer et al., in prep.). Probably this defense could protect against relatively slow attackers (e.g., ram-feeding juvenile fish) rather than against high-speed attacks (e.g., strikes of *Chaoborus* larvae). After perceiving the predator, the prey has an opportunity to escape, and a faster swimming rate is advantageous in escaping predators. Link (1996) reports that larger cladocerans (and copepods) are captured less frequently by lake herring than smaller cladocerans, and suggests the advantage may be due to the greater swimming strength of the larger animals.

ADAPTIVE DEFENSES OR BY-PRODUCTS?

In assessing whether a trait is a defensive adaptation, it must first be shown to be effective and second to be heritable. Genotype differences within and between populations have been shown for morphological defenses (Havel 1985a; Parejko and Dodson 1991; Schwartz 1991; Spitze 1992), for behavioral defenses (De Meester et al. 1995), and for life history adaptations (Weider and Pijanowska 1993).

Chemical Cues

Cladocerans live in an "olfactory sea" (fig. 10.5A). The typical pelagic cladoceran receives and "reads" chemical cues from a variety of sources including food, conspecifics, predators, and abiotic factors (Larsson and Dodson 1993). Chemicals are a good source of information in turbid aquatic environments. Chemicals can transmit information over long temporal and spatial scales. The ability of an individual cladoceran to respond to a chemical signal probably depends on genetics, the intensity of several environmental factors (fig. 10.5B), and the internal state of the individual (fig. 10.5C).

Environmental factors include turbulence, which stimulates expression of helmets in *Daphnia* (Brooks 1947); light intensity, which affects *Daphnia* response to fish kairomones (Loose 1993); and photoperiod, which affects production of male *Daphnia* in response to a chemical signal produced in crowded cultures. Food level also affects inducibility in cladocerans (e.g., van Gool and Ringelberg 1995; Ramcharan et al. 1992). Probably pH and water chemistry will modify kairomone effectiveness.

Internal state depends on memory, learning, and motivation, which are the result of past experience over probably more than one timescale. For example, Ringelberg and van Gool (1995) proposed that *Daphnia* learn to respond to specific light and kairomone cues. There is also the possibility that signals are filtered and prioritized, and that internal allocations of energy and rates (of development) are controlled by complex switching mechanisms.

Morphological defenses, life history defenses, and many behaviors in Cladocera are induced by chemical substances released by the predators. These so-called kairomones (Brown et al. 1970) are characterized as chemicals in interspecific signal transmission which are exclusively advantageous to the receiving organism in the specific context (Nordlund and Lewis 1976;

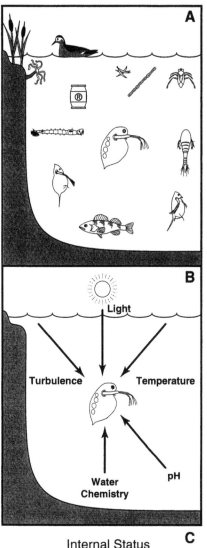
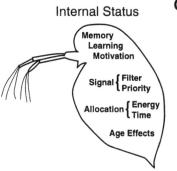

Fig. 10.5. (A) *The olfactory sea.* The organisms in this figure are either known or suspected producers of chemicals that influence *Daphnia* behavior or development. (B) *Environmental factors.* These environmental factors affect either the distribution, perception, or persistence of chemical signals. The barrel represents anthropogenic chemicals, such as pesticides, that are known to mimic chemical signals of predators. (C) Internal state of a *Daphnia*.

Dicke and Sabelis 1988a). The chemical might be advantageous to the emitter in a different context, e.g., as a pheromone in intraspecific communication. Thus the questions arise, what characteristics render the kairomone a good indicator of predation threat, and why is emission of the kairomone not suppressed by the predator?

Chaoborus *Kairomone*

The kairomone is a heat-stable, low-molecular-weight (<500 Dalton), water-soluble nonprotein, with a chemical structure that includes hydroxyl and carboxyl groups (Hebert and Grewe 1985; Parejko and Dodson 1990; Tollrian and von Elert 1994).

Actively feeding larvae (but not starved larvae) release the kairomone (Krueger and Dodson 1981). The signal is not an alarm substance. Larvae feeding on copepods release the kairomone, while crushed *Daphnia* do not (Walls and Ketola 1989). Release of the kairomone is directly related to prey uptake. The kairomone is released during egestion of indigestible prey remnants from the pharynx, and it is quickly degraded by bacteria. Thus the kairomone directly signals a potential danger by the predator (Tollrian, submitted).

If a pelagic predator like *Chaoborus* is present, it will generally not disappear within a short time. Therefore, a readily water-soluble cue ensures that the information is spread rapidly, but it stays present in the water column for a certain span of time. A cue indicating danger by diel vertical migrating predators should persist longer than the absence period of the predator to keep the prey alert before the predator returns. However, reactions to kairomones are concentration-dependent (Parejko and Dodson 1990; Tollrian 1993). At low predator densities there is a threshold for the reaction. It might well be that vertically migrating predators decrease defenses of their prey by releasing only part of their kairomones in areas where their prey lives.

All chaoborid species tested so far release the kairomone. This result together with the finding that the kairomone is involved in the feeding process of the predator indicates that coevolution and suppression of the kairomone release is constrained on the side of the predator.

Experiments with the kairomone are hampered by the difficulty of enriching the kairomone without accumulating toxic metabolites or substances that facilitate bacterial growth. Tollrian and von Elert (1994) introduced a method to purify the kairomone from rearing water by solid phase extraction, which allows work with a purer kairomone.

Chemical identity has not been proven for *Chaoborus* kairomones inducing behavioral, life history, and morphological defenses. Tollrian (1995b) used purified kairomones to induce neckteeth and life history shifts in *D.*

pulex. The purified kairomone was active in both traits, indicating at least chemical similarity.

Hanazato and Dodson (1993) found that helmets and neckteeth increased in several cyclomorphic species after addition of insecticides in sublethal concentrations. This result may provide insight into the developmental mechanism of induction. Many insecticides are cholinesterase inhibitors. It is possible that the nervous system of the daphnids is activated similar to chemoreceptor-mediated kairomone perception.

Fish Kairomones

Loose et al. (1993) developed a behavioral bioassay to study fish kairomones. Von Elert and Loose (1996) characterized the DVM-inducing kairomone as a nonolefinic, low-molecular weight (< 500 Dalton) anion of intermediate lipophilicity. A hydroxyl group is essential for activity, while the presence of amino groups in the molecule could be excluded. Separation by HPLC yielded only one active fraction. So far there is no evidence that different fish release different DVM-inducing kairomones. The benthivore Crucian carp (*Carassius carassius*), the plankton-eating roach (*Rutilus rutilus*), and the piscivorous pike (*Esox lucius*) all induce DVM (Loose et al. 1993). The origin of the substance is not yet known. Release of the kairomone is not related to diet or nutritional state. Fish starved for 5 days released the kairomone as well as fish that had been fed with various types of food including a nonplanktonic diet. Activity occurs both in fish feces and in mucus (Loose et al. 1993).

A dose-response relationship exists for morphological reactions in *D. lumholtzi* (Tollrian, unpubl.), behavioral reactions in *D. hyalina x galeata* (Loose 1993), and life history reactions in *D. hyalina* (Reede 1995) and *D. magna* (Stibor 1995). *Daphnia* may even respond to the absence of kairomone. For example, fish present at low densities (which are possibly under the detection threshold for cladocerans) reduce *Chaoborus* numbers and are a threat to large *Daphnia* (Stenson 1980). It is possible that the absence of *Chaoborus* kairomone induces an anti-fish morphology in these *Daphnia*.

Costs and Trade-Offs

Defense against predation is generally assumed to incur costs (Maynard Smith 1982). By forming inducible defenses only when they are beneficial, part of the costs can be saved. Costs can be estimated from individual growth rates, but the number of surviving offspring in future generations should be the currency in which costs are paid.

The occurrence of morphological defenses combined with life history changes highlights the dilemma of cost assessments. Is a longer time until maturity a cost for neckteeth formation or a trade-off for life-history changes? Studies aiming to assess costs of morphological defenses, even within a single species, have led to contradictory results. However, most studies report fitness disadvantages for induced morphs in the absence of predation (e.g., Havel and Dodson 1987; Riessen and Sprules 1990; Black and Dodson 1990; Walls et al. 1991). Spitze (1992) compared several clones from different populations and reported clonal differences in body size, growth rate, fecundity, and age at first reproduction.

Although clonal differences might explain some of the variation between studies, part of the variation may also be explained by experimental conditions (reviewed in Tollrian 1995b). Chemicals released by *Chaoborus* in addition to the kairomone, or bacteria growing in the presence of *Chaoborus* can be harmful to *Daphnia*. Ketola and Vuorinen (1989) found reduced growth in the robust species *D. magna* (which does not form neckteeth), when reared in *Chaoborus*-conditioned water. Survival was drastically reduced for induced daphnids in several studies (e.g., Ketola and Vuorinen 1989; Black 1993), which were possibly indications of deleterious effects rather than "adaptive" responses. Other reasons for possible overestimations of costs include co-occurring behavioral adaptations that could result in different food availability (Dodson 1988b; Ramcharan et al. 1992).

Separation of "Costs" and Life History Shifts

The dilemma of addressing costs can be overcome in *D. pulex*, because changes in life history parameters (e.g., size and age at maturity, clutch size) and morphological changes (neckteeth syndrome) can be separated by timing of induction. Tollrian (1995b) showed that life history shifts were induced after the induction of neckteeth formation and thus could be uncoupled from neckteeth formation. The observed changes (higher growth per instar, larger size at maturity, higher fecundity, and longer instar duration) occurred without neckteeth formation, while neckteeth formation occurred without these life history shifts under the experimental conditions. Therefore, the longer time to maturity was not the "cost" of neckteeth formation but a trade-off for the life history shifts. Data by Lüning (1994) and Repka et al. (1994) also indicate that neckteeth and life history shifts are independent defenses.

So far there is only one complete cost-benefit analysis for a single clone of *D. pulex* (Riessen 1992). The variable success in measuring costs and the reported clonal differences indicate that the results of this study are not applicable for all clones.

The above-cited likelihood that costs have been overestimated should not imply that they are nonexistent. We have to assume disadvantages connected to the formation of defenses. If there are no disadvantages, a defense should be expressed permanently (Riessen 1984). Even very small costs that are hardly measurable in experimental conditions would influence population development.

It is likely that costs have been studied in a simplified context. In most studies energy expenditures are regarded as the main costs. However, other types of costs might be more important. The real magnitude of costs might rather be the result of a mixture of several costs. Therefore it is worthwhile to assess possible sources of costs for the different types of defenses.

Possible Costs of Defenses

Costs are classified in Tollrian and Harvell (chapter 17). Here we estimate the relevance of these costs for cladocerans. Studies so far cannot rule out that plasticity costs exist, that is, the ability to respond to predator chemicals could include a permanent cost either as a result of negative pleiotropic effects, or for providing and maintaining the facilities (e.g., chemosensors, genetic mechanisms, hormones). These costs would act on inducible morphological, life history, and behavioral defenses.

Allocation costs are due to changes in energy allocation associated with morphological defenses. Higher energy and material expenditures for the synthesis of extra tissue are likely costs for some induced traits but not for all. Helmets in *Daphnia cucullata* or *Daphnia galeata mendotae* are larger than neckteeth. Nevertheless, Jacobs (1967) estimated the mass of a fully developed helmet in *D. galeata mendotae* to equal just one-seventh of a parthenogenetic egg, and neckteeth might even be less. Lynch (1989) calculated for small *D. pulex* that only 6% of the dry weight is lost as total molt, and differences between morphs due to the neckteeth should be much smaller. In contrast, the huge crests produced by *Daphnia carinata* (fig. 10.1) as defense against notonectids (Grant and Bayly 1981) might require a significant proportion of the energy budget (Barry 1994).

Energy demand related to feeding, grooming filters, and perhaps swimming can be a significant part of the *Daphnia* energy budget (e.g., Porter et al. 1982). Energy demand for swimming increases exponentially with increasing speed. The high metabolic cost of fast swimming (escape response speeds) limits the duration of fast swimming in copepods (Alcaraz and Strickler 1988) and probably in cladocerans as well. However, regular swimming might not be too costly. Dawidowicz and Loose (1992a) measured the costs of swimming and found energetic costs only at very low food conditions.

For helmeted *Daphnia* species a reduced maximum clutch size due to a reduced brood chamber volume caused by a more slender body has been proposed (Jacobs 1967). Field data on *D. retrocurva* supported this hypothesis (Riessen 1984). However, such an opportunity cost could not be found in *D. cucullata*, neither in the field nor in the laboratory (Tollrian 1991).

Environmental costs could result from the interaction of all types of defenses with the environment: morphological changes can alter the hydrodynamics of zooplankton and probably result in higher energy requirements for swimming. Jacobs (1967) reported that helmeted *D. galeata mendotae* sink faster, and Stenson (1987) proposed that a larger capsule size in *Holopedium gibberum* should increase the drag or lower food uptake. Also, large crests in *D. carinata* should alter hydrodynamics.

Living in a swarm can incur the cost of living in a food-depleted environment. Loose and Dawidowicz (1994) found that the main cost of DVM is a reduced population growth rate due to lower temperature in lower strata in temperate lakes, independent of food conditions. Reduced activity could be associated with a reduced ability to detect food patches and with a reduced food uptake.

Other environmental costs result from the constraint that adaptations to one predator can prevent adaptation to changing environments. For example, the production of more but smaller offspring under fish predation will have the disadvantage of reduced starvation resistance. Thus, the population growth rate would be reduced if the environment shifts to a state of food limitation.

Intermediate growth rate might be a compromise to the presence of a mixture of predators with different size preferences. The actual threat by each predator indicated by the relative amount of chemical substances in the water may shift the hierarchy of reaction toward or close to an optimal life history in a range of constraints and limitations evolved in the clonal history. When kairomones are absent, life history might be switched to a "general purpose phenotype."

The assumption that a disadvantage of defenses in cladocerans exists in different environments is supported by theoretical studies. Taylor and Gabriel (1992) calculated that the optimal strategy against invertebrate predation will be the worst strategy under fish predation. In an expansion of their model to seasonal environments, they concluded that the fitness reduction due to wrong adaptations to selective predation can be large (Taylor and Gabriel 1993).

Tollrian (in prep.) conducted an experiment in large mesocosms (Plankton Towers in Plön; description in Lampert and Loose 1992) and tested for reactions of a single clone of *D. hyalina* to multipredator environments: no predator, *Chaoborus*, fish, and *Chaoborus* + fish conditions. The daphnids reacted to fish with diel vertical migration (night up, day down) compared to control conditions (permanently up) and a smaller SFR. For *Chaoborus*, the reaction was reverse: a larger SFR and a reverse migration (day up, night down). Furthermore, neckteeth were formed also at later instars (early instars had neckteeth also in control conditions in this clone). In the mixed predator environment, life history was switched to a fish defense (small SFR), neckteeth were formed against *Chaoborus*, and behavior was mixed against both predators (permanent down). Thus one reason for the evolution of different types of defenses in one species might be that it allows a compromise against variable environments with different predators.

Conclusions

We identified five factors that possibly favor the evolution of inducible defenses in cladocerans.

1. Each of the inducing predators can have a strong, seasonally variable impact on zooplankton communities. Many cladoceran predators have a seasonal, predictable timing. The actual impact of each predator is not fully predictable.

2. The predators release reliable chemical cues that can be used by cladocerans to evaluate the potential risk of predation from specific predators.

3. Cladocerans can form effective defenses against many predators within their developmental and environmental constraints. Prey body size in relation to predator size selectivity is a key for the evolution of effective defense mechanisms in different species.

4. Costs are not in each case identified. For some defenses, direct energetic costs (allocation costs) of development are measurable. For other defenses, indirect costs (environmental costs, opportunity costs) might be important.

5. Inducible defenses allow cladocerans to adapt phenotypically to multipredator regimes. Although effects of misplaced strategies could be classified under cost savings, we judge multipredator environments especially important for the evolution of different types of inducible defenses in cladocerans. Morphological defenses, life history shifts, and behavioral defenses work together to form effective defenses against single predators and in concert with reversibility allow the existence of single clones under contrasting predation pressures.

The kairomones are still only partly characterized. Identification of the kairomones will be a major task for the future. Knowledge of the kairomones is not only crucial for analyzing sensory or developmental pathways or genetic mechanisms, but even for cost assessments. Clearly detailed analysis of trade-offs for the defenses are needed. Allocation costs are not always found and probably do not exist in various inducible defense systems. Environmental costs, e.g., disadvantages caused by misplaced defenses and variability created by heterogeneous environments, offer alternatives to allocation-cost-based models. Selection should generally force the evolution of inducible defenses toward a high degree of protection at low costs.

We don't have to be prophets to propose that many new prey species with inducible defenses and many new types of defenses will be found within the next years. However, this should not lead to the impression that we expect all prey to evolve inducible defenses. It still seems that species with inducible defenses represent the minority. This view might change with the increasing number of species forming cryptic defenses (e.g., resource allocation shifts). The variability in responses to predators, especially in heterogeneous environments, might allow for different strategies that lead to the coexistence of species.

11

Predator-Induced Defense in Crucian Carp

CHRISTER BRÖNMARK, LARS B. PETTERSSON,
AND P. ANDERS NILSSON

Abstract

Crucian carp (*Carassius carassius*) develops a deeper-bodied morphology when exposed to piscivorous fishes, and we have suggested that this is an example of an inducible morphological defense against predation by gape-limited piscivores. Here, we review our work on factors that have promoted the evolution of an inducible defense in crucian carp. First, stochastic disturbances, such as a winterkill event that only crucian carp survive, coupled with re-colonization from nearby systems may create the variability in predation pressure necessary for an induced defense to evolve. Crucian carp use waterborne, chemical cues to detect the presence of a predator. Cues from predators feeding on crucian carp elicited a change in body morphology, whereas no reaction was found in response to predators feeding on invertebrates or on fish species lacking alarm substance cells. Crucian carp have a benefit of being deep bodied when exposed to gape-limited predators such as northern pike (*Esox lucius*). Handling time of pike feeding on crucian carp increased as a function of body depth. An increased handling time may affect pike negatively through increasing risk of cannibalism and kleptoparasitism. When exposed to both deep-bodied and shallow-bodied crucian carp, pike fed on shallow-bodied individuals at a higher rate. However, a deeper morphology should also incur some costs, possibly through an increase of drag, increasing the energy expenditure when swimming. A field experiment showed that growth of deep-bodied crucian carp was reduced compared to shallow-bodied conspecifics, but only at higher population densities when resources were limited. This demonstrates the importance of relating defense costs to levels of resource availability.

Introduction

Most examples of inducible defense adaptations involve aquatic invertebrates that respond to a waterborne chemical cue from the predator (Adler and Harvell 1990). Only recently have examples of inducible morphological defenses been shown in vertebrates, i.e., in the crucian carp (*Carassius carassius*; Brönmark and Miner 1992) and in tree-frog tadpoles (Smith and Van Buskirk 1995; McCollum and Van Buskirk 1996). Crucian carp develop a deep-bodied morphology in habitats with piscivorous fish, whereas tadpoles acquire larger and more brightly colored tail fins in response to the presence of predators. Here, we will review our recent work with crucian carp.

Crucian carp is a cyprinid fish, common in ponds, lakes, and slow-flowing rivers in large parts of Europe and Central Asia (e.g., Maitland and Campbell 1992). It is a generalist forager, feeding on a range of zooplankton and benthic macroinvertebrates. Surveys of fish assemblages in lakes and ponds have shown that though crucian carp populations are dense and stunted in monospecific pond populations, they are generally sparse and dominated by large individuals in larger lakes where abiotic conditions allow a more diverse fish assemblage (Piironen and Holopainen 1988; Brönmark et al. 1995). Experimental studies have shown that smaller crucian carp are very vulnerable to predation and that the size structure with a dominance of large crucian carp found in lakes and ponds with piscivores is due to size-selective predation (Tonn et al. 1989; Brönmark and Miner 1992).

More interesting in this context is that crucian carp exists in two different body morphs. Individuals from dense pond populations where piscivores are absent are small and have a shallow, fusiform body shape, whereas crucian carp that coexist with piscivores are typically large (>15 cm) and have a hump-backed body shape (see, e.g., Poleo et al. 1995). The two morphs, the shallow bodied and the deep bodied, are so different in size and shape that they were described as two separate species, *Cyprinus carassius* and *Cyprinus gibelio*. It was not until the early 1800s that it was realized that both morphs belonged to one species. Large, deep-bodied fish were moved to a small pond and developed a population of small, shallow-bodied fish (Ekström 1838). The difference in body morphology was generally ascribed to differences in resource levels, shallow-bodied morphs developing in situations of severe resource limitation.

A survey of south Swedish ponds showed crucian carp to be an important predator on snails and other macroinvertebrates in these systems (Brönmark, unpubl.). In order to evaluate the strength of direct and indirect interaction in benthic food chains involving crucian carp, Brönmark and Miner (1992 and unpubl.) performed a series of manipulations of fish assemblage structure in south Swedish ponds. Ponds were divided into two sections with a plastic curtain, and fish assemblage structure was manipulated in one of the sections, whereas the other was kept as a control. Two of the ponds had high densities of crucian carp, the only fish species present. In mid-June, pike were introduced to one of the sections in each pond. A test fishing in September revealed that pike had had a dramatic effect on crucian carp popula-

Fig. 11.1. Deep-bodied (*upper*) and shallow-bodied (*lower*) crucian carp (*Carassius carassius*) from ponds with and without piscivores, respectively.

tions. The density of crucian carp was significantly reduced and individual size was skewed toward larger individuals in sections with pike (Brönmark and Miner, unpubl.). Further, in the presence of pike, crucian carp had changed in body morphology resulting in a more humpbacked, deeper body shape (fig. 11.1).

Three different hypotheses that could explain the change in body morphology were suggested. First, the crucian carp populations may be genetically polymorphic with regard to body morphology, i.e., both morphs may have coexisted in the ponds already before pike were introduced. Introducing a gape-limited predator that feeds selectively on a shallow-bodied morph would then result in the pattern found at the end of the experiment. However, the low variance in body depth for a given body length and an absence of overlap in relative body depth between treatments (fig. 11.2) suggested that there was no polymorphism with regard to this trait in the original populations. Second, as population densities of crucian carp decreased as a consequence of predation, the per capita food resource levels should have increased. Thus, differences between treatments may have been due to differences in resources. Third, the change in body morphology may be a predator-induced defense adaptation against predation by a gape-limited pis-

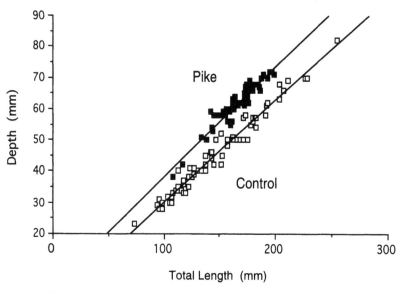

Fig. 11.2. The relation between total body length and body depth of crucian carp in experimental sections with or without pike in a south Swedish pond. Regressions are significantly different (ANOVA, $p < 0.001$). (Excerpted with permission from Brönmark and Miner 1992, *Science* 258: 1348–1350. Copyright 1992, American Association for the Advancement of Science.)

civore. The two last hypotheses were tested in a laboratory experiment (Brönmark and Miner 1992), where changes in body shape of crucian carp in the presence and absence of pike and at low and high food rations were monitored. Crucian carp that were held at high resource levels showed a small but significant increase in body depth when compared to control fish held at low resource levels. This is in accordance with studies on crucian carp in Finland, suggesting that resource level is an important determinant of body morphology. Holopainen and Pitkänen (1985) eliminated crucian carp from a pond with a monospecific, high-density (29,000/ha) population of shallow-bodied individuals using rotenone. The resulting fishless pond was then stocked with 280 crucian carp (187/ha). The stocked carp were initially shallow bodied but increased in body depth over time, and this was coupled to rapid growth and high resource levels. Similarly, Tonn et al. (1994) found that crucian carp from low-density enclosures were deeper-bodied than fish from high-density enclosures. However, when shallow-bodied crucian carp from a south Swedish pond without piscivores were transplanted to a fishless pond with high resource levels, there was no change in body morphology (Brönmark and Miner 1992). Further, although food level produced a significant effect in the laboratory experiment, the absolute change in relative body depth was small. On the other hand, at low resource levels but in the presence of pike there was a much larger increase in relative body depth. Accordingly, Brönmark and Miner (1992) suggested that the predator-induced change in morphology in crucian carp was an example of an inducible morphological defense.

Certain characteristics of predator and prey organisms promote the evolution of predator-induced defense adaptations (see chapter 17 and references therein). These include variability in predation pressure, i.e., the predator should not always be present in time or space. Prey, on the other hand, should have reliable cues of detecting the presence of the predator, and the induced morphology should increase the probability of survival. There should also be a cost associated with the defended morphology in the absence of predators. Below we discuss these different factors applied to the piscivore-crucian carp system.

Variability in Predation Pressure

As described above, crucian carp populations show a dichotomy in population density, body morphology, and size structure in different ponds and lakes, depending on the presence/absence of piscivores. We suggest that the variability in predation pressure necessary for the evolution of an inducible defense in crucian carp arises through a shift between these two states. A stochastic environmental disturbance acting selectively on piscivores may shift a system from the piscivore state to the monospecific crucian carp state. The most likely such disturbance is winterkill, i.e. low oxygen levels, or even anoxic conditions, under ice cover during winter. Crucian carp have physiological adaptations to survive anoxic conditions. During fall it builds up a glycogen storage in the liver, which can be used for anaerobic metabolism at extreme oxygen conditions. Pike and most other fish species are very sensitive to oxygen depletion, and, thus, during a winterkill event a lake or a pond may change from having a multispecies fish assemblage to a monospecific assemblage with crucian carp only. Crucian carp has a large reproductive potential, and it has been shown that populations may increase from a few hundred individuals to $>$ 25,000 in just a few years (Piironen and Holopainen 1988). At these high densities, resources become limiting and intraspecific competition is an important structuring force (Tonn et al. 1994). It may be less clear how a system changes from a monospecific to a multispecies state, i.e., how do piscivores colonize isolated ponds and lakes? In modern times, humans have surely been an important vector for the transport of fish between isolated ponds and lakes, either as part of fish management programs or as more spontaneous introductions. Nevertheless, over evolutionary time, transport by man is probably not very important. However, older maps show that the landscape has undergone drastic changes in large parts of the distribution range of the crucian carp over the last 100–150 years. Ponds and wetlands have been drained, and efficient drainage systems have reduced the frequency and magnitude of flooding events. Lakes and ponds that are now isolated may have been connected historically, at least during flooding periods, and it is well known that pike, for example, migrate up into smaller water bodies and flooded areas to spawn (Jude and Pappas 1992). Thus, waterbodies that had lost piscivores due to winterkill may well have been recolonized by piscivores from populations in nearby, larger systems, unaffected by winterkill. In summary, the variability in predation pressure experienced by crucian carp may be due to catastrophic disturbance events followed by recolonization of the piscivore.

The temporal scale of the variability in predation pressure, especially in relation to generation time of prey organisms and time lags in the development of a defense structure, is important for the evolution of inducible defenses (e.g., Padilla and Adolph 1996). Padilla and Adolph found that plasticity is generally favored when time lags are short. In crucian carp the period from a change in piscivore abundance to the full development of a deeper body is long compared with other aquatic organisms with inducible defenses (weeks rather than hours or days). However, it is important to relate the time lag to the timescale of the organism in question, and the previous studies involved short-lived invertebrates (e.g., cladoceran zooplankton, bryozoans). From a perspective of generation time and frequency of environmental change, the time lag in crucian carp is short. Crucian carp may live up to 10 years on average (Maitland and Campbell 1992), and the period between environmental changes may be several generations.

Cues to Detect Piscivores

Prey organisms that have predator-induced defense systems must have reliable cues to detect when the predator is present. Studies on aquatic macroinvertebrates with inducible defenses have shown that they generally rely on waterborne chemical cues to detect the presence of a predator. In the aquaria used by Brönmark and Miner (1992) pike and crucian carp were separated by a large-mesh plastic screen partition, i.e., crucian carp could use both visual and chemical cues to detect the presence of pike. To determine if chemical cues alone could trigger the change in morphology, aquaria with an opaque plastic wall that separated crucian carp and pike compartments were used (Brönmark and Pettersson 1994). Crucian carp could not see pike, but water was circulated between compartments and thus any chemical cues emitted by pike were available for crucian carp olfaction. There was no difference in the change in body morphology between the treatment where crucian carp could both see and smell the pike and the treatment where carp

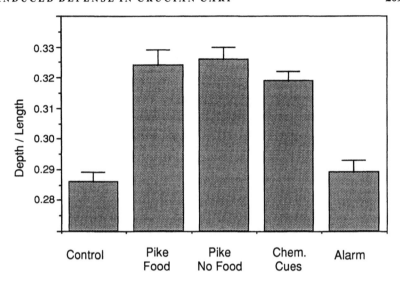

Fig. 11.3. The body-depth to total-length ratio of crucian carp grown without pike (Control), with pike feeding in the experimental aquaria (Pike Food), with pike feeding in holding tanks (Pike No Food), with pike behind an opaque PVC partitioning (Chem. Cues), and with the addition of crucian carp alarm substances (Alarm). Error bars denote one standard error. (Data from Brönmark and Pettersson 1994.)

could use chemical cues only. Thus, crucian carp can use chemical cues to detect the presence of a predator (fig. 11.3). This is another in a growing number of studies that have shown the importance of chemoreception in fishes (Hara 1992). For example, salmonids use chemoreception when returning to their home stream for spawning (e.g., Stabell 1992), other fish use chemoreception to locate food sources (e.g. Jones 1992), recognize kin (Olsén 1992), or to detect the presence of predators (e.g., Keefe 1992).

It is well known that cyprinid fishes release an alarm substance when attacked by a piscivore (see review in R. Smith 1992). The alarm substances are produced in epidermal club cells and are released by even a minor injury. Nearby conspecific fish react to the release of alarm substances with a fright reaction, including sudden flight, rapid swimming, cover seeking, increasing shoal cohesion, or immobility. In the experiments above, the pike were fed crucian carp in the experimental aquaria, therefore the chemical cues could either have originated in pike or they may have been released by injured crucian carp (alarm substances). To determine the origin of the chemical cue, a treatment was added where pike were not allowed to feed in the experimental aquaria, but instead were replaced weekly with pike kept and fed in a holding tank. The effect of alarm substance was also tested directly by adding crucian carp alarm substance to aquaria without pike. Alarm substance in itself had no effect on crucian carp body morphology, whereas carp in aquaria with nonfeeding pike increased in body depth (fig.11.3). Thus, it was concluded that crucian carp are able to recognize the presence of a predator based on chemical cues associated with the predator itself, not with an actual attack event between predator and other crucian carp. However, Stabell and Lwin (1997) found that crucian carp increased in body depth after exposure to homogenized skin extract of crucian carp. Extract from Arctic charr, a species that lacks alarm substances, did not elicit any change in crucian carp body morphology. Their results suggest that crucian carp indeed responds to alarm substances released by conspecifics. The contrasting results among experiments may be due to differences in experimental design, e.g., length and strength of exposure to the chemical stimuli.

Pike is an important predator that coexists with crucian carp, but other piscivores may also be important, including, for example, perch and zander. In order to assess if the cue is specific to pike or if it is a cue associated with piscivores in general, Brönmark and Pettersson (1994) did an experiment where they used perch (*Perca fluviatilis*) as a predator. In addition, studies on behavioral fright responses to chemical cues from predators had indicated that predator diet is important, and by feeding perch macroinvertebrates or fish, the effect of predator diet on crucian carp morphology could be evaluated. The experiment showed that perch that were feeding on chironomids had no effect on crucian carp body shape; but as soon as perch started on a piscivorous diet, crucian carp responded and started to increase in body depth (Brönmark and Pettersson 1994). Thus, crucian carp can recognize a chemical cue that is associated with a predator's piscivorous diet. This is in accordance with recent studies on behavioral responses to chemical cues in fish. Prey fish have been shown to exhibit avoidance and fright behavior in response to chemical cues from piscivores, whereas nonpredatory fish elicited no change in behavior (e.g., Rehnberg and Schreck 1987; Mathis et al. 1993). Keefe (1992) further showed that avoidance behavior in brook trout was directly related to the predator's piscivorous diet. A change in diet should affect metabolites or metabolic by-products released by mucus, urine, or feces and thus allow prey to recognize a potential predator by chemoreception. Bryant and Atema (1987) noticed a change in the composition of free amino acids in the urine of bullhead catfish following a change in diet, whereas Stabell and Lwin (1997) found that pike feeding on crucian carp (with alarm cells) elicited an increase in body depth, in contrast to pike fed Arctic charr (without alarm substance cells), indicating that crucian carp react to residues of alarm substances from digested fish. Further, Mathis and Smith (1993a) similarly found a change in behavior in minnows when these were exposed to odor from pike-fed fish with alarm substance cells, but there was no reaction when pike were fed fish without alarm substance cells.

It would be very interesting to determine the chemical nature of the cue in further detail, but these substances are probably highly labile. Thus, long-term monitoring of responses in crucian carp morphology to specific chemical substances would be very difficult. A shortcut may be to monitor behavioral changes in crucian carp when they are exposed to different chemical cues. Smith and coworkers have very successfully applied this method to study interactions between piscivores and prey (e.g., Mathis and Smith 1993b; Chivers and Smith 1995). Preliminary studies on crucian carp antipredator behavior show that crucian carp immediately decrease activity and change microhabitat when exposed to chemical cues from piscivorous fish (Pettersson, Nilsson, and Brönmark, unpubl.).

It might be argued that the experimental conditions were highly artificial, with the low volume of the aquaria resulting in concentrations of chemical cues that are never experienced by crucian carp in nature. However, studies using electro-olfactograms (EOG), where nerve pulses from chemosensory cells are monitored directly in response to chemical cues, have shown that fish respond to extremely low concentrations of the active substance (e.g., Hara 1993). In a preliminary EOG study using water from an aquarium that had contained pike for 12 hours as a stimulus, it was shown that pike water diluted 300,000 times still elicited an EOG response in crucian carp (Bjerselius and Nilsson, pers. comm.), indicating a very refined chemosensory ability in crucian carp.

Benefits of a Change in Body Morphology

Naturally, an induced morphological defense structure should benefit the prey organism through a decrease in predation rates. Crucian carp are normally very vulnerable to predation, but an increase in body depth should be an efficient defense against predation by gape-limited piscivores. When a piscivore, such as northern pike, attacks a prey, it makes a short-distance rush and strikes toward the midbody of the prey (Webb and Skadsen 1980; Webb 1986). If the prey is too large to swallow directly, the pike handles the prey so it can be swallowed head or tail first (Hart and Connellan 1984; Hoyle and Keast 1987). The widest part of the prey during swallowing is then its body depth, as measured dorsoventrally, and given that piscivore prey choice is limited by the width of its gape, the prey body depth relative to predator gape width should be an important morphological character affecting piscivore-prey interactions (Gillen et al. 1981; Hoyle and Keast 1987; Hambright 1991; Hambright et al. 1991). Comparisons among species have shown that deep-bodied prey fish species enjoy a morphological refuge from predation at shorter lengths than shallow-bodied species (Hart and

Hamrin 1988; Hambright et al. 1991). Introduction of pike to two monospecific crucian carp ponds resulted in a shift from a population consisting of small, shallow-bodied individuals that were all vulnerable to predation by pike, to a population with large, deeper-bodied individuals that had reached an absolute size refuge and were invulnerable to predation (fig. 11.2; Brönmark and Miner 1992). More anecdotal suggestions that show a benefit in being deep-bodied was provided by the observation of deep-bodied crucian carp with wounds and scars, indicating successful escapes after being attacked by a piscivore (Brönmark and Miner, pers. obs.).

A crucian carp that increases in body depth and reaches an absolute size refuge from predation by gape-limited piscivores obviously has benefited from the change in body shape. However, smaller fish that have not yet reached an absolute size refuge may still benefit from body depth increases. Piscivores that strike deeper-bodied prey fish have been shown to redirect the strike caudally, which increases the probability of escape (Webb 1986). An increase in body depth should also increase handling time, and given that profitability decreases with increasing handling time, shallow-bodied prey should be preferred. The effect of crucian carp body morphology on piscivore handling times and preference was tested in a number of laboratory experiments (Nilsson et al. 1995). We found that the handling time of pike when feeding on crucian carp with different body shapes increased with increasing carp body depth when the effect of total length was controlled for. There was no effect of crucian carp body length on pike handling times when the effect of body depth was controlled for; thus, within the size ranges used in the experiment, body depth was the most important factor affecting pike handling times. Further, in a predation experiment where we offered pike a choice of deep- and shallow-bodied crucian carp of similar lengths, pike fed more frequently on the shallow-bodied individuals (fig. 11.4). This is in accordance with other studies on prey choice in pike and other piscivores (e.g., Gillen et al. 1981; Hart and Hamrin 1988; Hambright 1991; Tonn et al. 1991). Generally, piscivores prefer to feed on prey that are smaller than the maximum size possible or as predicted from foraging models.

Why then do pike choose to feed on shallow-bodied crucian carp rather than deep-bodied ones? Foraging models predict that profitability decreases with increasing handling times, i.e., pike should choose prey that incurs the lowest handling times, which is in accordance with our results. However, although there is a substantial increase in handling time with increasing body depth of crucian carp, the longest handling time measured in our experiments was still only a couple of minutes. What difference would a couple of minutes make to a predator that successfully attacks a prey maybe once a day or at even longer intervals? First of all, it must be emphasized that we used crucian carp with body depths well below maximum depths possible

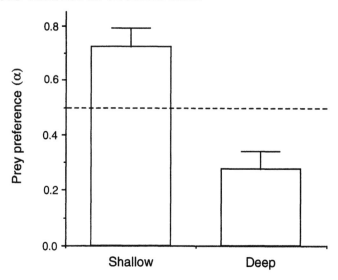

Fig. 11.4. Prey preference, expressed as the Chesson-Manly alpha, of pike given a choice between shallow-bodied and deep-bodied crucian carp. Error bars denote one standard error, and the broken line shows the value for no preference ($a = 0.5$). (From Nilsson et al. 1995.)

for the pike to feed on (Nilsson, unpubl.). The effect of handling time should increase as prey size approaches the absolute size refuge. Besides decreasing profitability, a longer handling time should give the prey more time for possible escape. Scars and wounds on deep-bodied crucian carp indeed indicate that successful escapes by no means are rare. Furthermore, cannibalism is an important mortality source in pike populations (e.g., Raat 1988; Craig 1996; Grimm and Klinge 1996). Pike are often found in the interface between the vegetated littoral zone, characterized by high habitat complexity, and the more open water habitat. Inhabiting a complex macrophyte habitat should increase foraging success for an ambush predator such as pike, and in addition, provide a refuge from predation by larger conspecifics. A pike that leaves its refuge in the complex macrophyte habitat to strike a prey should minimize the time spent in the more open habitat in order to reduce its exposure to other piscivores, i.e., it should attack prey that incurs a minimum handling time. Further, a long handling time may increase the probability of kleptoparasitism. In the laboratory, it has been observed that pike handling their prey may lose it to another pike that sneaks up from behind and takes over the prey (Grimm and Klinge 1996; Nilsson, pers. obs.). Thus, to minimize the probability of kleptoparasitism or cannibalism, a pike should decrease its handling time, which in this case is achieved by a preference for small, shallow-bodied crucian carp.

In addition to increasing survival by reducing piscivore efficiency, a deeper body may benefit crucian carp by improving fast-start and fast-turn performance (Webb 1984). Fast starts are crucial in escaping piscivore attacks, and Webb (1986) showed that largemouth bass aborted their strikes more often when encountering prey fish with high acceleration rates. A deeper body is also generally associated with a high maneuverability, which may be beneficial in complex habitats (Webb 1984).

Costs Associated with a Deep-Bodied Morphology

The fact that the defensive, deep-bodied morphology is inducible suggests that it involves a trade-off in which advantages in the presence of predators are balanced by disadvantages when predators are absent, otherwise a permanent defense would have been favored. Several direct disadvantages of inducible defenses are possible, e.g., a defense can be costly to produce, its maintenance may cause increased energy expenditure, and even the ability to activate the defense may divert resources from growth and reproduction. Further, it also seems likely that indirect costs can be important in the evolution of inducible defenses (Taylor and Gabriel 1993; Tollrian 1995b), for example, the developmental canalization into one phenotype may constrain possibilities for other options (Tollrian 1995b).

An increased energy expenditure when swimming has been suggested as a direct cost of being deep bodied in crucian carp (Brönmark and Miner 1992). A deeper body results in a larger surface area, which in turn increases the total drag, causing a hydrodynamic disadvantage (Webb 1975). The energy spent for movement is directly proportional to total drag and, hence, deep-bodied crucian carp should expend more energy for their propulsion than shallow-bodied conspecifics. For example, the theoretical total drag for a deep-bodied crucian carp (depth : length ratio = 0.36) is 33% higher than for a shallow-bodied individual (ratio = 0.29). Thus, these theoretical estimates suggest that maintenance of the defense incurs an energetic cost for deep-bodied crucian carp.

Costly defense structures should be resorbed or reallocated if predation pressure is relaxed or if the prey organisms grow and reach a size refuge where they are no longer susceptible to predation (e.g., Havel and Dodson 1987). Deep-bodied crucian carp that have reached an absolute size refuge should allocate energy toward growing longer, thus decreasing drag. Interestingly, Holopainen and Pitkänen (1985) found that in populations coexisting with piscivores, the crucian carp body depth to length ratio decreased for individuals exceeding 150 mm, possibly indicating that these fish had reached a size refuge and changed their growth pattern toward growing more fusiform. Another indication that a deep body can be disadvantageous to crucian carp in the absence of piscivores was found by Brönmark and Pettersson (1994). Chemical cues from pike induced crucian carp to grow deep bodied, but when the cue was removed, deep-bodied fish decreased in depth. Hence, this suggests that the well-established hydrodynamic disadvantage associated with a deep-bodied morphology, so far only shown in comparisons between species (e.g., Facey and Grossman 1990), may be at work in this system.

In addition, there are other possible disadvantages, e.g., that permanent structural investments divert resources from reproduction (Harvell 1990a; Clark and Harvell 1992), or that the difference in body shape interacts with foraging strategies, making pelagic feeding relatively less profitable for deep-bodied individuals (Schluter 1995; Robinson et al. 1996). Also, there may be costs of the plasticity itself, i.e., resource needs for maintaining the ability to activate the defense (Harvell 1990a). However, these costs are likely to be of minor importance in this system, given the marked drag effect suffered by deep-bodied fish.

Several studies have demonstrated costs associated with inducible morphological defense structures. However, most of these studies are performed in the laboratory and few have been performed in actual field situations (Harvell 1992; Lively 1986b). Further, it is important to test possible costs of inducible structures at different resource levels (cf. Stemberger 1988) as costs may only appear at limiting resource levels. To explicitly investigate the performance of deep-bodied and shallow-bodied crucian carp in absence of predators and to relate this to normal levels of competition, we performed a field experiment where we monitored the growth rate of crucian carp of different morphologies at different densities (Pettersson and Brönmark 1997). We assigned groups of either deep-bodied or shallow-bodied fish into cages (12 m^2) with either low (0.5 individuals/m^2) or high crucian carp population density (1.5 individuals/m^2) and monitored these fish for 4 months. Differences in population density were achieved by adding appropriate numbers of shallow-bodied crucian carp to the cages. We found that high densities of crucian carp reduced food levels (cladoceran zooplankton) and resulted in reduced growth in both morphs. At low density, there was no difference in performance between deep-bodied and shallow-bodied individuals. However, at the high density, shallow-bodied fish gained body mass significantly faster than deep-bodied fish (fig. 11.5), demonstrating that induced morphological defense incurs a density-dependent cost manifested as reduced growth rate. Thus, as fitness in fishes is closely related to body size (Wootton 1985), deep-bodied crucian carp should suffer a fitness cost when competing with more shallow-bodied conspecifics in absence of piscivores.

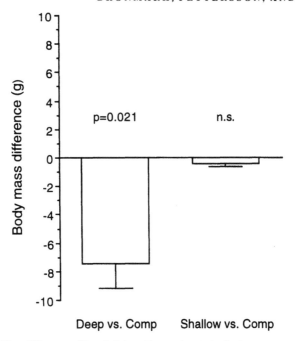

Fig. 11.5. The difference (Focal fish—Competitors) in body mass gain for deep-bodied (Deep) and shallow-bodied (Shallow) focal fish grown for 4 months in enclosures together with shallow-bodied competitors (Comp). The difference in body mass gain between deep-bodied individuals and their within-cage competitors represents a 45% reduction in growth. Probability values indicated are from pairwise t-tests of focal fish vs. within-enclosure competitors using enclosure means; n.s. denotes $p > 0.05$. N = 4 in both treatments. Error bars denote one standard error. (Data from Pettersson and Brönmark 1997.)

Conclusions

We have shown that crucian carp change in body morphology in response to waterborne chemical cues associated with a piscivorous diet of the predator, and suggest that this is an example of an inducible morphological defense. A deeper body provides a benefit by reducing predation rates in gape-limited piscivores, but it also incurs a cost manifested as reduced growth rate at the high population densities typical of crucian carp populations in lakes and ponds without piscivores.

It is worth noting that overall resource abundance also affects crucian carp morphology, albeit more slowly (Brönmark and Miner 1992; Tonn et al. 1994). This apparent contradiction, with an increase in body depth even in the absence of chemical cues from predators, resolves when the allocation problem faced by these fish is seen not in terms of absolute investments but as a state-dependent strategy (sensu Clark and Harvell 1992). If a defense is effective, relative to, e.g., the increased maintenance costs it causes, then this can favor investments in the defense even at very low risks of predator invasion (Clark and Harvell 1992). Clearly, the relative importance and interactions of food levels and predation pressure for determining body morphology in crucian carp deserves further study.

In addition, more work is needed to determine if there is a genetic variability in induction characteristics between different populations of crucian carp. Differences in reaction norms may be expected when comparing populations that vary in the frequency of changes in predation pressure. For example, crucian carp from northern-latitude ponds that always winterkill should respond differently to predator cues than carp from more southern ponds that winterkill during extreme winters only; furthermore, crucian carp from small ponds should differ from carp in large lakes with permanent populations of piscivores.

Phenotypic change in morphology is, of course, a very interesting phenomenon in itself, but it also gives us an excellent tool for testing theories on costs and benefits of different morphologies without being constrained by phylogeny. For example, when evaluating the effects of body depth by comparing predation success on different prey fish species differing in body shapes, other factors, such as fin rays, spines, behavior, etc., that vary among species will affect the outcome. By using the crucian carp we have been able to show that body depth in itself is beneficial to prey fish. At present, we are testing predictions from hydrodynamic theory on the effect of body shape on swimming energetics.

12

Density-Dependent Consequences of Induced Behavior

BRADLEY R. ANHOLT AND EARL E. WERNER

Abstract

Acquiring resources requires movement that often increases the encounter rate of animals with their predators. The widespread observation of reduced activity in the presence of predators is evidence of the generality of this trade-off. A collection of optimization models that address this trade-off predict that foragers should reduce foraging activity when predation risk increases or when the availability of resources increases. We have tested these predictions and their consequences in a series of artificial pond experiments with larval frogs and their larval dragonfly predators. In the first experiment we found that when we increase food availability, larval bullfrogs reduced activity and had lower mortality rates in the presence of a single predator. In the second experiment we manipulated food availability to wood frog larvae in the presence and absence of additional caged predators to manipulate predator density. As in the first experiment we found lower predation mortality of tadpoles at higher food levels and when predator density was augmented by caged predators. In the third experiment we examined whether changes in behavior due to the presence of predators could affect competitive interactions among tadpoles. We found that small vulnerable tadpoles were lighter in the presence of predators. In contrast, large invulnerable tadpoles were heavier in the presence of predators. We interpret this as the result of competitive release for large tadpoles when small tadpoles reduce their foraging. The widespread observation of antipredator behavior suggests that similar trait-mediated indirect effects will be equally widespread. We may be unable to adequately predict food web dynamics unless we incorporate these indirect effects into our models.

Introduction

In the presence of predators, animals very often alter their behavior (reviewed in Sih 1987a; Lima and Dill 1990; Werner and Anholt 1993). This is a change in the expression of the phenotype (behavior) that depends on the environment (presence of the predators). Although this change may be short-lived, it can be treated in the same theoretical context as other forms of phenotypic plasticity associated with predation risk. The presumed benefit is reduced risk of predation mortality. There must also be some cost associated with this change in phenotype, otherwise we expect the behavior pattern to be fixed (Lively 1986a; Harvell 1990a; N. Moran 1992). Indeed, animals can lack behavioral plasticity where predators are always absent. For example, stream-dwelling insects that normally drift at night show no diel periodicity in streams lacking predators (Flecker 1992), and copepods that normally show daily vertical migrations in the presence of chaoborid predators lack this ability where predators are absent (Neill 1990, 1992).

Plasticity of behavior has important potential consequences for population dynamics and community structure. In traditional formulations of food web dynamics, the characteristics of the members of a food web are independent of the web itself. For example, the foraging rate of a consumer does not depend on the presence of its own predators. If this is true, the measured effect of an individual consumer on its resources will be the same whether predators are present or not. It would be convenient if this were true; knowing the strength of the pairwise interaction terms between predator and consumer, and between consumer and resources, would allow us to predict the indirect effect of predators on the resources (e.g., Bender et al. 1984). However, as soon as the consumer alters its foraging behavior in the presence of the predator, pairwise estimates of interaction strengths are no longer sufficient to allow us to make predictions about the behavior of the system as a whole (Abrams 1984, 1991a, 1992a, 1993a,b, 1995; Matsuda et al. 1995).

The Trade-Off between Resource Acquisition and Predation Risk

The adjustment of foraging in the face of predation risk is a consequence of the fundamental trade-off between the acquisition of resources and encounter rates with predators. The faster an animal moves, the more likely it is to encounter a randomly located point in space (Gerritsen and Strickler 1977). If that point is a predator, encounter rates with predators will increase with movement speed. If that point is food, encounter rates with food will increase with movement speed. Similar arguments can be proposed for the amount of time active. The more time spent foraging, the more likely an animal is to encounter its predators. Even for an animal that does not move from place to place, such as a barnacle, the risk of predation can be a function of foraging effort because it is most at risk of predation when it is filtering food from the water column. Because it is unlikely that the place where, or time when, food is most available is also the place and time where predation risk is lowest, an animal cannot simultaneously maximize its rate of food intake and minimize its predation risk.

Proposed Resolutions of the Trade-Off

A collection of optimization models have proposed resolutions to this trade-off (Milinski and Heller 1978; Abrams 1982, 1984, 1990, 1991b, 1992a, 1993a; Houston et al. 1993; Werner and Anholt 1993; McNamara and Houston 1994). The models commonly predict that not only should animals reduce foraging activity when the world is more dangerous, but that they should also reduce activity when the world is richer in resources. There are some important exceptions. First, this prediction depends on the benefits of acquiring additional food showing diminishing returns (Abrams 1991b) and not being too low (Abrams 1982, 1984). However, if fitness is concave upwards with the acquisition of additional resources (i.e., has a positive second derivative), we expect increased foraging activity with additional resources. We can easily imagine this being the case where the probability of successful metamorphosis or successfully holding a mating territory increases sharply with body size. Second, this prediction applies to long-term increases in resource availability; a short-term increase in resources is predicted to increase foraging activity (Abrams 1991b; McNamara and Houston 1994). Mass flights of insects can sometimes provide rich, ephemeral resources that are capitalized on by predators. For example, lizards have been reported to increase their foraging in response to the nuptial flight of termites (Huey and Pianka 1981), and trout are famous for their increase in foraging activity when aquatic insects emerge en masse.

Larval Anurans as a Model System

Larval anurans appear to fit the assumptions of these models rather well. Both interspecific and intraspecific comparisons have shown that more active animals grow faster: *Pseudacris crucifer* is relatively inactive and is a poor competitor (Morin 1983). *Rana pipiens* is more active and grows faster than *R. sylvatica* (Werner 1992a), which is translated into competitive superiority. Similarly, *R. catesbeiana* is more active than *R. clamitans* in the presence of predators and is also competitively superior under these conditions (Werner 1991). *Scaphiopus couchi* is a temporary pond species that is more active than two permanent pond species, *R. catesbeiana* and *R. pipiens,* and is competitively superior to both (Woodward 1982). Presumably, activity is related to food gathering ability and this is translated into growth.

However, more-active species tend to experience higher predation mortality (Woodward 1983; Lawler 1989; Azevedo-Ramos et al. 1992; Werner and

McPeek 1994), and, intraspecifically, more active individuals have higher mortality rates due to predators (Skelly 1994).

Thus, the antipredator behavior of larval anurans provides us with a powerful and convenient model system for investigating the ecological consequences of antipredator behavior. Our approach has been to develop a simple model that allows us to predict the change in behavior as a function of presumed benefits and costs (Werner and Anholt 1993), test whether the behavior matches the predictions of the models, and then ask how this adaptive variation in behavior might affect the major demographic variables and community-level phenomena such as competition.

Measurements of Induced Behavior

Although animals are capable of varying several aspects of their foraging effort, including both the proportion of time active and movement speed while foraging, the majority of data have been collected on the proportion of time active. This reflects the relative ease of collecting such data rather than their relative importance.

Reduced activity induced by the presence of predators has been demonstrated in the larvae of *Bufo americanus* (Skelly and Werner 1990; Anholt et al. 1996); *B. woodhousei* (Lawler 1989); *B. calamita* (Tejedo 1993); *Rana catesbeiana* and *R. clamitans* (Werner 1991); *R. lessonae* and *R. esculenta* (Stauffer and Semlitsch 1993; Horat and Semlitsch 1994); *R. sylvatica* and *R. pipiens* (Anholt and Werner, unpubl. data); *Hyla versicolor, H. andersonii,* and *H. crucifer* (Lawler 1989); *Pseudacris triseriata* and *P. crucifer* (Skelly 1995); and *Ascaphus truei* (Feminella and Hawkins 1994). Decreases in activity are not universal: Hews and Blaustein (1985) and Hews (1988) found increases in activity level in the presence of predators, while Tejedo (1993) found no effect in *Pelobates cultripes*.

Studies of how activity levels vary with food levels are somewhat rarer. The available results, however, generally support the predictions of the model. Reduced activity levels at higher food levels (or, alternatively, higher activity levels at higher hunger levels) have been reported for *Bufo americanus* (Anholt et al. 1996); *Rana sylvatica* and *R. pipiens* (Werner 1992a); *R. clamitans* (Werner 1992b); and in the European waterfrogs *R. lessonae* and *R. esculenta* (Horat and Semlitsch 1994). Video image analysis of *R. catesbeiana* raised at two different food levels (Anholt and Werner 1995) showed that these tadpoles reduce the proportion of time active at high food levels, and that when they are active at the higher food level they swim more slowly (fig. 12.1B).

This simultaneous adjustment of both proportion of time active by *R. catesbeiana* and movement speed while active highlights one of the limitations

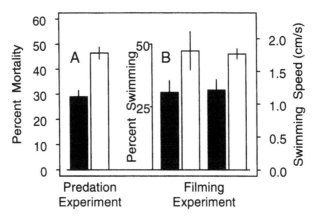

Fig. 12.1. Mortality and activity of bullfrog tadpoles under low food (0.68 g/sq m; clear bars) and high food (13.7 g/sq m; shaded bars) treatments. (A) Percentage of animals killed by one *Tramea lacerata* larva in 48 hours. (B) Percentage of animals that were observed to be swimming at any moment in time when kept under identical food conditions with a caged dragonfly larva, and tadpole speed while swimming. All values are presented with their standard errors.

of Werner and Anholt (1993). We generated separate predictions for each of these aspects of activity, but, clearly, animals are capable of simultaneously adjusting both the frequency of activity and its intensity. When we substitute the speed equations into the proportion of time active we are unable to solve for both simultaneously. In their current formulation there is no solution in the real plane (P. Taylor and Anholt, unpubl. results). If we incorporate some estimates of the metabolic costs of activity, there is a solution but we have been unable to solve it analytically.

Population Consequences of the Trade-Off

We have conducted a collection of experiments (Anholt and Werner 1995 and in press; Werner and Anholt 1996; Peacor and Werner 1997) that demonstrate that these behavioral adjustments to the availability of food and the risk of predation have consequences for the fundamental demographic processes of survival and growth. Each experiment was motivated by the relationships implied by an optimality solution to the trade-off between the acquisition of resources and encountering predators.

The Effect of Food Availability

Models that explicitly incorporate the trade-off between resource acquisition and mortality risk (see above) generally predict that at lower food availability, animals should increase their level of foraging activity. If increasing activity leads to higher encounter rates with predators and therefore to higher mortality rates, then at lower food levels animals should experience higher predation mortality.

In this experiment (Anholt and Werner 1995) we kept one hundred bullfrog tadpoles (*R. catesbeiana*) in the presence of a single larval dragonfly (*Tramea lacerata*) at two food levels. Both predator and prey were collected from nearby ponds on the George Reserve of the University of Michigan. Each replicate was a flexible wading pool containing 75 liters of water and 48 meters of polypropylene rope to provide some structure to the environment. The rope was weighted down by a pair of wooden dowels held in place with a clay brick. We had ten replicates at a low food level of 0.68 g/m^2 of pulverized Purina Rabbit Chow, and ten replicates at a much higher food level of 13.7 g/m^2. The initial 24 hours was an acclimatization period for the tadpoles, during which the dragonfly larva was restricted to a cage. Previous laboratory experiments had demonstrated that the tadpoles reduce their activity in the presence of caged predators. After the 24-hour acclimatization, the dragonfly larva was released from its cage and allowed freely to hunt tadpoles. At the same time we also filmed the behavior of animals in the laboratory that were being kept under the same food regime as the tadpoles subjected to predation. In each container there was also a caged dragonfly.

We found that the tadpoles adjusted their behavior in general agreement with the predictions of the models: tadpoles increased the amount of time active from 30.7% at the high food level to 47.3% at the low food level. When the tadpoles were swimming, they increased their swimming speed from 1.21 cm/s at the high food level to 1.77 cm/s at the low food level (fig. 12.1B). As predicted, we found that at the low food level, where the tadpoles were more active, they had higher mortality rates. On average, more than forty-six tadpoles were killed by the predator in the low food treatment, and only twenty-nine were killed in the high food treatment (fig. 12.1a). It seems clear from this experiment that there is a mortality risk associated with either moving more frequently or more quickly, and tadpoles accept this risk when there is less food available. Although tadpoles reduce activity in the presence of predators, the level of activity also depends on the availability of resources.

The Effect of Predator Density (and Food Availability)

We tested the predictions associated with the risk of predation in a similar experiment where we manipulated both the density of predators and the availability of food in a two-way design (Anholt and Werner, in press). We used a different species pair in these experiments, partly because of availability but also partly to convince ourselves that the results were not simply the fortuitous result of having chosen the right pair in the first experiment. In this second experiment, one hundred wood frog (*Rana sylvatica*) tadpoles were the prey, and the larval dragonfly predator was *Anax junius*. As in the previous experiment, these species can be collected from the same ponds in southern Michigan. The experimental arenas were the same as before. In every experimental container, a single larval dragonfly was free to forage 24 hours after the tadpoles had been allowed to acclimatize to the experimental arena. We manipulated predator density by adding seven empty cages to the low-predator treatment, and seven cages that each contained another dragonfly larva to the high-predator treatment. The world therefore seemed more dangerous in the high-predator treatment, but was not. We predicted that the tadpoles should reduce activity at the higher predator density and that this would result in lower mortality. The predator treatment was crossed with five food treatments where we manipulated food in a geometric series from a low food level of 0.37 g/m^2 of pulverized Purina Rabbit Chow, up to a maximum of 11.9 g/m^2. Each treatment combination was replicated four times. In previous laboratory experiments we had demonstrated that *R. sylvatica* tadpoles reduce their activity at higher predator density and at higher food availability (Werner 1992a; Anholt and Werner, unpubl. data).

Once again, with increasing food availability the survival of the tadpoles increased (fig. 12.2A). In addition, more tadpoles survived when additional caged predators were present than in the low-density treatment (fig. 12.2B). All of these results are consistent with the predictions. We expected animals to reduce their activity when food availability increased and to suffer lower mortality as a result of encountering their predators less frequently. By the same logic, reduced mortality when additional caged predators were present is also the expected result of reduced activity with increasing (in this case apparent) predation risk.

Density-Dependent Predator/Prey Interactions

These results lead us to infer that induced variation in the behavior of prey is capable of generating density-dependence in the interaction between predator and prey as suggested by Abrams (1984), Sih (1987b), and Ruxton (1995). Tadpoles are capable of depleting their resources, and have been repeatedly shown to compete (e.g., Morin 1983; Wilbur 1987; Semlitsch and Reyer 1992; Werner 1991, 1994). At high densities of tadpoles, resource levels should be depleted more quickly and to lower levels. Once resources are depleted, our results suggest that tadpoles will be more active in the search for resources and suffer a higher predation rate as a result.

Our two experiments did not manipulate tadpole density, but rather generated the effect of higher tadpole density, lower food resources. In a more

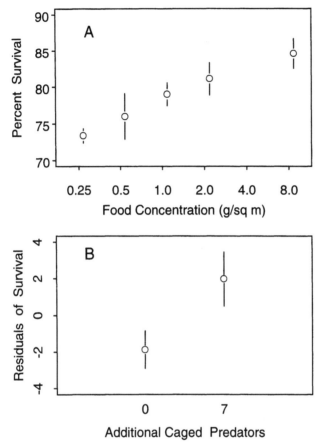

Fig. 12.2. (A) Number of wood frog tadpoles surviving out of 100 (± S.E.) in the presence of one unrestrained *Anax* larva under a range of food concentrations. (B) Residuals of survival (± S.E.) from (A) as a function of the density of additional caged predators. Tadpoles have lower predation mortality at higher food levels and in the presence of additional caged predators. Both factors reduce the activity of tadpoles and therefore their encounter rate with predators.

refined experiment, Peacor and Werner (1997) added large, invulnerable bullfrog tadpoles to reduce the availability of resources to small vulnerable tadpoles with essentially the same results. In the presence of large tadpole competitors, small tadpoles were more active and suffered higher predation mortality.

Previous explanations of density-dependent predation have been focused on changes in the behavior of the predator. For example, predators exhibit area-restricted search, accumulate in areas of high prey density, and learn to handle prey more efficiently (Murdoch and Oaten 1975; Curio 1976; Hassell

1978). Each of these mechanisms may well be true or contribute to density dependence in individual cases. However, our results suggest that induced variation in behavior in response to the activity mediated trade-off between resource acquisition and predation mortality provides a more general mechanism for density dependence that is applicable even in situations where predators are immobile, or do not learn.

Our results also show that the predation rate per predator declines with predator density. This is usually interpreted as interference competition, again based on the behavior of predators rather than prey. The possibility that prey might adjust their behavior in the face of higher predator density and thus reduce predation rates was recognized by Charnov et al. (1976), who called it "resource depression." However, the more usual explanation of this result is changes in predator behavior: the time that predators waste interacting with each other reduces their foraging success (Beddington 1975). There is no doubt that this form of interference does occur, but attempts to directly manipulate interference competition as an alternative to estimating the interference coefficient have had mixed results (Anholt 1990; MacIsaac and Gilbert 1991).

Thus, induced variation in behavior can generate density dependence in both halves of the predator/prey relationship. Moreover, because this density dependence is generated on a short behavioral timescale, it is likely to be stabilizing (May 1981). Population-dynamic models that incorporate antipredator behavior often find enhanced stability properties (Abrams 1984; Sih 1987b; Ives and Dobson 1987; Ruxton 1995). However, if the effect is too strong, it can also be destabilizing (Abrams 1992b).

Being able to predict induced variation in behavior with the optimality approach of behavioral ecology offers the promise that from relatively simple premises we can make reasonable predictions about how behavior should vary and what the consequences of that are likely to be at the population level. Simple and general models may be unable to predict the consequences of induced variation behavior as well as detailed models of individual behavior. However, we should be able to make more accurate predictions using these general models than if we ignored variation in behavior.

Strong tests of the model and its population consequences in individual cases will require estimates of both the instantaneous probability of mortality and the functional response as functions of behavior. Because mortality risk is dependent on the environment, estimates made in the laboratory will have to be treated with some caution. They are unlikely to yield accurate quantitative predictions of how animals will alter behavior in the field. Similarly, functional responses measured in the laboratory may lack precision because of differences in the resources available. On the other hand, precise estimates of mortality risk in the field as a function of activity may well be impossible to attain.

There are some additional possibilities for testing the real-world relevance of the general predictions. None of the possibilities will be completely convincing by themselves, but if several of these approaches yield concordant results, then our confidence in their validity will increase. We can use the comparative method and ask whether different species, or genotypes, differ in their activity level and their mortality risk and functional responses. This is essentially the approach of Lawler (1989), Azevedo-Ramos et al. (1992), and Werner (1991, 1992a). We can also manipulate behavior pharmacologically or hormonally and generate neophenotypes (sensu Ketterson and Nolan 1994), which is the approach taken by Skelly (1994). Each of these approaches has a set of confounding variables, but with luck not the same set (Anholt 1997).

Community Consequences

The results of these first two experiments encouraged us to think that in a highly interactive system where competition occurs frequently, we might see the effect of variation in behavior transmitted to other members of the community through the indirect connections of the food web. Therefore, we conducted an experiment to examine the effect of the presence of caged predators on competitive interactions between classes of tadpoles that differed in their vulnerability to predation (Werner and Anholt 1996). We used small bullfrog tadpoles and small green frog tadpoles that were vulnerable to *Anax* predation and large second-season bullfrog tadpoles that were invulnerable to predation. Our prediction was that reductions in activity by the small vulnerable animals in the presence of predators would reduce the competitive effect of small tadpoles on large ones. We expected that large tadpoles, because of their invulnerability to predation, would not respond to the presence of the predators and experience a competitive release due to the reduced foraging activity of small, vulnerable size classes. The experiment had five density levels in a target-neighbor competition design (table 12.1) crossed with the presence and absence of predators for a total of ten treatments. The experiment was replicated in five blocks.

When we compare the mass of small bullfrogs and small green frogs raised under the same density conditions in the presence and absence of predators, we see that in almost every case these animals were heavier in the absence of predators than in their presence (fig. 12.3). In contrast, the large bullfrogs were heavier in the presence of the predators, consistent with our prediction of competitive release. The lower growth rates of the small tadpoles in the presence of the predators is strong evidence for the cost of the antipredator behavior. Indeed, we also found that these smaller animals also had higher mortality rates in the presence of the predators. This mortality

Table 12.1
Density Treatments Used to Investigate Behavioral Indirect Food Web Effects Generated by Induced Antipredator Behavior in Bullfrog and Green Frog Tadpoles

	Treatment				
	I	II	III	IV	V
Large bullfrog	8	8	8	16	32
Small bullfrog	100	50	25	25	25
Small green frog	30	30	30	30	30

Notes: Values are numbers of each class of tadpoles stocked in a 1300-liter tank. Small green frog tadpoles are always a target of competition. Large bullfrogs are the target of competition in the first three treatments, while small bullfrog tadpoles are the target in the last three treatments. Density treatments were crossed with the presence and absence of caged predators making ten treatment combinations. The entire design was replicated in five spatial blocks.

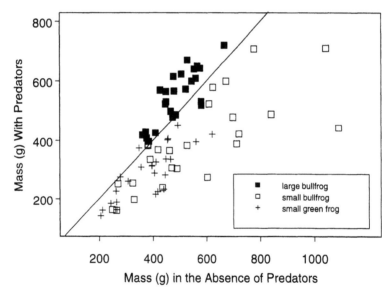

Fig. 12.3. Comparison of the mass of tadpoles raised in the presence and absence of predators. Each plotted point is the harvested mass of tadpoles at the end of the experiment from two experimental units with the same density treatment from the same experimental block. The *x*-value is the mass of animals from the tank without predators, and the *y*-value the mass of animals from the tank with predators. The diagonal line is the line of equal mass under the two conditions. Small green frog larvae (+), and small bullfrog larvae (open squares) are lighter in the presence of predators, while large bullfrog larvae (filled squares) are heavier in the presence of predators. Values for large bullfrog larvae have been divided by 10 for presentation.

could not be directly due to predation because the predators were caged. We suggest that the reduction in growth rate is to the point where these animals are occasionally starving to death.

The exploitative competition exhibited by tadpole larvae is an indirect interaction mediated by the density of resources. Density-mediated indirect interactions are common in nature (reviewed in Schoener 1993; Menge 1994; Wooton 1994), and the measured strengths of the indirect effects can be greater than the direct effects (Schoener 1993). The widespread occurrence of induced antipredator behavior argues that trait-mediated indirect effects will also be common (Abrams 1993a, 1995). When the presence of a predator reduces the foraging rate of a consumer, predator density is indirectly affecting the interaction strength between consumer and resource. Theoretical investigations of trait-mediated indirect effects also show that they can be larger than the direct effects (Abrams 1984, 1991a,b, 1992a). It remains to be demonstrated that induced antipredator behavior has community consequences in the field (Anholt 1997).

Induced Behavior versus Induced Morphology

Just as animals are capable of adjusting different aspects of their behavior (habitat choice, proportion of time active, speed of movement) in the face of predation risk, there are additional kinds of phenotypic plasticity that can be induced by predators. Predator-induced morphological changes have long been recognized in invertebrates, but recent evidence suggests that this may occur among vertebrates as well (Brönmark and Miner 1992).

Larval anurans also have marked developmental changes in the presence of predators. Hylids develop deeper and shorter tails than animals raised in the absence of predators (McCollum 1993; Smith and Van Buskirk 1995; McCollum and Van Buskirk 1996). Deep tails may facilitate fast starts, suggesting that these animals are more efficient at evading predators once they are attacked. This is a different tactic than the reduction of activity, which is for avoiding encounters with predators in the first place. We suspect that the relative costs and benefits of these two tactics depend in part on the temporal overlap of animals with their predators. A morphological defense has long-term costs, sometimes for the lifetime of the organism, but a behavioral defense can be turned on and off as necessary. Thus costs and benefits need to be considered on the appropriate timescale. When prey are long lived but predators appear and disappear seasonally, it may be more profitable to use behavioral defenses whose costs are only paid when they are used (and useful). When predators are likely to be present for long periods of time, then morphological defenses may be favored. This implies that behavioral defenses have higher operating costs than morphological ones, but that morphological defenses have higher fixed costs. As far as we know there are no measures of the relative costs of these two forms of defense as a function of their efficacy.

Conclusions

Understanding inducible variation in behavior by animals foraging under the threat of predation is vital for assessing interactions in food webs. The widespread observation that animals change their foraging activity in the face of predation (Sih 1987a; Lima and Dill 1990) makes it clear that we cannot assume that the interaction strengths of links between members of the web are insensitive to other species that are present. However, the generality of the trade-off between acquiring resources required for growth and reproduction and the additional mortality risk that this entails holds out the promise that incorporating realistic behavioral variation in our models will allow us to make successful predictions about the indirect consequences of changing some features of the web.

13

Complex Biotic Environments, Coloniality, and Heritable Variation for Inducible Defenses

C. DREW HARVELL

Abstract

Colonial marine invertebrates are notorious for rapidly changing their forms with shifts in the biotic and abiotic environment. For sessile marine organisms growing on hard substrates, the unpredictable biotic environment of simultaneously encroaching competitors, predators, and pathogens provides intense selection for inducible responses, localized in time and space. Inducible defenses are especially favorable when individuals have a cue to distinguish among several alternative states. For example, the bryozoan *Membranipora membranacea* produces zooids with straight spines against predators, branched spines in some hydrodynamic regimes, and stolons against competitors.

Here I address two questions common to the more general study of the evolution of inducible defenses:

(1) *How widespread are inducible defenses in colonial invertebrates, and to what extent does inducibility promote regionalization of defense within colonies?* In addition to *Membranipora*, another bryozoan (*Tegella robertsonae*) shows a range of induced morphological responses to the biotic environment, and several other cheilostomes produce stolons against competitors; the remainder of bryozoans have not been comprehensively investigated. Similarly within the Cnidaria, although several species of hydrozoans and scleractinian and gorgonian corals have been shown to change morphology in response to cues from competitors, the extent of this induced variation is not yet documented. Gorgonian corals change the form of their sclerites in response to mechanical damage and attacks by consumers. Corals produce sweeper tentacles in response to proximity of competitors. All of these responses are regionalized within the colony and produced adjacent to a biotic threat, thus a benefit of the inducibility is allowing precise spatial partitioning of defenses within colonies.

(2) *Can we, as a next step in evaluating the microevolutionary potential of these responses, document the range of inducible phenotypes in natural pop-*

ulations and the factors that maintain variation in the inducible response? Previous studies with *Membranipora* reveal high levels of heritable variation in the inducible spine response and a genetic polymorphism within populations, producing a mix of inducible, constitutively spined and unspined individuals. Costs of defense, as well as frequency- and density-dependent effects of both predation and competition, probably contributes to the maintenance of genetic polymorphism type (constitutive, inducible, or undefended) within a population and the maintenance of quantitative variation within a type.

Introduction

Colonial marine invertebrates represent a unique system for studying inducible defenses, combining attributes of both plants and animals. Like plants, most colonies are sessile and substrate bound and have a modular, iterated organization. But they are animals and, as such, have an integrated nervous and resource-sharing system, and rely on chemical cues to mediate morphology (Mackie 1986; Harvell 1991). A final attribute of colonial invertebrates that is shared with many plants is that the genotype can be larger than the biotic grain of the environment, and inducible responses can be localized to certain regions within an individual genotype. This introduces a complex decision about where and when to distribute inducible defenses, thus the normal time-course issue of when to turn on and off a defense is complicated in colonies by a further decision of where to put different defense types. Although colonial invertebrates are present in freshwater and marine habitats, work with inducible defenses has only been conducted in the ocean, where colonial invertebrates such as corals, sponges, bryozoans, and ascidians are of ecological significance in establishing the structure of coral reefs and dominating many subtidal communities.

Due to the sessile, substrate-encrusting growth habit of many colonial invertebrates and the inability to flee threats, interactions with both predators and competitors can provide a lethal biotic environment. Predators often feed like herbivores, removing modules without killing the entire genotype; thus an attack provides a reliable, nonlethal cue for activating a defense (Harvell 1984b, 1986). Competitors kill by overtopping colonies, smothering and taking over their space (Buss and Jackson 1979). Competitive interactions can be mediated by chemicals, stinging cells, or strangling stolons. Nothing is known in any colonial invertebrates about mechanisms of resistance to pathogens (but see Kim et al. 1997; Harvell et al. 1997), although death by disease is well known for corals (Rutzler et al. 1983; Guzmán and Cortés 1984; Nagelkerken et al. 1997; G. Smith et al. 1996). Inducible morphological defenses to both predators and competitors appear common in colonial invertebrates: bryozoans make stolons against competitors and spines against predators; hydrozoans make stolons against competitors; individuals in two orders of corals make sweeper tentacles against competitors; and at least two species of gorgonians change the density and size of sclerites in response to damage and predators (table 13.1). Although this review suggests the possibility that inducible morphological responses are common in some taxa of marine colonies, in no taxon have species been comprehensively assayed to generate estimates of the frequency of these inducible structures.

In reviewing what is known of inducible defenses of colonial marine invertebrates, I will begin with a well-studied system, the inducible defenses of *Membranipora membranacea*, then proceed to other examples that demonstrate the diversity of response. To date, inducible defenses are only described in marine bryozoans and two orders of cnidarians. Yet there remain several phyla with colonial members (Rotifera, Hemichordata, Urochordata) in which inducible defenses have not been demonstrated but might be expected. *Membranipora membranacea* has been an unusually good study system for addressing, in nature, issues about the evolution of inducible defenses because they are an abundant, fast-growing, subannual species. I will review our work with costs of defense and causes of variation in inducible defenses, with particuar emphasis on factors maintaining the variation in inducible defenses and the potential for evolution of inducible types.

How Widespread Are Inducible Defenses in Marine Colonies

The Membranipora-Doridella Interaction

Membranipora membranacea is a circumglobally distributed and abundant cheilostome bryozoan that grows on laminarian kelps. Bryozoans are colonial invertebrates that not only reproduce sexually, through the production of a motile larval stage, but also propagate asexually through the production of daughter zooids. The zooids remain connected (and are also nutritionally and electrically integrated) in a circular colony. In the San Juan Archipelago, Washington, the sexually produced planktonic larvae of *Membranipora* first recruit onto kelps during May. Larvae settle and metamorphose in high numbers, creating dense populations of the bryozoan on kelp blades. Recruitment continues until September, when the adult colonies die (Harvell, Caswell, and Simpson 1990). Like some other cheilostome bryozoans, *M. membranacea* is a protandrous hermaphrodite: individual colonies progress from a prereproductive stage through a male stage, to a transitional stage, when both sperm and oocytes are produced. This sequence of reproductive stages appears to be developmentally set; however, the time spent in any one stage is not well correlated with colony size, and transitions can be accelerated by

TABLE 13.1
Inducible Responses of Marine Colonies to Biotic Factors

Induced Taxon	Inducer Predators	Inducer Competitors	Induced Structure	Reference
Cnidaria				
Gorgonacea				
Erythropodium caribaeorum		Other corals	Sweeper tentacles	Sebens and Miles 1988
Briareum asbestinum	Damage		Sclerite size	West 1997
Plexaurella sp.	*Cyphoma gibbosum*, *Cyphoma signatum*		Sclerite density	Nowlis, West, and May, in prep.; Nowlis 1994
Scleractinia				
Montastrea cavernosa		Other corals	Sweeper tentacles	denHartog 1977; Chornseky and Williams 1983
Agaricia agaricites		Other corals	Sweeper tentacles	Lang and Chornesky 1990
Agaricia agaricites		Other corals	Sweeper tentacles	Chornesky 1983
Pocillopora damicornis		Other corals	Sweeper tentacles	Wellington 1980
Galaxea fascicularis		Other corals	Sweeper tentacles	Hidaka and Yamazato 1984
Antipatharia				
Antipathes fiordensis		Other corals	Sweepers	Goldberg et al. 1990
Hydrozoa				
Hydractinia echinata		*H. echinata*	Stolons	Ivker 1972
Podocoryne sp.		Podocoryne	Stolons	McFadden et al. 1984
Millepora sp.		Gorgonians	"Fingers"	Wahle 1980
Bryozoa				
Membranipora membranacea	*Doridella steinbergae*		Spines	Harvell 1984a
Membranipora membranacea		*M. membranacea*	Stolons	Harvell and Padilla 1990
Tegella robertsonae	*Doridella steinbergae*	*T. robertsonae*	Spines, stolons	Harvell, this chapter

TABLE 13.1 (Continued)

Induced Taxon	Inducer		Induced Structure	Reference
	Predators	Competitors		
Celleporaria sp.	Bryozoans		Stolons	Osborne 1984
Cleidochasma bassleri	Bryozoans		Stolons	Osborne 1984
Hipopodina feegeenis	Bryozoans		Stolons	Osborne 1984
Stylopoma duboisii	Bryozoans		Stolons	Osborne 1984
Thalamoporella tubifera	Bryozoans		Stolons	Osborne 1984

crowding (Harvell and Grosberg 1988; Harvell and Helling 1993). Crowded colonies reproduce at an early age and small size. Colonies release long-distance dispersing larvae that live and feed in the plankton for 2–4 weeks (Yoshioka 1982).

Colonies of *Membranipora membranacea* produce spines on the surface of the frontal membrane and at the corners of the zooecia within 48 hours of exposure to chemical cues from the nudibranch *Doridella steinbergae* (Harvell 1984a; fig. 13.1A). Colonies produce spines in response to the chemical cue without damage or physical contact by the predator (Harvell 1986), and can be induced by exposing colonies in culture to filtered seawater that has contained actively feeding nudibranchs (Harvell 1998). Our experiments were usually conducted with individual, uncrowded colonies cut from their algal blade or grown on lucite panels. Colonies are maintained in 4-liter vessels on oscillating paddles that keep the algal food well stirred and facilitate active feeding by the bryozoans. In all induction experiments, colonies were fed once or twice daily at 5000–10,000 cells/ml (*Dunaliella turtox* or *Rhodomonas* sp.), depending on the experiment.

Preliminary studies revealed that 2–3 species of nudibranchs other than *Doridella* will trigger spine production, but neither as reliably nor with such large spines as triggered by *Doridella steinbergae* (Iyengar, unpubl.). *Doridella* feeds by creating suction on the frontal membrane and then extracting the zooid. The spines of the bryozoan appear to disrupt the establishment of the predator's suction. It remains likely that other predators could also induce spines. Similarly, some abiotic conditions will trigger spination. In California, small colonies grown in still water produce long, forked spines (Harvell and Trager, unpubl.). However, forked spines rarely occur further north in Friday Harbor, Washington (FHL), and induction studies with northern populations in slow and no water flow did not result in the production of spines (Harvell, unpubl.).

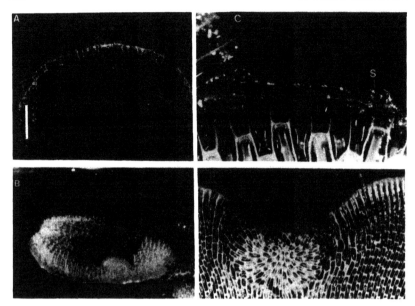

Fig. 13.1. Inducible spines and stolons of marine Bryozoa. (A) *Membranipora membranacea*, spines induced by exposure to *Doridella steinbergae*. Scale = 100 microns. (B) *Tegella robertsonae*, spines induced by exposure to *Doridella steinbergae*. Spines approximately 1.2 mm high. (C) *Membranipora membranacea*, two stolons induced on a large colony in contact with a small competitor. Each stolon approximately 3 mm long. (D) *Membranipora membranacea*, close-up of a large stolon induced at the junction between two colonies. Stolon approximately 3 mm long.

For most invertebrates with inducible defenses, nothing is known about the heritability of the response or the magnitude of genetic variation in the trait. However, a recent study has revealed high heritable variation among colonies of *Membranipora membranacea* in the magnitude of inducible response (Harvell 1998; fig. 13.2). The clonal nature of these colonies allowed this experiment to be conducted as a norm of reaction to reveal variation among genotypes. Colonies were collected from a single kelp blade and each colony was divided into four replicates; each clonal replicate was then independently assayed for inducible spine length in a separate vessel. All colonies were subjected to the same concentration of inducer. The resulting variation is a measure of genetic variation in inducible spine length among colonies. The small standard errors indicate that genotypes responded uniformly, indicating high broadsense heritability (Harvell 1998).

Even more rare than studies to measure heritable differences among inducible types are attempts to map the distribution of different phenotypes in nature. Data for two years (1993, 1995) indicate that there are three different phenotypes in the population at FHL, inducible (approximately 80%), consti-

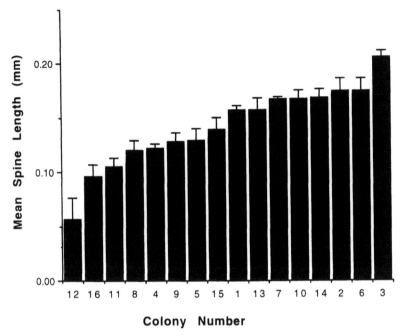

Fig. 13.2. Variation in inducible spine length of *Membranipora membranacea*. Each colony is assayed at the same level of inducer. Mean and standard error of 4 clonal replicates per colony.

tutively spined (5–10%, depending upon the year) and constitutively unspined (5–10%, depending on the year) (Harvell, 1998). These experiments were conducted by culturing newly metamorphosed recruits (N = 129, 187 colonies, depending on the year) in a common lab environment for 30 days, prior to applying the inducer. Colonies were assayed 4 days after applying the inducer. The constitutive colonies in 1993 were unusually spinose, with all zooids producing long, dense spines. Even among the inducible colonies, there is large variation among genotypes in the length and type of spines induced (fig. 13.2; Harvell, 1998). This is one of the first studies not only to examine the range in variation of the inducible response, but also to detect the presence of a polymorphism for inducibility within a population. In general, very little is known in any organism about the levels of variation in inducible response or the factors maintaining variation in the inducible genotypes.

Studies showing the degree of genetic variation in inducible defenses have also been conducted with *Daphnia*. Parejko and Dodson (1991) showed heritable variation among clones for inducible neckteeth formation. Spitze and Sadler (1996) showed clone-specific variation for eight inducible characters of *Daphnia*; four of these characters affected vulnerability to a predator. De

Meester's work with induced vertical migration of *Daphnia* shows genetic polymorphisms in inducible type (De Meester et al. 1995; De Meester 1996a) and suggests that polymorphism in vertical migration behavior is maintained by variation in predation regime (De Meester, this volume). For *Membranipora*, the polymorphism in defensive type could well be maintained by spatial and temporal variance in predictability of predator attack. In low predator sites or years, fast-growing undefended colonies may escape predation. Furthermore, since colonies are not killed in an attack (nudibranchs kill individual zooids, but the colony can regenerate from even the most severe attack), the consequences of forgoing defense may not be fatal. However, both the undefended and constitutively defended types appear rare in the population (5–10%), suggesting the overall favorability of the inducible strategy in the FHL populations.

Costs of defense are theoretically expected to balance the benefit of the inducible defense (Clark and Harvell 1992) and thus could affect the maintenance of variation (Lively 1986b,c). I examined whether growth rates of experimentally induced colonies were reduced. Spines induced weekly over the lifetime of colonies are correlated with reduced growth rate and a reduction in life span, but spined colonies reproduced sooner than unspined ones (Harvell 1992). The reduction in growth rate of isolated colonies grown in the field was relatively small. However, biologists rarely consider the compounding effects of multiple biotic challenges. Since these colonies suffer from intense intraspecific competition that drastically affects eventual colony size (Harvell, Caswell, and Simpson 1990), the effects of reduced growth should be measured in this same competitive milieu. If this experiment were conducted with the target colony in its natural habitat of a high intraspecific density, even a small slowdown in growth would likely have an even larger effect on the final size a colony would reach. These density-dependent "opportunity costs" have been considered earlier in plant studies (Gulmon and Mooney 1986), but are usually neglected. For some organisms like *Membranipora membranacea*, which typically live in high population densities and require bare substrata, the opportunity costs may far exceed the base-measured metabolic costs of defense. Grünbaum (1997) also detected a cost of producing spines and suggested the cost was caused by spines interfering with colony feeding, since the magnitude of the cost varied with flow regime. To date, we have not investigated differences in growth rate among the three polymorphs, but at least some constitutively spined colonies appear to grow more slowly than unspined colonies, suggesting genotype-specific negative correlations between spine type and fitness (Harvell, unpubl.). It is plausible that variations in growth rate among colonies of varying spine type, in this keenly competitive environment, help balance selection for the benefits of spines in deterring predators. Furthermore, high variation in the density of competitors may contribute a stochastic component in the intensity of density-dependent selection. This variance in the importance of density dependence could also maintain polymorphism in defensive types. Finally, high variance in the probability of attack by predators could also contribute to the maintenance of variation, by allowing some undefended fast-growing colonies to succeed in the absence of predators at low-predator times and locations. Taken together, the cost of defense, coupled with variance in the probability of predator and competitor attack, likely contribute to maintenance of variation.

In many species with inducible defenses, the effects of predation should not be evaluated without simultaneous consideration of the effects of competition. The interwoven influences of a complex biotic environment are immediately apparent in examining the multiple inducible responses of some colonies to their attackers. *Membranipora membranacea* simultaneously produces spines against predators and stolons against competitors (fig. 13.1C,D). The presence of competitors does increase the opportunity costs of defenses because a slowly growing colony will be more quickly surrounded by competitors. What we have yet to learn is how simultaneous challenges by predators and competitors affect the priority of response. For example, does a stolon-making colony produce the same length spines as a nonstolon-making colony? Does a spined colony make stolons with the same probability as a nonspined colony?

Other Inducible Responses of Bryozoans: A New Example

Tegella robertsonae is also a cheilostome bryozoan, but instead of broadcasting its newly fertilized eggs like *Membranipora*, it broods them and releases larvae that are competent to metamorphose. It is likely that life history differences (brooding vs. broadcasting progeny) affect the types of costs incurred by inducible defenses. For example, in *Membranipora* costs of spines include reduced growth, reduced feeding efficiency (Grünbaum 1997) and an accelerated investment in reproduction. I was unable to measure a decrease in per zooid investment (zooids can contain over forty oocytes at one time), although slowed growth resulted in smaller final colony size and thus lower numbers of reproductive zooids. The results in a brooding species, with a much higher per embryo investment (one embryo brooded per zooid from fertilization to larval release), could be even more extreme or at least easier to detect. In an induction experiment with *Tegella*, spines were produced on colonies exposed to nudibranch (*Doridella*) water, but not on control colonies. Spines on treatment colonies were significantly longer than controls 3 days after the experiment began, and increased further in length by day 5 (figs. 13.1B, 13.3). The mean difference in spine length from day 1 to day 5 for the nudibranch treatment was 0.11 ± 0.01 (se) from the initial pretreat-

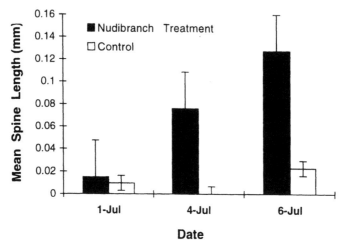

Fig. 13.3. Induction of spines on *Tegella robertsonae*. Mean spine length and 95% confidence intervals are plotted for control and treatment colonies at the beginning and 3 and 5 days post exposure.

ment size and 0.01 ± 0.01 for the control treatment (t test, $t = 5.86$, $p = .0001$). *Tegella* also produces stolons against conspecific and heterospecific competitors. This is the first evidence that bryozoans other than *Membranipora* produce stolons as well as spines against competitors. These multiple, specifically cued defenses in a single organism require complex allocation decisions in timing and placing morphologies.

Inducible Responses of Corals

Gorgonian and scleractinian corals are similar to bryozoans in being colonial, and are also similar to plants in being sessile and even autotrophic through the symbiotic algae housed in their tissues. Many gorgonians rely completely on carbon fixed by their algae and may not be carnivorous like other groups of corals. Scleractinian hard corals (reviewed in Lang and Chornesky 1990) and gorgonian soft corals have shown inducible responses against competitors in the form of nematocyst-laden tentacles. Thus at least five species of scleractinians respond to encroachment of heterospecific competitors with the production of sweeper tentacles (table 13.1). The sweepers of scleractinian corals can exceed 10 cm in length and develop along small sections of corals in proximity to competitors. The induction of sweeper tentacles takes several weeks, depending upon the species, and is usually accompanied by shifts in the relative proportions of nematocyst type from spirocysts for food capture to stinging cells. The black coral (*Antipathes fiordensis*) also produces elongated (15 mm) sweeper tentacles filled with a distinctive stinging cnidal form and none of the normal spirocysts used in food capture (Goldberg et al. 1990). The sweeper tentacles of gorgonians are somewhat distinctive; surprisingly they are almost identical morphologically to scleractinian sweepers, as elongate, smooth nematocyst-laden tentacles, although typical gorgonian tentacles have the characteristic pinnate structure of octocoral tentacles (Sebens and Miles 1988).

The structural agents, called sclerites, which are embedded as skeletal elements in the soft tissue of gorgonian colonies, can also be induced to change form by damage or predators. West (1997) found that the sclerites of *Briareum asbestinum* increased in size and density in response to mechanical damage. The increased density of sclerites in artificial foods functioned as a deterrent to the corallivorous mollusc *Cyphoma gibbosum*. Nowlis, West, and May found that the density of sclerites increased in colonies of *Plexaurella* sp. in response to actual attacks by *Cyphoma gibbosum*, and then returned to normal densities when the consumer was removed (Nowlis et al., in prep.; Nowlis 1994). Together these studies support the hypothesis that gorgonian corals are capable of mounting inducible, morphological responses against their macroconsumers. These are the first studies to provide any information on inducible defenses of corals to their consumers. Kim et al. (1997) have shown that sea fan sclerites change from pale to dark purple in response to a fungal pathogen. Thus in the Gorgonacea, induced changes to predators, pathogens (sclerites), and competitors (sweeper tentacles) have been identified, and even these are not necessarily simultaneous responses in the same species.

Gorgonians are also notorious for producing a diverse complement of secondary metabolites, rivaling the types of terpenes produced by plants against a range of herbivores and pathogens (Rodriguez 1995). Some of these compounds are deterrent to fishes (Pawlik et al. 1987; Harvell, Fenical, and Greene 1988; Pawlik and Fenical 1989; Harvell et al. 1993; Harvell, West, and Griggs 1996) and invertebrates (Paul and Van Alstyne 1992), and inhibitory to bacteria (Jensen et al. 1996) and fungi (Merkel et al. 1997). No studies have yet demonstrated inducibility of these secondary compounds to biotic agents, but Harvell et al. (1997) did show elevated anti-fungal activity to fungal lesions on gorgonian sea fans. Harvell et al. (1993) showed that the levels of secondary compounds changed in colonies transplanted between two depths, suggesting the ability to regulate these compounds in different ecological contexts. In fact, induction of chemical defenses against biotic agents has not been shown in any colonial marine invertebrate, despite the ubiquity of high levels of bioactive secondary compounds (Paul 1992) in these groups. Although Hay (1996) suggests that the rarity of chemical induction in marine systems is due to an actual rarity of response, I am more inclined to think for the marine colonies that researchers have not implemented the appropriate experiments yet. I am expecting that gorgonians will eventually be very useful for examining the relative importance of inducible chemical vs. morphological defenses.

Multiple Biotic Agents and Spatial Localization of Inducible Defenses

It has been suggested previously that spatial and temporal variation in predation regime is likely a primary attribute favoring inducible defenses (Harvell 1990a; Dodson 1989b) or phenotypic plasticity in general (Levins 1968); less attention has been paid to the interactive effects of a complex of predators and competitors in inducing defensive responses. Specifically, an emerging hypothesis emphasized in this book is that unpredictable shifts in a complex biotic regime require inducible defenses, whether they be the immune system or inducible morphological defenses. My work with bryozoans shows that being caught between multiple predators and competitors necessitates the need for not only an inducible response to changing biotic regimes, but also the ability to localize the response within different regions of a colony. Bryozoans (and other colonial invertebrates such as corals), unlike solitary invertebrates like *Daphnia*, can partition responses within the colony.

In *Membranipora*, the phenotypic possibilities for inducible responses of a colony are many and hierarchical. First, single zooids can choose between antipredator or anticompetitor morphologies. *Membranipora membranacea* (and some other bryozoans; Osborne 1984; Harvell, unpubl.) produces stolons in response to attacks by conspecifics. Colonies live in a space-limited monoculture and fight actively for space by deploying stolons at contacts with non-clonemates (Harvell and Padilla 1990; Padilla et al. 1996). The production of stolons is highly context dependent: stolons are usually produced by large colonies interacting with smaller heteroclonal competitors (Padilla et al. 1996); thus colonies appear to respond to both absolute and relative size as well as genetic identity (Harvell and Padilla 1990). The stolons quite effectively slow or stop the local growth of competitors at the contact point. However, for the individual zooid, developing into a competitive form (stolon) makes it more vulnerable to predation, because the nudibranchs may selectively feed on the stolons (DeGroat and Harvell, unpubl.). It is not yet known how the production of spines affects subsequent competitive ability, except that it slows the rate at which a colony can fill space (Harvell 1992; Grünbaum 1997). Secondly, the arrangement of zooids within colonies can vary. A predation-resistant colony deploys spines around the edge; this appears to be a favorable configuration since predators approach at the edge (Harvell 1991). This predation-resistant configuration may also increase competitive ability through resistance to overgrowth, but at a reduction in final size due to reduced growth rate. This cost of defense will be density dependent because growing slowly in a high-density population will produce especially small colonies (Harvell et al. 1989). Third, colonies can spatially partition inducible defenses to predators and competitors:

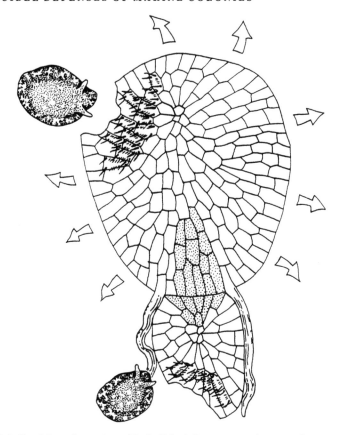

Fig. 13.4. Spatial configuration of inducible defenses to predators and competitors in a marine bryozoan. In asymmetrically sized two-colony interaction, stolons are produced by the large colony at the border of the small colony (Harvell and Padilla 1990). Both colonies reproduce locally, adjacent to the interaction, as represented by stippling (Harvell and Helling 1993). Nudibranchs feed on the induced stolons and also induce spines by their proximity. The other side of the colony can be involved in producing spines locally against an attacking nudibranch and show localized growth reductions (Harvell 1991). Undisturbed regions of the colony continue to grow rapidly, as represented by arrows (Harvell and Helling 1993).

zooids on one side of the colony can be producing stolons, and on the other side, spines (fig. 13.4). Spines and stolons are also produced by different types of zooids—spines are produced on young zooids adjacent to but not on the growing edge; stolons are produced only by growing buds at the colony edge.

The benefits of inducibility to a modular organism in a complex biotic regime are therefore threefold. First, it allows appropriate timing; second, there is the option for a colony to select among alternative phenotypes; and third, the spatial configuration of both predation and competition-resistant morphologies can potentially be optimized.

Another example of the need to consider the structure of the entire biotic regime in the evolution of inducible defenses is seen in the induced vertical migration behavior of *Daphnia*. Visual vertebrate predators (fish) inhabit the surface waters during the day and force their prey, *Chaoborus*, to migrate (up at night, down in the day) as an inducible defense (Neill 1990). In turn, the *Daphnia* are preyed upon by both vertebrate and invertebrate predators (*Chaoborus*), and only an inducible response allows them to tailor their migration strategy to the exact combination of biotic threat (DeMeester et al. 1995; Tollrian and Dodson, chapter 10). *Daphnia* exposed to fish show diel migration, and those exposed to *Chaoborus* show reverse diel migration. The picture could become even more complex if there were inducible responses to competitors at the same time. This focus on complex biotic environments suggests an overriding benefit of inducibility that perhaps favors these inducible responses, irrespective of their fitness costs.

Conclusions

The sessile nature of colonial invertebrates and their propensity to respond morphologically to a changing biotic environment provides an unusual opportunity to visualize the dynamics of inducible responses to predators and competitors. Thus, colonies in two genera of bryozoans localize the distribution of inducible defenses within the colony, with stolons produced on the growing edge adjacent to competitors and spines in a ring or cluster nearest predator cues. Although localized inducible competitive structures are well known in corals, less is known about combinations of inducible defenses, since inducible morphologies against predators and pathogens have been reported only in gorgonaceans.

The detection of significant genetic variability in inducible responses among colonies of *Membranipora* (Bryozoa) raises the issue of what environmental factors may interact to maintain the observed variation in types of defenses. What maintains both inducible and nondefended types in the same population? Large costs of defense may be one factor that offsets the benefit of constitutive spines and allows undefended colonies to persist in the population. Other possibilities include negative correlations between the rate or magnitude of defenses against predators and competitors.

14

Developmental Strategies in Spatially Variable Environments: Barnacle Shell Dimorphism and Strategic Models of Selection

CURTIS M. LIVELY

Abstract

Predator-induced defense in an intertidal barnacle (*Chthamalus anisopoma*) is reviewed, and strategy models of development are presented in an effort to understand the general conditions under which induced defense is evolutionarily stable. The default developmental form of the barnacle has the conic shape that is characteristic of barnacles, and it is more fecund and faster growing than the induced form. The induced form is more resistant to attack by a specialized gastropod predator, and it is most common near rock crevices that serve as refuges for the predator during high tide. The models suggest that induced defense is evolutionarily stable for a wide range of parameters, even when the cues used for induction are less than perfect predictors of the future environment. These conditions for evolutionary stability of induced defense are also very sensitive to the structural and material costs of defense, as well as the cost of plasticity per se (i.e., the cost of mechanisms used for sensing the environment). The results also suggest that the conditions for genetic polymorphism of canalized morphs are very narrow, and that under some conditions mixtures of canalized and inducible individuals can be evolutionarily stable. This latter result may explain the apparent mixtures of inducible and noninducible individuals in barnacles, Daphnia, and bryozoans.

Introduction

Discrete phenotypic variation in natural populations is of interest because it suggests a trade-off among morphs in their abilities to survive and/or reproduce in the different patches of a heterogeneous environment (Levene 1953; Levins 1962, 1963, 1968; Lloyd 1984). This is true whether the morphs are under strict genetic control, or whether they result from environmentally induced switches among alternative developmental programs. The challenges that such polymorphisms present to ecologists are to determine whether the environment is indeed heterogeneous in a relevant way, whether there are patch-dependent trade-offs among morphs as suggested by theory, and whether the development of individuals is canalized or environmentally induced.

Environmental induction of alternative forms has gained theoretical attention as an important evolutionary strategy (Lloyd 1984; Via and Lande 1985; Lively 1986c; Bull 1987; Hazel et al. 1990; N. Moran 1992; see Travis 1994 for a thorough review), and recent empirical studies suggest that it may be a taxonomically widespread phenomenon in animals (compare Hazel and West 1979; Krueger and Dodson 1981; Grant and Bayly 1981; Yoshioka 1982; Gilbert 1966, 1980; Gilbert and Stemberger 1984; Eberhard 1982; Collins and Cheek 1983; Harvell 1984a, 1986; Hebert and Grewe 1985; Lively 1986a; Etter 1988; Crespi 1988; Pfennig 1990; reviews in Harvell 1990a; Travis 1994; Gotthard and Sören 1995). Most of these later studies have been conducted in the last 15 years and have uncovered induced adaptive responses that might have been inconceivable in the 1970s. In what follows, I first review predator-induced defense in an intertidal barnacle. Then, using barnacles as a "handle," I present strategy models of selection designed to determine the general conditions under which an induced defense is evolutionarily stable.

Induced Defense in a Barnacle

The intertidal barnacle *Chthamalus anisopoma* is dimorphic for shell shape in the northern Gulf of California (fig. 14.1), where it overlaps in distribution with the predatory snail, *Acanthina angelica* (Lively 1986a,b). The undefended form of the barnacle has the conic, volcano shape that is characteristic of acorn barnacles (fig. 14.2), and it is less resistant to attack by *Acanthina*. This snail is a specialized barnacle predator (Paine 1966; Malusa 1985; Perry 1985; Lively 1986a; Dungan 1987), which uses a labial spine to gain access to barnacle prey (see Keen 1971 and Brusca 1980 for photos of *Acanthina*). In a typical attack sequence, the predator first wraps the front margins of its foot around the barnacle's base. The snail then positions the spine over the barnacle's operculum and rams the spine down. About 45% of the time, the spine is successfully pushed through the operculum of undefended barnacles (Lively 1986a); the spine is then withdrawn and the barnacle is eaten.

The defended morph of the barnacle (fig. 14.2) is more resistant to attack of this kind, because the aperture is oriented in a plane that is perpendicular to the substratum (Lively 1986a). This 90° shift in the plane of the aperture

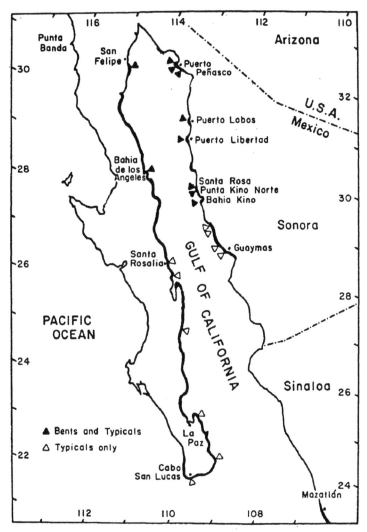

Fig. 14.1. Map of the Gulf of California, showing the known distributions of the bent form of the barnacle *Chthamalus anisopoma* and the predatory snail *Acanthina angelica* (closed triangles). The conic form of the barnacle is distributed throughout the Gulf of California (open and closed triangles).

results from shortening the lateral plates on one side while lengthening the lateral plates on the other. Apparently, one side simply stops growing while the other side continues to grow, pushing the aperture upright. This differential growth makes the barnacle appear bent over, resembling a hood. During the attack sequence, when the predator's spine is rammed down, it usually misses the aperture of the bent morph altogether.

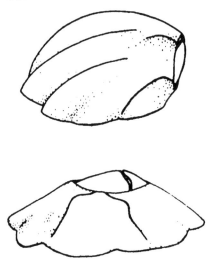

Fig. 14.2. Line drawings of the "conic" and "bent" morphs of the intertidal barnacle *Chthamalus anisopoma*. The bent form (*top*) of the barnacle is induced by presence of the gastropod snail *Acanthina angelica*, and it is more resistant to attack by this predator. The conic form (*bottom*) grows faster and is more fecund per unit size. The basal diameter of adult *C. anisopoma* ranges from 2 to 7 mm (Brusca 1980; Lively 1986b.)

This defense has two direct costs: (1) the bent form grows more slowly, and (2) it is less fecund per unit size (Lively 1986b). Both of these costs result from the structure of the bent form. There is less room for brooding larvae in the bent shell; hence, overall fecundity is reduced. In addition, growth is slower in the bent morph, since it is restricted to one side. To my surprise, I found no competitive disadvantage of the bent morph (Lively 1986b). The survivorship of bent individuals in competition with conics was not less than that of conics when similarly grown in competition with conics. Hence there is no competitive asymmetry between morphs that would be analogous to the well-known competitive asymmetries between different species of barnacles that inhabit adjacent zones in the intertidal (Connell 1961; Dungan 1985).

Typically, the snail predators emerge from rock crevices or from under boulders as the tide recedes. They tend to move quickly, while the rocks are still wet, and begin foraging. Once they engage in attack, they move very little, presumably to minimize their water loss while crawling over the rocks, which are sometimes quite hot (Lively and Raimondi 1987). Hence, for the induced defense to be of any value, the bent morph must be able to withstand repeated attack during a low tide. The bent form of the barnacle is sometimes killed during repeated attacks but generally is much more likely to survive than the typical, conic form of the barnacle (Lively 1986a). When the tide returns and wets the foraging snails, they tend to move back to the rock crevices. This back and forth movement of the predators creates a distinct area where predation intensity is high. In adjacent areas, only a few centimeters farther from crevices, predation intensity is low or absent (Lively 1986a; Lively et al. 1993). Hence there are two discrete patches for predation risk: high risk and low risk (see Menge 1978, Moran 1985, and Garrity and Levings 1981; see Fairweather 1988 for similar reports of "crevice effects").

This is the kind of heterogeneous environment that would seem to favor a developmental switch, and indeed the bent form of the barnacle is induced by the presence of *Acanthina*. The details of this induction are not clear, but juvenile barnacles between the ages of 1 and 5 days old were induced to develop the bent form by placing *Acanthina* directly on them just prior to inundation by the incoming tide (Lively 1986a). In this experiment, where individuals were followed over time, 37% of the barnacles exposed to the predator developed the bent morphology within 30 days (Lively 1986a). Surprisingly, the rest of the barnacles did not respond to contact with the predator, and all became the default, conic form. It is not presently known whether the cue is chemical or mechanical, but the addition of herbivorous snails in one experiment did not induce the bent form, suggesting that mechanical stimulation by a snail's foot is not sufficient to induce the defended form (Lively 1986a).

Two curious twists to these main results are of theoretical as well as empirical interest. One twist is that some juvenile barnacles exposed to the predator were not induced to develop the attack-resistant form. The second is that the cue seems to be a poor indicator of the future environment, at least during certain seasons. The reason for this is that the snails estivate during the summer (Lively 1986a), when most of the barnacle settlement occurs (Raimondi 1990). Hence most barnacles that settle in the high-risk zone during the summer do not receive the cue; they therefore develop as conics and are susceptible to predation when the snails come out of estivation in the early autumn (Lively et al. 1993). This raises the question: how reliable does the cue have to be in order for selection to favor induced defense over constitutive defense? I have studied this question using strategy models (Lively 1986c), which are extended here in two ways. First, costs of inducibility are added (following Van Tienderen 1991); and second, I assume that there is spatial autocorrelation in the environment, such that inducible individuals make the same developmental choice as their neighbors.

Strategy Models of Selection

Selection on Two Canalized Developmental Strategies

I begin by considering two canalized alternatives, defended and undefended. Along with these pure unconditional strategies, I also consider a stochastic switch between morphs (a mixed strategy) where individuals develop as the undefended morph with an average probability (q), which is independent of the patch in which they settle. In the context of the barnacles, then, when would we expect to find (1) unconditional development of the bent form, (2) unconditional development of the conic form, (3) an evolutionarily stable mixed strategy, or (4) an evolutionarily stable state of the population composed of individuals showing either canalized defense or canalized development of the undefended conic form? Further, we want to know which of these four alternatives to expect as the frequency of patch types varies. I use game-theoretic models, which (strictly speaking) assume clonal reproduction. However, sexual populations will evolve toward the evolutionarily stable (ES) strategy or ES state in population genetic (Maynard Smith 1982) and quantitative genetic models of selection (P. Taylor 1996).

Barnacle larvae are planktonic and likely to become widely distributed. They do, however, use cues (such as the presence of conspecifics) for settlement, which increases the likelihood of attachment and metamorphosis in the "right" intertidal zone (Raimondi 1988). Within this zone, I assume that the barnacles settle more or less at random. Let the probability of attachment in the predation-free, low-risk patch be p; thus $(1 - p)$ gives the probability of settling in the high-risk patch where the probability of contact with the predator is high. Similarly, let k be the relative fitness of the defended morph, where $k < 1$ reflects structural or material costs of investment in defense. Finally, let e_{ij} be the effect of competition on the ith morph when rare in a population of the jth morph. For example, e_{du} gives the relative fitness of a rare, defended individual in a population composed almost entirely of undefended individuals.

In what follows, I assume that the patch frequencies are constant over time, and that the undefended morph cannot survive in the high-risk patch. Now consider a resident, mixed strategy such that all individuals in a large population develop the undefended form with probability q and the defended form with probability $(1 - q)$. Using standard evolutionary game theory (Maynard Smith 1982; Bulmer 1994), we consider the fate of a rare individual in this population. Let q_i be the probability of developing as an undefended morph by this rare individual, and let $(1 - q_i)$ be the probability of developing as the defended form. Assuming that fitness is a linear function of the frequencies of the two morphs in the population, individual fitness is estimated as

$$W_i = q_i p[q e_{uu} + (1 - q) e_{ud}] + (1 - q_i)(1 - p)[q + (1 - q) e_{dd}]k \\ + (1 - q_i) p[q e_{du} + (1 - q) e_{dd}]k. \tag{1}$$

A mutant strategy cannot increase under selection when the resident population is at a local maximum for fitness, which is when

DEVELOPMENTAL STRATEGIES

$$\frac{\partial W_i}{\partial q_i} = q[k(e_{dd} - pe_{du}) + p(e_{uu} - e_{ud}) - k(1 - p)] - (ke_{dd} - pe_{ud}) = 0, \quad (2)$$

provided the second derivative is less than or equal to zero, which is indeed the case for eq. (1). Hence, the evolutionarily stable strategy, represented by the variable q^*, is

$$q^* = \frac{ke_{dd} - pe_{ud}}{k(e_{dd} - pe_{du}) + p(e_{uu} - e_{ud}) - k(1 - p)}. \quad (3)$$

A mixed strategy at q^* is also continuously stable (Eshel 1983; Christiansen 1991) when

$$\left[\frac{\partial^2 W_i}{\partial q_i^2} + \frac{\partial^2 W_i}{\partial q_i \, \partial q}\right]_{q_i = q = q^*} < 0, \quad (4)$$

which is satisfied when

$$\frac{(e_{uu} - e_{ud}) + k(1 - e_{du})}{k(1 - p)(1 - e_{dd})} < \frac{(1 - p)}{p}. \quad (5)$$

Note that the equilibrium attained at q^* may be achieved by either a mixed strategy, or a population in which q^* of the individuals show canalized development of the undefended form (see Maynard Smith 1982). The latter population would represent a genetic polymorphism in a sexual population.

The parameter space for which a mixed strategy or genetic polymorphism might be expected (i.e., $0 < q^* < 1$) is given graphically in figure 14.3. This region of parameter space is surprisingly narrow (see also Maynard Smith and Hoekstra 1980), and requires that the probability of the low-risk patch (p) be on the interval,

$$\frac{ke_{dd}}{e_{ud}} < p < \frac{k}{e_{uu} + k(1 - e_{du})}. \quad (6)$$

Canalized development of the defended morph will be favored by selection when $p < ke_{dd}/e_{ud}$, and canalized development of the undefended morph will be favored by selection when $p > k/[e_{uu} + k(1 - e_{du})]$.

Introduction of a Developmental Switch to the Strategy Set

I now consider the fate of a rare mutant that responds to cues in the local environment that induce a developmental switch to the defended morph in

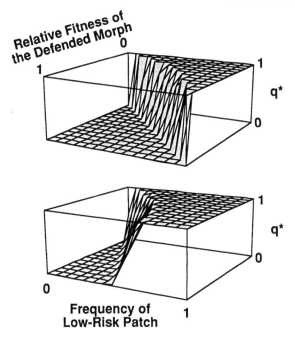

Fig. 14.3. Conditions for genetic polymorphism. Both figures give the equilibrium values of a constitutively undefended morph, q^*, as a function of (1) the relative fitness of the defended morph, and (2) the frequency of the low-risk patch where predation is a rare event. In the top figure, there are no competitive asymmetries, thus the interaction coefficients are all set to one-half (i.e., all $e_{ij} = 0.5$). Note the steplike transition between selection for fixation of the defended form and selection for fixation of the undefended form as the frequency of the low-risk patch increases. In the bottom figure, the interaction coefficients are set to reflect the most extreme type of competitive asymmetry, where the undefended morph excludes the defended morph ($e_{du} = 0$), while the defended morph has no effect on the fitness of the undefended morph ($e_{ud} = 1$). Both morphs decrease the fitness of an individual of the same kind by 50% ($e_{uu} = 0.5$; $e_{dd} = 0.5$). Note that the transition between fixed states is less steep than in the top figure, but the transition is nonetheless quite abrupt. In both figures, the equilibrium at q^* meets the conditions of eqs. (4) and (5) and is continuously stable.

the high-risk patch. I retain the useful idea of allowing patch frequencies to vary, but also include some probability of making the "wrong" choices in both the low-risk and high-risk patches. Biologically this means getting the cue and becoming the defended morph in the low-risk patch (with probability: $1 - F$), or not getting the cue, and becoming the undefended morph in the high risk patch (with probability: $1 - G$).

These assumptions are the same as those in a previous model of selection in spatially variable environments (Lively 1986c). The present model also

TABLE 14.1
Payoff Matrix for a Game between Canalized and Conditional Morphological Strategies

Fitness of Rare Developmental Strategy	Resident Developmental Strategy		
	Undefended ($q = 1$)	Defended ($q = 0$)	Conditional
Undefended ($q = 1$)	pe_{uu}	pe_{ud}	$p(Fe_{uu} + (1 - F)e_{ud})$
Defended ($q = 0$)	$k[pe_{du} + 1 - p]$	ke_{dd}	$k\{p[Fe_{du} + (1 - F)e_{dd}] + (1 - p)[Ge_{dd} + 1 - G]\}$
Conditional	$c\{p[Fe_{uu} + (1 - F)ke_{du}] + (1 - p)Gk\}$	$c\{p[Fe_{ud} + (1 - F)ke_{dd}] + (1 - p)Gke_{dd}\}$	$c\{p[Fe_{uu} + (1 - F)ke_{dd}] + (1 - p)(Gke_{dd})\}$

Notes: "Undefended" represents a strategy for canalized development of an undefended morph. "Defended" represents a strategy for canalized development of a defended morph whose fitness is scaled by the cost variable, k. "Conditional" represents a strategy for conditional development of a defended morph in a high-predation-risk patch, and an undefended morph in a low-predation-risk patch. The frequency of the low-risk patch is p. The probability of making the "wrong" choice in the high-risk patch is $(1 - F)$, and the probability of making the "wrong" choice in the low-risk patch is $(1 - G)$. The variable, c, gives the cost of plasticity, i.e., the cost of maintaining the developmental machinery for sensing and responding to the environment.

includes some additions and changes that seem warranted. One is that I have included a cost to plasticity per se. By this I mean a fitness cost to developing and maintaining any machinery for sensing the environment and flipping the developmental switch. Let the relative fitness of inducible individuals be c (where $0 < c < 1$). Note that the cost of plasticity (c) is independent of the cost of having the defended structure (k). In addition, my previous model allowed mistakes to be made on an individual-by-individual basis, which seems unreasonable given the diffusive nature of most chemical cues. Thus, in the present model, all conditional individuals make the same choice ("right" or "wrong"). So, in a population containing only conditional strategists, defended individuals compete for resources only with other defended individuals, and undefended individuals compete only with other undefended individuals. The payoff matrix is given in table 14.1.

To complete the model, we must also know when a conditional strategy cannot increase in a population of individuals playing a mixed strategy such that the probability of development as an undefended individual is at the ESS (q^*). Setting $q_i = q$ in eq. (1), and then substituting q^* from eq. (3) for q, the fitness of a mixed strategist when the mixed strategy is common is

$$W_{q^*q^*} = \frac{kp[(1 - p)e_{ud} + pe_{du}e_{ud} - e_{dd}e_{uu}]}{k(1 - e_{dd}) + p(e_{ud} - e_{uu}) - kp(1 - e_{du})}. \quad (7)$$

The fitness of a rare conditional mutant in a population of mixed strategists, W_{cq^*}, is closely approximated by:

$$W_{cq^*} = c\{qp[Fe_{uu} + (1 - F)ke_{du}] + (1 - q)p[Fe_{ud} + (1 - F)ke_{dd}]$$
$$+ q(1 - p)Gk + (1 - q)(1 - p)Gke_{dd}\}. \quad (8)$$

The rare conditional strategy cannot increase in a population of mixed strategists when $W_{q^*q^*} > W_{cq^*}$. Similarly, the mixed strategy at q^* cannot increase when rare when $W_{cc} > W_{q^*c}$, where W_{cc} is given in table 14.1, and

$$W_{q^*c} = p[q(Fe_{uu} + (1 - F)e_{ud}) + (1 - q)k(Fe_{du} + (1 - F)e_{dd}) \quad (9)$$
$$+ (1 - p)(1 - q)k(Ge_{dd} + 1 - G)].$$

A feeling for the parameter space for which each developmental strategy is evolutionarily stable and can increase when rare can be gained by numerical substitution. I have chosen to examine this space in terms of the frequency of the low-risk patch (p), and the probability of making the wrong choice in the high-risk patch ($1 - G$). This was done by substituting values for the relative fitness of the conditional strategy (c), the relative fitness of the defended form (k), and the probability of making the wrong choice in the low-risk patch ($1 - F$). To reflect the situation for the barnacles, I have set the interaction coefficients equal to one-half (all $e_{ij} = 0.5$), since there was no competitive asymmetry between morphs, but the presence of conspecifics did reduce the size (and hence the fecundity) of both forms. Following these substitutions, I determined the areas of parameter space for which each of the different developmental strategies would resist invasion by rare mutants having any one of the alternative strategies. Induced defense, for example, is evolutionarily stable if an individual having conditional development has a higher expected fitness in a population of conditional individuals than any alternative developmental strategy (defended, undefended, or mixed) when rare. The equations for these calculations are given in table 14.1 and eqs. (7–9) above.

The results of the numerical substitutions are interesting for several reasons. For example, they show that a mixed developmental strategy is evolutionarily stable for only a narrow set of parameters (fig. 14.4); in other words, development of the undefended morph with probability q^* (such that $0 < q^* < 1$) is subject to invasion by canalized mutants or conditional development over most of the parameter space. When there is no cost to plasticity ($c = 1$), a conditional strategy can increase when rare, and fix in a population composed of constitutive strategists whenever the average probability of making the right choice is greater than one-half (Lively 1986c). This result means that, if the probability of making the right choice is 80% in the low-risk patch (as presented in fig. 14.4), a conditional strategy can still increase when rare over a narrow region of parameter space when the probability of making the right choice in the high-risk patch is marginally

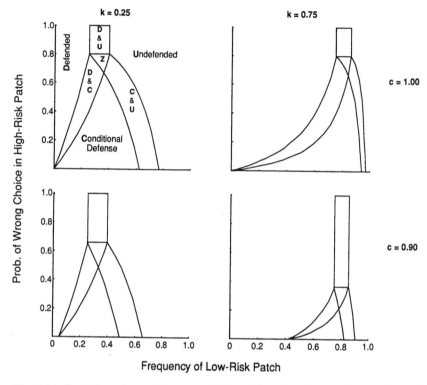

Fig. 14.4. Conditions for evolutionary stability of three developmental strategies and mixtures of these same strategies. The "torso" is the parameter space where constitutively defended and constitutively undefended developmental strategies can both increase when rare and reach a continuously stable equilibrium. The right "leg" is the region where constitutively undefended development and conditional defense can both increase when rare; and the left leg is where constitutively defended development and conditional defense can increase when rare (conditions for continuous stability were not analyzed for either leg). In the region marked "Z," all three strategies have a rare advantage. For the calculation of these figures, the probability of making the right choice in the low-risk patch was set at 80% ($F = 0.80$), and the interaction coefficients were all set at one-half (i.e., all e_{ij} 0.50). The left-hand figures assume that the defended morph (however produced) has only 25% of the reproductive output of the undefended morph; the right-hand figures assume that the defended morph has a relative fitness of 75%. Similarly, in the top two figures there is no cost to plasticity per se ($c = 1$), but in the bottom two figures, plastic individuals have a relative fitness of 90% ($c = 0.90$). Note that the parameter space under which the conditional strategy is an ESS is sharply reduced by a small cost to plasticity. Note also that changes in patch frequency or the probability of mistakes will lead to changes in the frequencies of the two morphs that may or may not be genetically based.

greater than 20%. The region of parameter space for which a conditional strategy can increase when rare increases dramatically as the probability of making the right choice in the high-risk patch increases (fig. 14.4). The overall result is that a large fraction of the parameter space contains some frequency of conditional strategists at the ESS, provided there is no cost to the plastic response.

If there is some cost to plasticity, the region of parameter space for which a conditional strategy is either stable or can increase when rare is greatly reduced (fig. 14.4). Surprisingly, a 10% cost of plasticity (i.e., $c = 0.9$) reduces this parameter space by about 50% or more, depending on the cost of the defended form (k) (see also Van Tienderen 1991 and León 1993). Thus it would seem that costs of developing and maintaining the "machinery" for sensing and responding to the environment have the potential to prevent the spread of rare conditional development, especially if the cost of defense is high. Nonetheless, once established, selection would be expected to fix any mutations that reduce the cost of plasticity, thereby increasing the range of conditions that favor induced defense.

The results also show that there are regions of parameter space under which conditional and canalized strategies can both increase when rare ("control polymorphisms," Lively 1986c). These regions reflect earlier results, but add one striking observation: some regions of parameter space support all three strategies. Hence, constitutively defended, constitutively undefended, and conditionally defended strategies can all increase when rare under some conditions. The coexistence of three strategies would seem to be unlikely in the wild, though it is known for male morphs of isopods (Schuster and Wade 1991) and lizards (Sinervo and Lively 1996). Recently, Harvell (1998) has shown the co-occurrence of three developmental types (inducible, constitutive defense, and undefended) in a marine bryozoan.

Discussion

Several points raised by the model seem worth fleshing out. The three most important are: (1) the conditions for a genetically determined polymorphism are very narrow; (2) the conditions for the existence of conditional development are relatively broad; and (3) mixtures of conditional and canalized development can be evolutionarily stable for some combinations of parameters.

Regarding the first point, it is of interest to see that the conditions for the evolutionary maintenance of canalized developmental morphs (one defended and one undefended) are very sensitive to patch frequencies and the magnitude of the trade-off between morphs, at least under the assumptions of the present model (fig. 14.3). Similar results under different assumptions were gained by Maynard Smith and Hoekstra (1980) in their genetic models of selection in spatially variable environments, and thus it seems that multiple niche polymorphism (Levene 1953; Levins 1963) may be expected to be rare in nature. These results have implications for one of the primary hypotheses for the maintenance of sexual reproduction, which relies on soft selection in spatially heterogeneous environments (i.e., the Tangled Bank Hypothesis; see Bell 1982). The idea behind the Tangled Bank hypothesis is that selection for genetic variation might favor the production of sexual offspring over asexual offspring even if there is a two-fold cost to cross-fertilization. The hypothesis, however, loses some of its force if the conditions for genetic polymorphism are narrow. In any case, the results of the present study suggest that conditional developmental strategies, such as induced defense, could increase when rare and fix in a genetically polymorphic population under a wide array of conditions (fig. 14.4).

The second major point is that induced defense is stable over a wide range of conditions, especially if the cost to plasticity is low and juveniles make the right choice regarding their future environment in at least one of the patches. These conditions are further increased if the cost of defense is intermediate, such that the fitness of defended individuals is about half that of undefended individuals. These results suggest that when observing multiple discrete morphs in nature, chances are very good that there is at least partial environmental determination of development. I also found that induced defense can be evolutionarily stable, even when the probability of making a mistake and developing as the "wrong" morph is reasonably high. This may explain one anomaly in the intertidal barnacles reviewed here. A large fraction of the individuals settle in the summer during a period of estivation by the predator that would otherwise attack them. This results in a decoupling of the high-risk patch with its cue. Nonetheless, the results suggest that induced defense can be stable to replacement by constitutive defense, even if the inducing cue is a poor predictor of the future.

The third point, and the qualification ("at least partial") in the previous paragraph, stems from the finding that mixtures of canalized and conditional development exist over part of the parameter space (see also Lively 1986c). This result might explain the situation for the barnacles as well as for other similar examples of "control polymorphisms" in table 14.2. In field experiments, I found that I could induce only about 45% of the barnacles, even though I "inundated" juvenile barnacles with *Acanthina*. The barnacles in the northern Gulf of California may therefore be a mixture of conditionally defended and constitutively undefended individuals. Alternatively, the mixture of individuals may result from gene flow from the southern Gulf of California, where the predatory snail is absent. In either case, the main point is that mixture of developmental control is a possibility in these barnacles. A plausible quantitative genetic basis to such a mixture is given by Hazel et al. (1990) and Hazel and Smock (1993).

TABLE 14.2
Examples of Developmental Conversion Where Less than 100% of the Individuals Tested Could Be Induced to Develop the Alternative Form

Water fleas:	30–79% developed neckteeth when exposed to predator (Krueger and Dodson 1981)
Rotifer:	13–55% developed spines when exposed to predator (Stemberger and Gilbert 1984)
Barnacle:	28–46% developed bent form when exposed to predator (Lively 1986a)
Salamander:	7% developed as cannibalistic morph at high density (Collins and Cheek 1983)
Nematode:	6–16% developed as males under high density (Clark 1978)
Bryozoans:	80% developed spines when exposed to predator (Harvell 1998)

Note: The remaining animals did not develop the alternative form, and may have canalized development.

Finally, the model can be used to consider the effects of gradual changes in the environment over time. For example, consider the effects of changing the frequency of the high-risk patch from close to zero to close to one in figure 14.4. Intuitively, we would expect an increase in the frequency of the defended, bent morph of the barnacle; and if we were examining the change in the fossil record, we might be inclined to declare saltational evolution of the bent form, because the shell shapes are so different (see Levinton 1988, p. 377). But, depending on the reliability of cues associated with the two patches, no such change is expected. An increase in the frequency of the high-risk patch can increase the occurrence of the defended morph in two important ways: (1) replacement of a population composed entirely of constitutively undefended individuals by a population composed entirely of constitutively defended individuals; or (2) a change in a conditional population from mostly undefended phenotypes to mostly defended phenotypes. Hence inferences regarding evolutionary change based on the replacement of one discrete form by another in the fossil record should be made cautiously (Palmer 1985). In addition, changes in the reliability of cues, without necessarily changing the frequency of patches, can also lead evolutionary change. For example, increasing the probability of making the wrong choice in the high-risk patch can lead to selection for canalization of the defended morph when the high-risk patch dominates the environment. Such selection would lead to genetic assimilation (in the sense of Waddington 1957) of the defended phenotype.

15

Evolution of Forager Responses to Inducible Defenses

FREDERICK R. ADLER AND DANIEL GRÜNBAUM

Abstract

Most theoretical and empirical investigations of inducible defenses have focused on identifying conditions that favor the evolution of inducibility by the prey species. These analyses outline the essential consequences of frequency-dependent benefits of deploying the defense, degrees of predictability of future predation, and existence of multiple predators. However, they say little about the ecological effects of inducible defenses on the foragers that create the prey's selective environment, or how long-term coevolutionary dynamics might shape foragers' strategies. Because of the tight coupling between forager and prey, such a forager-based perspective is essential both to understand the evolution of inducible defense systems and to assess the community-level effects of inducible defenses.

In this chapter, we develop a simplified theoretical framework with which we can examine jointly the forager and prey strategic perspectives. Our framework is centered on an environment subdivided into patches, in which defended and undefended prey phenotypes are periodically attacked by predators. Our principal focus is to look for evolutionarily stable strategies for predators moving into and out of patches, for emission of cues by predators while foraging, and for the degree of prey responsiveness to those cues. Our quantitative approach allows us to examine interactions between forager behavior and frequency distributions of defended and undefended phenotypes, and to distinguish some of the underlying evolutionary dynamics in inducible defense systems.

Introduction

Inducible defenses are phenotypic expressions of defensive traits that occur preferentially or exclusively in the presence of predators (Havel 1987; Adler and Harvell 1990; Harvell 1990a). These defenses take a wide variety of forms, including morphological changes, behavioral avoidance, changes in life history, and accumulation of allelochemicals. These selectively deployed inducible defenses give the prey a wider array of options than nonselectively deployed constitutive defenses; tactics that would be prohibitively costly if adopted continuously may be easily affordable if they are invoked only when needed. Inducible defenses provide a mechanism through which specialization can partially insulate prey from predation; this in turn provides predators with an opportunity to specialize in overcoming the prey defense. Inducible defenses thus have the potential to alter both the short-term dynamics and long-term evolution of predator-prey systems (Harvell 1990a).

Most theoretical and empirical investigations of inducible defenses have focused on identifying the conditions favoring the evolution of inducibility by the prey species, generally in comparison with constitutive defenses. Inducible defenses are thought to evolve under two general scenarios: "cost-benefit" and "moving target." The cost-benefit model requires unpredictable variation in predation risk, a fitness cost of employing the defense unnecessarily, and the availability of nonfatal predictive cues of future attack (Lively 1986; Edelstein-Keshet and Rauscher 1989; Harvell 1990a; Clark and Harvell 1992; Riessen 1992). The prey trade off the risk of deploying the defense too late or not at all when attacked against the cost of the defense in an unpredictable world with ambiguous cues.

The trade-off in the "moving target" scenario does not depend on a cost of defense, but instead on an array of predator types which respond differently to different prey phenotypes (Karban and Myers 1989; Adler and Karban 1994). In this case, prey that are attacked while deploying one type of defense are induced to switch to another defense. By switching defenses, prey attempt to find an effective deterrent to their particular attacker by trial and error. Furthermore, switching may make prey different from their neighbors, reducing their risk from predators specializing in the locally predominant defense phenotype.

These evolutionary scenarios focus on the strategy set available to the prey but say little about inducible defenses from a predator's point of view. How might a predator evolve a foraging strategy to best contend with the array of defensive strategies it confronts? Predator responses to inducibly defended prey can include physiological responses, such as detoxification or sequestration of defensive compounds; behavioral responses, such as foraging movement strategies (Lima and Dill 1990); and adjustments in life history characteristics. In many ways, these responses reflect the same types of strategies available to the prey. As responses to other organisms, inducible defenses always exist in an environment with the potential for coevolution, whether it be tightly coupled or diffuse (Futuyma and Slatkin 1983; S. Levin et al. 1990).

Several aspects of information transfer are essential to evolution in inducible defense systems (Harvell 1990a). At a minimum, prey must be able to recognize that an attack is occurring before succumbing. For example, in many aquatic systems, prey can detect waterborne chemical cues of their predators before an attack begins, and can prepare a specific response (Havel 1987). The consequences of delays in responding to such cues have been discussed by Clark and Harvell (1992) and Padilla and Adolph (1996). The question of whether prey should signal their level of defense has been recently studied with game-theoretic models (Augner 1994). However, in many examples of inducible defenses, additional traits such as the production of cues by predators are likely also under strong selection, and have dynamic, reciprocal influences on the selection for type and level of response by prey. Even with the spate of interest in signaling and signal selection (Guilford 1995), this aspect of inducible defenses has received little attention from empiricists or theorists.

In this chapter, we present a framework for simultaneously analyzing the evolutionary perspectives of the prey and the predator in an inducible defense system. We first discuss this framework in general terms, and then work out a simplified concrete example demonstrating the potential for coevolution of foraging, signaling, and signal response. We predict the strategies that might result from evolution by looking for evolutionarily stable strategies (ESSs) for the prey and predator species, both separately and simultaneously (Maynard Smith 1982). Finally, we look for cases in which there is no ESS, where evolution can maintain a diversity of predator and/or prey strategies (Ellner and Hairston 1994).

Our overall aim is to call attention to three aspects of inducible defense systems generally overlooked by empiricists and theorists: predator behavior, predator evolution, and the evolution of cues or signals. We develop a modeling framework to address these issues, and use it to demonstrate the possibility of counterintuitive and interesting evolutionary dynamics, even in a simple inducible defense scenario. The framework points toward several promising empirical angles for addressing these overlooked issues.

A Framework for Predator-Prey Dynamics with Inducible Defenses

In this section we present our framework for thinking about evolution in inducible defense systems. First, we summarize some of the factors affecting evolution of inducible defenses from the more familiar point of view of the prey (Harvell 1990a). Next, we summarize factors affecting responses of inducible defenses from the predator's point of view, considering both foraging strategy and signal production. We then show graphically how the various dynamical terms might affect populations of prey. This graphical treatment lays the groundwork for a more mathematical study of the evolution of prey foraging and signal behavior later in the chapter.

Inducible Defenses from the Prey's Perspective

Our view of the predator-prey population dynamics in an inducible defense system is summarized in figure 15.1a. This is clearly not a comprehensive view, because it omits many factors that may be important in particular examples of inducible defense systems in nature. In fact, a completely general treatment would likely be hopelessly complicated conceptually and intractable mathematically. However, this framework (combined with the following graphical and mathematical versions of the ecological dynamics) is sufficient to demonstrate how our approach can be used to model aspects of inducible defense evolution that would be difficult to understand without a quantitative analysis, and to suggest how it might be adapted to specific examples from nature.

In our model, the prey habitat is divided into a large number of discrete patches. Every patch is characterized by its densities of defended and undefended prey, and by whether or not it is under attack by a predator. (We could further discriminate between patches with different numbers of predators, but here we restrict our attention to patches that are small enough and predator densities that are low enough that we can neglect patches with more than one predator.) We refer to this set of characteristics describing a patch as the patch's "state."

What are the dynamical factors that change patch states? To begin with, these dynamics are dependent on whether a predator is present or absent (fig. 15.1a). Consider a patch at the instant a predator arrives. The patch changes in state from "recovery" to "attack" (i.e., from the right to the left box along the leftward-pointing arrow). While under attack, the predator is consuming the prey (maybe preferentially attacking the undefended phenotype). Simultaneously, the undefended prey phenotype is responding to cues of the predator's presence by deploying the defense. Both consumption and deployment reduce the quality of the patch for the predator by reducing the available pool of undefended prey. At some point, the predator abandons the patch (or dies, matures, or otherwise stops foraging).

When the patch is abandoned by the predator (moving back to the right box along the rightward-pointing arrow), prey in the patch no longer suffer predation, and the cue disappears. The prey population can then recover (there might be a reproductive advantage for the undefended phenotype). Defended prey may also revert over time to the undefended state. This recovery continues until the next predator attack, at which time the cycle starts again.

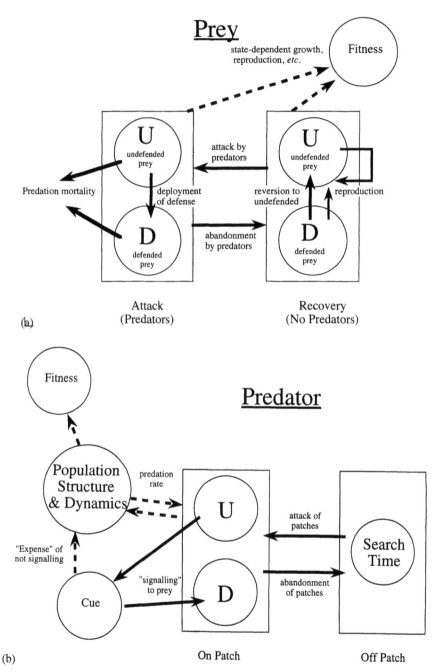

Fig. 15.1. Schematic diagrams of the differing evolutionary perspective of prey and predators with respect to inducible defenses.

The prey's strategic choice in this model is the rate at which they deploy their defense in response to predator cue (this includes the option of not deploying at all). The selective regime experienced by the prey when they use a particular strategy is determined by the frequency with which patches are found in various stages of the attack/recovery cycle, and the number of defended and undefended prey present at each stage. However, this frequency distribution of patch states depends not only on the prey's strategy, but also on the number of predators and the strategy they are employing. Thus, to estimate the fitness of a particular deployment strategy, we need to know the strategies, as well as the density, of the predators and of other prey in the patches, and how these translate into patch state distributions.

Calculating all these distributions and payoffs simultaneously is the technical hurdle that our mathematical analysis will address. By making quantitative assumptions about the demography and predator-prey dynamics, computing the resulting state distributions, and looking for a single response rate that, if adopted by most of the population, would exclude any alternative response rates, we will compute the evolutionarily favored (or ESS) prey strategy.

Inducible Defenses from the Predator's Perspective

A predator's experience of this inducible system is summarized in figure 15.1b. In the presence of a predator, a patch's quality decreases, both through depletion of prey and through conversion of prey into the unpalatable defended type. A predator attacking a patch keeps consuming prey until the patch's quality gets too poor, at which time it leaves. After the predator abandons the old patch, it must spend time searching for a new patch. Because the predator does not eat while searching, travel time is costly. For this reason, the predator should not abandon a patch before its quality is poor enough to justify paying that cost.

Just what does "poor enough" mean? The state or quality at which the predator should abandon the patch depends on how good it can expect the next patch to be. A succinct and quantitative way to express the patch quality at which the predator should depart is provided by the Marginal Value Theorem (Charnov 1976; Stephens and Krebs 1986). This theorem states that the optimal time to leave a patch occurs when the predator's rate of return on the patch is equal to the average rate of return it would get by leaving.

In addition to abandonment, there is another element to the predator's strategy: the cue. While a predator is in a patch, prey that have not yet been consumed can detect its presence, perhaps through odors or some other nonfatal, short-range signal. We assume in our model that the strength of the predator's signal is under selection. If the predator could reduce its signal without paying a cost, it would almost always be advantageous to do so. However, we expect that suppressing the cue is costly in most inducible defense systems. There could be a metabolic cost in not emitting an odor, or an efficiency cost of being visually inconspicuous. Alternatively, the signal used by the prey as a cue could have other important roles for the predators. For example, prey could cue on a predator's reproductive pheromone. In that case, the predator could not reduce or change its signal without directly reducing its own fitness. In this chapter, however, we focus on a simpler scenario, in which there is an energy cost to reducing the emission of cues.

The predator's choice of foraging and cue production strategies, and the prey's strategy of how to respond to the cue, jointly determine the structure of the prey state distribution. The prey state distribution then in turn determines the fitness accruing to the predator by using its particular strategy. Thus, as was the case with the prey, in order to estimate the fitness of a predator's strategy we need to know simultaneously the strategies and abundances both of the prey and of the other predators, and how they are distributed in patches of various states. And, again, our mathematical analysis will allow us to estimate the predator's fitness by calculating patch state distributions resulting from particular prey and predator strategies.

Describing the Evolutionary Dynamics

We have found phase plane diagrams to be a convenient way to conceptualize and graphically display these very complicated dynamics (fig. 15.2). Phase plane diagrams are *parametric* portraits of how a patch's state changes in time. In these plots, the horizontal axis is the population density of undefended prey, U, in a patch, and the vertical axis is the population density of defended prey, D. The vectors represent rates of increase or decrease in the defended and undefended prey densities. For example, a vector pointing toward the upper right in these plots means that both defended and undefended prey densities are increasing. In contrast, a vector pointing to the upper left means that undefended prey are decreasing, while defended prey are increasing. The trajectory of a patch's state over time is found by locating the patch's state on the plot and then moving it in the direction indicated by the nearby vectors. The size of the vector is proportional to the magnitude of the increase or decrease—large vectors represent rapid changes, and small vectors represent slower changes. Thus, phase plane diagrams are an intuitive way to describe quantitative dynamics without resorting to complicated mathematical expressions.

Although phase plane diagrams are unfamiliar to most biologists, they have two important features that make them potentially very useful. First,

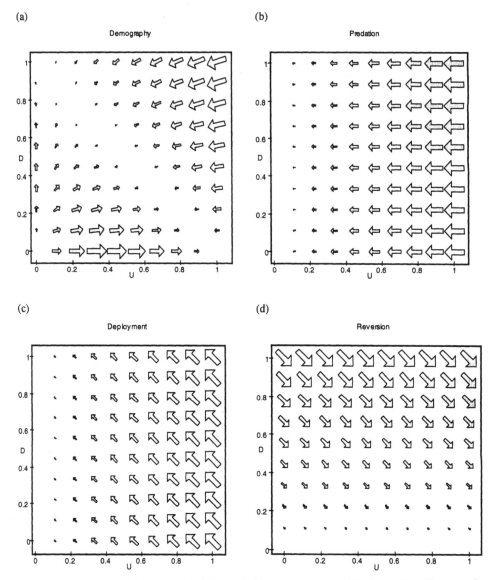

Fig. 15.2. Predator-prey dynamics and defense deployment represented in phase plane diagrams. In these plots, a given position represents the state of a patch, i.e., the density of undefended prey (U-axis) and defended prey (D-axis). The vectors represent rates at which patches change in state, due to (a) demography (growth and mortality in the absence of predators) of undefended and defended prey; (b) predation of undefended and defended prey; (c) deployment of the defense when under attack; and (d) reversion to the undefended condition when predators are absent.

they show at a glance how patches will change over time, for the entire range of possible initial states, in a format that is much clearer than equations would be. Second, in a complicated scenario such as our inducible defense model, in which many different factors affect patches, it is often more intuitive first to consider the factors in isolation. Then, the phase plot of the complete dynamics is found simply by adding the phase plots for the component factors, vector by vector. An example might be density-dependent demographics with a carrying capacity, coupled with a predation rate that saturates at high prey densities. Each of these dynamics is easier to understand when presented in separate diagrams, and the combined dynamics are thought of as the sum of the component dynamics.

We now give sample phase plane diagrams representing the various dynamical factors affecting patch states in our model.

Demography. Figure 15.2a gives an example of prey demography (in the absence of predators). Reproduction of prey tends to increase the density of prey (move the state toward the right and top). Mortality (e.g., from overcrowding) tends to decrease prey density (move the state toward the left and bottom). In this example, we have assumed a unit carrying capacity, so that demographic rates are zero whenever $U + D = 1$. This appears in the plot as zone of small or zero vectors running diagonally from the upper left to the lower right. Below this zone, densities are increasing; above it, densities are decreasing.

In the presence of a cost of defense, the undefended phenotype may have a reproductive advantage over the defended phenotype. In figure 15.2a, we assume that demographic rates are much more rapid for undefended than for defended prey. This is reflected in the plot by the vectors having a larger horizontal than vertical component, except when there are very many defended and very few undefended prey.

Predation. Figure 15.2b shows how predation moves the patch to lower densities (back toward the origin). If there is some advantage to being defended, predation on defended prey will be lower than on undefended prey. In this plot, we have assumed that there is no predation at all on defended prey. Therefore, the arrows point straight to the left. There is also some functional response on the part of the predator-to-prey density. We have assumed here that the predator consumes undefended prey strictly in proportion to their numbers.

Deployment of defense and reversion to undefended state. The transition of undefended to defended phenotypes and back does not create or destroy any prey; it is simply an exchange of one type for the other. Thus, the vectors representing these processes always point along a diagonal. However, the direction of these vectors (i.e., whether prey are deploying or reverting to undefended defenses) depends on whether a predator is present or not. Figure 15.2c shows how the population switches from undefended to defended when a predator is present in the patch. Figure 15.2d shows how the population reverts from defended to undefended in the absence of a predator.

Patch state distributions. Typically, a patch under attack will move along the deployment vectors (fig. 15.2c) together with the growth and mortality vectors (fig. 15.2a) and the predation vectors (fig. 15.2b), decreasing in quality until the predator is compelled to abandon it. When the predator is gone, the patch follows the growth and mortality arrows (fig. 15.2a) and reversion arrows (fig. 15.2d) until it is once again attacked.

These dynamics provide the information needed to compute optimal predator foraging strategies and ESS predator signal levels. To carry out this computation, we need to work with the mathematical formulas corresponding to the phase plane diagrams in figure 15.2, which we do in the next section. However, we emphasize again that the graphical and mathematical representations of our inducible defense system specify the same information in the same quantitative detail.

Models

In this section, we formalize mathematically the conceptual and graphical models presented in figures 15.1 and 15.2, and outline how we use this quantitative framework to compute ESS levels of predator cue production and patch departure and prey response strategies. The analysis proceeds in four basic steps:

1. Solve the population dynamic equations for a single prey type faced by a single type of predator.

2. Derive the fitness of predators that use an optimal departure time strategy computed with optimal foraging theory.

3. Find the ESS predator signal level as a function of various parameters.

4. Use the prey population dynamic equations to compute the prey's ESS signal response.

We make several simplifying assumptions to make the calculations tractable—we will describe these as they come up in our analysis. Additional mathematical details of the model's derivation can be found in the appendix at the end of the chapter. To make these details easier to follow, we number our equations continuously through the text and appendix, in the order that they appear in our derivation. We present the results of our model in the next section.

The Dynamical Equations for the Prey

In our model, prey can be in two discrete states, defended and undefended, and in two discrete environments, with and without a predator (fig. 15.1). Our basic equations track the population dynamics of the two types of prey in each environment (fig. 15.2). Let U and D represent the number of undefended and defended prey when a predator is present. The dynamics of these prey follow

$$\frac{dU}{dt} = -\begin{pmatrix}\text{rate} \\ \text{eaten}\end{pmatrix} \pm \begin{pmatrix}\text{other births} \\ \text{and deaths}\end{pmatrix} - \begin{pmatrix}\text{rate of} \\ \text{deployment}\end{pmatrix}$$

$$\frac{dD}{dt} = -\begin{pmatrix}\text{rate} \\ \text{eaten}\end{pmatrix} \pm \begin{pmatrix}\text{other births} \\ \text{and deaths}\end{pmatrix} - \begin{pmatrix}\text{rate of} \\ \text{deployment}\end{pmatrix}. \quad (15.1)$$

Similarly, let $\tilde{U}$ and $\tilde{D}$ represent the number of undefended and defended prey when no predator is present. In our model, these dynamics follow

$$\frac{d\tilde{U}}{dt} = \begin{pmatrix}\text{rate of switching} \\ \text{to undefended}\end{pmatrix} \pm \begin{pmatrix}\text{births and} \\ \text{deaths}\end{pmatrix}$$

$$\frac{d\tilde{D}}{dt} = -\begin{pmatrix}\text{rate of switching} \\ \text{to undefended}\end{pmatrix} \pm \begin{pmatrix}\text{births and} \\ \text{deaths}\end{pmatrix}. \quad (15.2)$$

Here we have assumed that prey switch to the defended state only when under attack, and revert to undefended only between attacks. We also will assume in the calculations to come that all births are into the undefended phenotype. Finally, in our calculations, we will assume that only one predator can occupy a patch at any given time, and that predators remain in patches for a relatively short time. These latter assumptions mean that few births and deaths occur while predators are present (eq. 15.1), relative to the large changes caused by consumption and deployment.

During predation, the various components of the model are parametrized as follows (see tables 15.1 and 15.2 for a list of variables and parameters):

$$\text{Undefended prey eaten} = c_u \lambda U$$
$$\text{Defended prey eaten} = c_d \lambda D$$
$$\text{Rate of deploying defense} = \epsilon \sigma U$$
$$\text{Births and deaths} = (\beta_u U + \beta_d D) g(U + D).$$

TABLE 15.1
The Prey Variables and Parameters

Symbol	Meaning
U	Undefended prey population in the presence of predators
D	Defended prey population in the presence of predators
$\tilde{U}$	Undefended prey population in the absence of predators
$\tilde{D}$	Defended prey population in the absence of predators
ϵ	Prey signal response
σ	Predator signal level
c_u	Eatability of undefended prey
c_d	Eatability of defended prey
λ	Encounter rate with prey
β_u	Reproduction of undefended prey
β_d	Reproduction rate of defended prey
ρ	Rate of switching to undefended

TABLE 15.2
The Predator Variables and Parameters

Symbol	Meaning
e_u	Food value of undefended prey
e_d	Food value of defended prey (assumed to be 0)
T	Average travel time between patches
z^*	Optimal number of undefended prey upon departure
θ	Optimal fraction of defended prey upon departure
D^*	Number of defended prey upon departure
U^*	Number of undefended prey upon arrival
t_m	Average time between predator visits to a patch
n_p	Number of predators per patch
$k(\sigma)$	Cost of signal level σ (decreasing)
k_0	Marginal cost of reducing signal level σ

(Again, we assume this latter term is negligible compared to the previous ones over the short time the predator remains on the patch.) Between predation events, the components are:

$$\text{Rate of switching to undefended} = \rho \tilde{D}$$
$$\text{Births and deaths} = (\beta_u \tilde{U} + \beta_d \tilde{D}) g(\tilde{U} + \tilde{D}) .$$

The function g describes the density-dependent demographics in the population. In the cases to be analyzed in detail, we make the additional simplifying assumption that defended prey are completely immune to capture ($c_d = 0$).

The dynamics during predation then follow

$$\frac{dU}{dt} = -(\epsilon\sigma + c_u \lambda) U ,$$

$$\frac{dD}{dt} = \epsilon \sigma U , \qquad (15.3)$$

while the dynamics without predation are given by

$$\frac{d\tilde{U}}{dt} = (\beta_u \tilde{U} + \beta_d \tilde{D}) g(\tilde{U} + \tilde{D}) + \rho \tilde{D}$$

$$\frac{d\tilde{D}}{dt} = -\rho \tilde{D} . \qquad (15.4)$$

Fitness and the Behavior of Predators

The fraction of time the prey spend fending off predators depends on the number of predators and their strategy. In particular, the length of a predator visit and the time between visits determines the experience of the prey. We use the Marginal Value Theorem (Charnov 1976; Stephens and Krebs 1986) to derive the optimal time for predators to leave a patch to maximize their fitness (see the appendix for additional details).

Predators should leave a patch when the instantaneous rate of return is equal to the average rate obtained, or

$$\text{instantaneous intake rate} = \frac{\text{average intake during visit}}{\text{average visit time} + \text{travel time}} \qquad (15.5)$$

In our case,

$$\text{instantaneous intake rate} = \lambda\, e_u\, U(t),$$

where e_u represents the food obtained from eating a single undefended prey item (recall that we have assumed no consumption of defended prey).

Let z^* represent the undefended prey population present at the optimal departure time. That is, z^* is the patch quality at which the predator should abandon patches in order to maximize its payoff if all other predators have the same strategy. We show how to calculate z^* in the appendix. A predator following this optimal departure strategy consumes food at the rate;

$$\text{optimal average intake rate} = \lambda\, e_u\, z^*.$$

This equation involves three components: the search efficiency or consumption rate of the predator λ, the food quality e_u, and the quality of the patch z^*. The last component, z^*, serves as a statistic that summarizes the patch state distribution. The procedure in the appendix determines the value of z^* that maximizes payoff of a strategy in response, if all predators use the same strategy. However, this strategy is also the evolutionarily stable strategy (ESS) because an invader that differed would leave a different fraction of the original undefended prey when it departs a patch (θ in the appendix). Such a strategy would fail to satisfy the Marginal Value Theorem condition, (15.5), and so would necessarily be inferior to the strategy of leaving at patch quality z^*.

Finding the ESS Predator Signaling Level

If it were possible, predators would reduce σ, the amount of signal produced. We envision three reasons why this might not occur or might not appear to occur. First, the predators might indeed always be reducing the amount of particular cues, but prey are simultaneously learning to recognize new cues, much as in some epidemiological situations where a virus constantly alters antigens (Pease 1987). Second, the prey may capitalize on a cue that predators cannot modify for other reasons, such as a mating pheromone. Finally, reducing cue production might have a direct energetic or foraging efficiency cost for the predators. Because we are unaware of data on this subject, we make the simple assumption that reducing signal level has an energetic cost. In particular, we assume that the payoff is modified as

$$payoff = \lambda\, e_u\, z^* - k(\sigma).$$

$k(\sigma)$ is a decreasing function, indicating it is *more costly* to produce a *lower* signal of level σ. This equation summarizes the trade-off for predators. Those with lower levels of signal can maintain a higher value of patch quality z^* by not inducing the prey defense, but must pay a cost to suppress cue production.

Our mathematical results (presented in the appendix) show that if $k''(\sigma) \leq 0$ (the cost function curves down), there is never an ESS at an intermediate level of cue. The ESS must then be at $\sigma = 0$ or at the maximum possible value of σ, i.e., the best strategy is either "all" or "none" cue emission. On the other hand, when the cost function is curved up, the ESS is an intermediate level of signaling in the cases we have examined. We used the costs of the form $k(\sigma) = e^{-k_0 \sigma}$ in all our calculations. The parameter k_0 describes how quickly the cost of suppressing cue decreases when a small amount of cue is produced, and can be thought of as the marginal cost of suppressing the last bit of cue. We give the formula for the ESS predator signaling level in the appendix.

Finding the ESS Prey Signal Response

Prey, too, face a trade-off. Those that defend rapidly (large ϵ) are protected from predators but may suffer lowered reproduction in their absence. Furthermore, this trade-off is frequency dependent: a slowly defending prey type (small ϵ) could be indirectly helped by rapidly defending types that induce a predator to leave. A rapidly defending prey type could be indirectly helped by slowly defending types that are severely reduced by predation and open up space for growth.

To determine the ESS signal response level ϵ, we must determine the fitness function for prey. Because the prey equations explicitly include population dynamics, we can compute fitness of an invader as its growth rate over a complete cycle of predation and recovery. In the cases we have looked at, the ESS response is always at $\epsilon = 0$ or $\epsilon = \epsilon_{max}$. That is, the prey should either deploy as fast as possible or ignore the cue completely, depending on the circumstances. This conclusion—that delays in deployment are costly— was arrived at by Clark and Harvell (1992) and by Padilla and Adolph (1996) in their studies of prey response strategies. Numerically, we find that the ESS is ϵ_{max} when $\sigma > 0$, i.e., whenever there is a cue it pays to respond to it as fast as possible. When $\sigma = 0$, the prey are indifferent to ϵ, because there is no cost in our model to responding to a cue that does not exist.

Results

We begin our discussion of model results by varying several parameters and plotting how the ESS signal level strategy of the predators (σ) responds. We then comment briefly on the implications of our model for prey responses.

ESS Predator Signal Levels

Figure 15.3a shows how signal level responds to the food quality of undefended prey, e_u. The predator's fitness increases roughly in proportion to the food value of the prey. The signaling level decreases, however, as the difference between the food values of undefended and defended prey increases (in our model, defended prey have zero food value) and the predators have more to gain by keeping the prey undefended.

Figure 15.3b shows how signal level responds to ϵ, the level of prey response to the predators' signal. As the prey become more and more adept at deploying the defense in response to the cue, predators are forced to signal less and less. The cost associated with lowered signaling drives the predators' fitness down. When ϵ becomes sufficiently large, predators do best by reducing the signal to zero.

Figure 15.3c shows how signal level responds to k_0, the marginal cost of reducing signal level. If k_0 is small, only predators that produce a huge amount of signal receive any energetic benefit. When k_0 is large, even a small increase in signal level above zero generates a dramatic reduction in energetic cost. The predator shows two types of ESS strategy in distinct ranges of k_0. For $k_0 < 1$, the marginal cost of signaling is small and predators do not signal at all. For $1 < k_0 < 2$, the ESS signal level increases rapidly; then, for even larger values of k_0, the ESS signal level decreases (because the cost function saturates). The increase in fitness with larger k_0 results from the fact that predators can avoid cue suppression costs at even a small level of the cue.

Figure 15.3d shows how signal level responds to λ, the search efficiency or attack rate of the predators. At very low λ, the predators have low energy intake, and the cost of signaling looms large. Therefore, predators are unable to pay the cost of suppressing signal, signal level is high, and predator fitness is low. With increasing attack rate, consumption of prey and fitness increase. At the same time, the relative cost of traveling to a new patch increases. It therefore pays for the predator to maintain patch quality by reducing signaling. Eventually, however, consumption rates become so high that undefended prey are consumed before they have time to deploy their defenses, so there is no profit in not signaling. Because all the other predators have realized this too, the world is full of heavily defended prey, and predator fitness decreases.

These same phenomena are apparent in figure 15.3e, which shows how signal level responds to n_p, the number of predators. Higher numbers of predators means that the world is populated by more heavily defended prey in lower-quality patches. This pushes predators to remain in patches longer, and to drive the quality in their own patches lower. Low signal levels can be

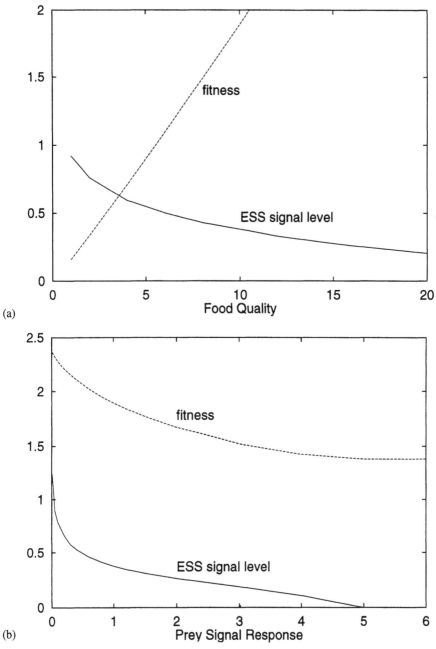

Fig. 15.3 continued on the next page

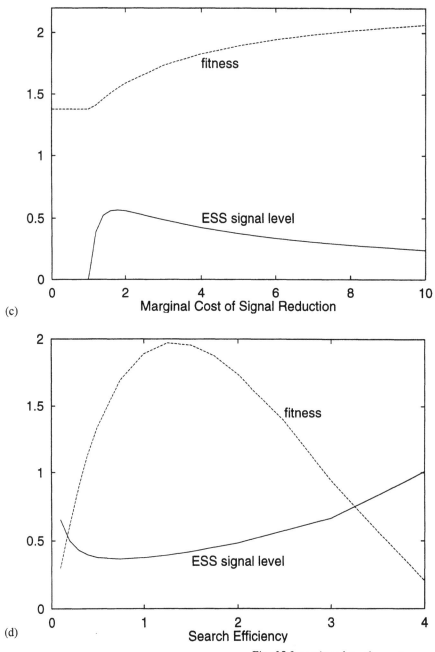

Fig. 15.3 continued on the next page

FORAGER RESPONSE TO INDUCIBLE DEFENSES 277

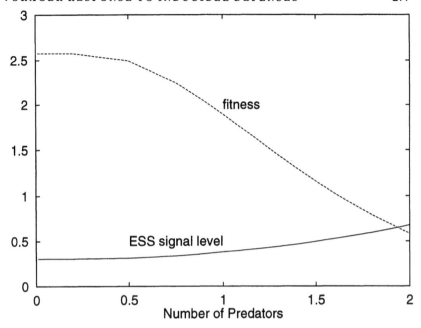

Fig. 15.3. (a) *ESS* predator signal level σ and resulting fitness as a function of the quality of food e_u. (b) *ESS* predator signal level σ and resulting fitness as a function of the prey signal response ϵ. (c) *ESS* predator signal level σ and resulting fitness as a function of the marginal cost of signal reduction k_0. (d) *ESS* predator signal level σ and resulting fitness as a function of the foraging efficiency or attack rate λ. (e) *ESS* predator signal level σ and resulting fitness as a function of the number of foragers n_p.

thought of in this context as a form of cooperation. Reducing your own signal is costly and will only benefit other individuals. Thus, overcrowding by predators may not only deplete patches, but also reduce the level of cooperative foraging. This sort of depressing feedback has been observed in fisheries (Clark 1990).

The Lack of a Joint ESS

We noted earlier that the best prey response to any positive signal level is a large value of ϵ. Therefore, according to figure 15.3b, the predators do best by reducing their signal to zero. In this circumstance, the fitness of the prey is then independent of ϵ. This suggests an interesting possibility for the long-term evolutionary dynamics: if there is any cost to maintaining the capacity to respond to predator cues, we might expect the prey's response to be rapidly lost when predators emit no cues. Even without a cost, we might expect the ability to respond to be lost by genetic drift. In the absence of prey response, the predators might then reduce their cost by resuming cue production, thus leading to another cycle of coevolution. However, these sorts of coevolutionary dynamics can be very subtle (Levin et al. 1990). The results depend on the relative timescales of the evolutionary responses by the predators and the prey, and could be affected in interesting ways by any spatial structure creating asynchrony among the populations. Of course, the cycles of coevolution suggested by our model are highly speculative, and it remains to be seen whether a more detailed analysis of a particular animal system would display these sorts of dynamics. Nonetheless, we find it surprising that such a range of complex dynamics can easily emerge from a relatively simplified treatment of inducible defenses such as ours.

Discussion

We have developed graphical and mathematical models to pose questions about three insufficiently studied aspects of inducible defense systems: predator behavior, predator evolution, and the evolution of cues or signals. Our framework includes the elements necessary to address these issues in a rudimentary way. We hope it can be the basis of more detailed inquiries about specific examples of inducible defenses.

Preliminary analyses of our simplified modeling framework have emphasized the importance of three elements of inducible defenses: (1) how predators respond behaviorally to inducible defenses; (2) how predator signal levels respond evolutionarily to inducible defenses; and (3) how prey signal responses respond evolutionarily to predator strategies. More particularly, our model suggests that predators should be more stealthy when stalking superior prey (fig. 15.3a) or more vigilant prey (fig. 15.3b). The response of predators to their cost function can be subtle (fig. 15.3c), and the details of optimal behavior seem to depend sensitively on the shape of the function. More interestingly, perhaps, we predict that predators do best with an intermediate level of search efficiency (fig. 15.3d), but note that this cannot be thought of as an evolutionarily stable strategy because an invader with superior foraging ability will gain a greater individual payoff. However, if predators are constrained to an intermediate search efficiency, a lower signal level is favored. Finally, greater crowding by predators can lead to a lack of cooperation in that higher ESS signal levels are more successful (fig. 15.3e). With our assumptions, however, we find no joint predator and prey ESS for signal production and response.

Of course, these results are dependent on the many simplifying assumptions we have made. Our analysis of cues has left out the very important question of unreliable cues or mistakes in identifying cues (N. Moran 1992). We suspect that the maximum cue response favored in our preliminary analysis may be an artifact of this omission. In addition, we ignored any temporal or spatial dynamics of the cue, assuming that a patch is instantly and uniformly saturated when a predator arrives. The time it takes cues to build up or decay is one factor that makes cues less reliable. It would be worthwhile to analyze how different cue diffusion and decay rate constants affect the results.

Our prey have also been simplified in many ways. Their strategic options are restricted to the speed of response to cue, but could realistically include different levels or types of defense. Our models assumed that the defense has the effect of reducing food quality, but it might be more realistic to assume an increase in handling time. The prey in our model are identical, although differences in size or apparency could create important opportunities for predators. The same differences could be important for entire patches. The selective forces acting on predators and prey could be altered by variability in patch size, patch location, or patch apparency.

The predators in this model interact only through their prey, but direct interactions between predators and the potential for defense of patches against other predators can have strong effects on the evolution of signaling. Even without such interactions, such factors as size structure within the predators could produce different abilities to handle defended prey or different interests (larger predators might be more interested in mating than in eating). We can only begin to speculate on the effects of multiple predator species (S. Levin et al. 1977).

Nonetheless, the models point the way toward several measurements that could clarify our thinking about predators and signaling. First, it would be interesting to test, by manipulating both the density and defense level of prey that they encounter, whether predators actually use optimal foraging in response to inducible defenses. Second, a basic assumption of the model is that reducing signal levels is costly. As with defenses themselves, testing for costs of reducing signaling can be difficult (Baldwin 1996), but it might be possible in those systems where the cue can be identified. In those same systems, it would be valuable to check for natural variability in signal level by predators or signal response level by prey. More speculatively, it would interesting to measure several variables across patches in the same system: defense level, degree of damage, cue response level, and cue production level. Additional modeling could seek robust predictions about the relation among these measurements as a function of ecological conditions.

Further consideration of these models might cast light on some larger issues in predator-prey interactions. Our model assumes a fairly tight relationship between predator and prey. Does a signal-mediated interaction between predator and prey favor specialists? Thought of as a signal detection game, what implications do our models have for the evolution of inducibility? Can these systems break down through the evolution of completely unreliable cues?

We hope that our model provides a useful tool for biologists trying to gain an intuitive understanding of how various parameters shape the evolution of predator and prey strategies in inducible defense systems. Intuition rather than quantitative prediction is the intent behind abstract models like ours. However, even though our results are not meant to be quantitatively accurate, we could not have approached this problem without a quantitative model because there are too many hidden and intertwined dynamics to guess at the solutions. Furthermore, the model is formulated with independently measurable parameters and could be made quantitatively accurate if those parameters were measured. In practice, this might prove difficult. Nonetheless, our framework outlines the types of information needed to make quantitative evolutionary arguments about these systems.

Our basic message is that we cannot assess the potential benefits to prey of deploying inducible defenses under an assumption of constant predator behavior. For the most part, studies of inducible defenses have focused on the prey's strategy set, with the implicit assumption that the predator has no strategic alternatives, or that the predator strategy set changes little or not at all with differences in prey type. Our simple model demonstrates that this point cannot be taken for granted. Coevolution of predator responses to prey defensive strategies can quantitatively and qualitatively change evolutionary outcomes in inducible defense systems. The additional evolutionary perspective of the predator certainly complicates the formulation of models and vastly increases the effort needed to experimentally characterize an inducible predator-prey system. However, the interactive dynamics make these systems yet more interesting and compelling as model ecological and evolutionary systems.

Appendix: Model Derivations

This appendix contains additional mathematical details that may be unnecessary for the average biological reader, but are necessary for the more mathematical reader to follow our analysis. In the interest of brevity, our format is fairly terse and is intended to accompany the verbal descriptions in the corresponding sections in the text.

The Dynamical Equations for the Prey

The only nonlinear equation is that for $\tilde{U}$. With initial conditions (U_0, D_0), the system of equations with predator present (15.3) has the solution

$$U(t) = U_0 e^{-(\epsilon\sigma + c_u\lambda)t} \tag{15.6}$$

FORAGER RESPONSE TO INDUCIBLE DEFENSES

$$D(t) = D_0 + \frac{\epsilon \sigma}{\epsilon \sigma + c_u \lambda} U_0 \left(1 - e^{-(\epsilon \sigma + c_u \lambda)t}\right). \tag{15.7}$$

With initial conditions (z,D), the predator-absent equations have solutions

$$\tilde{U}(t) = H(z,D,t) \tag{15.8}$$

$$\tilde{D}(t) = D e^{-\rho t} \tag{15.9}$$

where H is an as yet unknown nonlinear function.

Fitness and the Behavior of the Predator

Let $\phi(U)$ represent the probability density function giving the number of undefended prey when a predator encounters a patch, and $t_u(U)$ be the time it takes for that patch to be depleted to the optimal departure level z^*. Then (15.5) translates to

$$\lambda e_u z^* = \frac{\int_{z^*}^{\infty} \int_0^{t_u(u)} \lambda e_u U(t) dt \phi(u) du}{\int_{z^*}^{\infty} t_u(u) \phi(u) du + T} \tag{15.10}$$

where $U(t)$ represents the population of undefended prey starting from initial condition $U(0)$. Substituting in the solution for $U(t)$, (15.6), and simplifying, we find an implicit equation for z^*,

$$(\epsilon \sigma + c\lambda) T = \int_{z^*}^{\infty} \left(\frac{u}{z^*} - 1 - \log\left(\frac{u}{z^*}\right)\right) \phi(u) du. \tag{15.11}$$

How do we find $\phi(u)$, the probability density function of U, when predators enter patches? Even if all patches are abandoned with $U = z^*$, they will differ when next visited depending on the amount of time they have had to recover. In order to simplify the equations, we approximate this distribution of recovery times by the average recovery time t_m. Patches will then be encountered with the same (still unknown) number of undefended prey. Denote this number by U^*. The equation for z^*, (15.11), can be rewritten

$$(\epsilon\sigma + c\lambda) T = \frac{U^*}{z^*} - 1 - \log\left(\frac{U^*}{z^*}\right).$$

This expression involves only the ratio $\frac{U^*}{z^*}$, the fraction of the original undefended prey left when a forager departs, which we define as θ. The fundamental equation, (15.11), for z^* can be rewritten as

$$(\epsilon\sigma + c\lambda) T = \frac{1}{\theta} - 1 - \log\left(\frac{1}{\theta}\right). \tag{15.12}$$

This equation can be solved numerically for θ, the optimal fraction of undefended prey to leave. We use the equation for $U(t)$, (15.6), to find how long it takes to reach this level of depletion, solving for the optimal time in the patch t_u as

$$t_u = \frac{\log(\theta)}{(\epsilon\sigma + c\lambda)} \tag{15.13}$$

because patch quality declines exponentially during a visit.

To find the number of undefended prey upon arrival, U^*, we need to find the average time between visits, t_m. The average cycle from starting one visit to starting the next is $t_u + T$ for each predator, so the average time between arrivals at a patch is

$$t_m = \frac{t_u + T}{n_p}, \tag{15.14}$$

where n_p represents the number of predators per patch. As n_p becomes large, the time between visits becomes small. In a more complete analysis, t_m would have an exponential distribution with mean t_m. Preliminary simulations indicate that including this factor does not qualitatively change the results.

We can now use the population dynamics of the prey to write down the equations for z^*. The trick is to follow the dynamics through a full sequence of recovery and predation, requiring that patches end up exactly where they started. If predators leave patches when $U = z^*$ and $D = D^*$, then

$$\begin{pmatrix} z^* \\ D^* \end{pmatrix} \xrightarrow{\text{recovery}} \begin{pmatrix} H(z^*, D^*, t_m) \\ D^* e^{-\rho t_m} \end{pmatrix}$$

$$\xrightarrow{\text{predation}} \left(D^* e^{-\rho t_m} + \frac{\epsilon\sigma}{\epsilon\sigma + c\lambda} H(z^*, D^*, t_m)(1 - \theta) \right). \tag{15.15}$$

FORAGER RESPONSE TO INDUCIBLE DEFENSES

Recall that $H(z^*, D^*, t)$ gives the solution of the nonlinear recovery equations as a function of their initial conditions. Eq. (15.15) gives a pair of simultaneous equations that can be solved numerically for z^* and D^*.

Following the above procedure chooses the value of z^* that maximizes payoff of a strategy in response to itself. This strategy is evolutionarily stable because an invader that differed would have a different θ, and thus necessarily be inferior.

Finding the ESS Predator Signaling Level

Suppose a predator with signal level $\hat{\sigma}$ invades a population of predators with signal level σ. This invader can compute $\hat{\theta}$, the optimal fraction of prey to leave, exactly as above. This value will differ from θ (the fraction left by the resident type) because θ depends on the signal level σ, (15.12). If most predators use signal level σ, an invader will encounter patches with $U^* = \dfrac{z^*}{\theta}$ undefended prey, and will leave when $\dfrac{\hat{\theta} z^*}{\theta}$ undefended prey remain. Because $\hat{\theta}$ defines the optimal response, the intake rate at departure matches the long-term average. Therefore, the fitness of a type $\hat{\sigma}$ predator invading type σ is

$$W(\hat{\sigma}, \sigma) = \lambda\, e_u\, \hat{\theta}\, \frac{z^*}{\theta} - k(\hat{\sigma}). \tag{15.16}$$

The evolutionarily stable strategy, or ESS, can occur when

$$\frac{\partial W(\hat{\sigma}, \sigma)}{\partial \hat{\sigma}} = 0 \text{ at } \sigma = \hat{\sigma} > 0$$

or

$$\frac{\partial W(\hat{\sigma}, \sigma)}{\partial \hat{\sigma}} \leq 0 \text{ at } \hat{\sigma} = \sigma = 0.$$

We can compute the derivative from the implicit formula for θ, (15.12), finding that

$$\frac{\partial W(\hat{\sigma}, \sigma)}{\partial \hat{\sigma}} = \lambda\, e_u\, \frac{\partial \hat{\theta}}{\partial \hat{\sigma}} \frac{z^*}{\theta} - k'(\sigma)$$

$$= \lambda\, e_u\, \frac{\epsilon T \hat{\theta}^2}{\hat{\sigma} - 1} \frac{z^*}{\theta} - k'(\sigma).$$

Solving, a necessary condition for a positive ESS is

$$\frac{\lambda e_u \, z^* \, \epsilon T \hat{\sigma}}{\hat{\sigma} - 1} = k'(\sigma) . \tag{15.17}$$

The second derivative of W determines whether this critical point defines an ESS.

Finding the ESS Prey Signal Response

During predation, the invading prey follow

$$\frac{d\hat{U}}{dt} = -(\hat{\epsilon}\sigma + c\lambda) \hat{U}$$

$$\frac{d\hat{D}}{dt} = \hat{\epsilon}\sigma \, \hat{U}, \tag{15.18}$$

where $\hat{U}$ and $\hat{D}$ represent the populations of undefended and defended prey of type $\hat{e}$. These dynamics are coupled to those of the resident type only through the departure time t_u as determined by the predator response to the resident type, (15.13). With initial conditions $(\hat{U}^*, \hat{D}^*)$, the solution at time t_u can be written

$$\hat{U}(t_u) = \hat{U}^* \, e^{-(\hat{\epsilon}\sigma + c\lambda)t_u} = \theta \, \hat{U}^* \tag{15.19}$$

$$\hat{D}(t_u) = \hat{D}^* + \frac{\hat{\epsilon}\sigma}{(\hat{\epsilon}\sigma + c\lambda)} (1 - \theta^p) \, \hat{U}^*, \tag{15.20}$$

where

$$p = \frac{(\hat{\epsilon}\sigma + c\lambda)}{(\epsilon\sigma + c\lambda)} .$$

Between predation events, invading prey obey the equations

$$\frac{\partial \hat{\tilde{U}}}{\partial t} = (\beta_u \hat{\tilde{U}} + \beta_d \hat{\tilde{D}}) g(\tilde{U} + \tilde{D}) + \rho \hat{\tilde{D}}$$

$$\frac{\partial \hat{\tilde{D}}}{\partial t} = -\rho \hat{\tilde{D}}, \tag{15.21}$$

where $\hat{\tilde{U}}$ and $\hat{\tilde{D}}$ represent populations of undefended and defended prey.

These differential equations are linear in $\hat{U}$ and $\hat{D}$, with nonautonomous terms depending on the solutions $\tilde{U}$ and $\tilde{D}$ for the resident type. The solution has the form

$$\hat{U}(t_m) = H(\hat{z}^*, \hat{D}^*, t_m) = a\,\hat{z}^* + b\,\hat{D}^* \tag{15.22}$$

$$\hat{D}(t_m) = \hat{D}^* e^{-\rho t_m}, \tag{15.23}$$

with the constants a and b to be determined later. A full cycle of the dynamics follows the matrix equation

$$\begin{pmatrix} \hat{z}^* \\ \hat{D}^* \end{pmatrix} \xrightarrow{\text{recovery}} \begin{pmatrix} \hat{H}^*(\hat{z}^*, \hat{D}^*, t_m) \\ \hat{D}^* e^{-\rho t_m} \end{pmatrix}$$

$$\xrightarrow{\text{predation}} \begin{pmatrix} \hat{D}^* e^{-\rho t_m} + \dfrac{\hat{\epsilon}^*\sigma}{\hat{\epsilon}^*\sigma + c\lambda}\, \hat{H}^*(\hat{z}^*, \hat{D}^*, t_m)(1 - \theta^p) \\ \theta^p\, \hat{H}^*(\hat{z}^*, \hat{D}^*, t_m) \end{pmatrix} \tag{15.24}$$

$$= \begin{pmatrix} a\dfrac{\hat{\epsilon}^*\sigma}{\hat{\epsilon}^*\sigma + c\lambda}\theta^p & b\dfrac{\hat{\epsilon}^*\sigma}{\hat{\epsilon}^*\sigma + c\lambda}(1-\theta^p) + e^{-\rho t_m} \\ a\theta^p & b\theta^p \end{pmatrix} \begin{pmatrix} \hat{z}^* \\ \hat{D}^* \end{pmatrix}.$$

The invader will invade if the leading eigenvalue $\lambda(\hat{\epsilon}, \epsilon)$ of the matrix exceeds 1. The prey ESS signal response level occurs at a point where

$$\frac{\partial \lambda(\hat{\epsilon}, \epsilon)}{\partial \hat{\epsilon}} = 0, \tag{15.25}$$

when evaluated at $\hat{\epsilon} = \epsilon$, or it could be at an endpoint (ϵ is equal to zero or some maximum possible value $\epsilon_{\max}$).

The fitness of an invader depends on the values of a and b, which in turn depend on the unknown solution of the nonlinear equations for $\tilde{U}$ and $\tilde{D}$. One can be computed in terms of the other using the fact that $\lambda(\epsilon, \epsilon) = 1$, and the second must be computed numerically. We find both by simulating the differential equation (15.21) starting with initial condition (1,0) to find a and with the initial condition (0,1) to find b.

Although it is possible to find an analytical formula for the existence of a critical point in (15.25), such points are always minima rather than maxima in the cases we have examined (unless we add an explicit cost for larger ϵ).

16

Evolution of Reversible Plastic Responses: Inducible Defenses and Environmental Tolerance

WILFRIED GABRIEL

Abstract

This study is a modeling approach to evaluating the selective advantage of reversible plastic responses within the concept of environmental tolerance. In the model of Gabriel and Lynch (1992) a trait value can be modified by external triggers during development but remains fixed for the rest of the life span. This model is expanded to include more flexible traits, e.g., temporal modifications of behavioral traits or inducible defenses that are rebuilt if no longer advantageous, like predator-induced helmets in *Daphnia*.

Compared with an inflexible genotype, the selective advantage of reversible plasticity in shifting the mode of the tolerance curve is large. The advantage increases exponentially both with the fitness loss of nonplastic genotypes and with the difference $t_s - t_r$, where t_s denotes the time span for which a shift is advantageous and t_r the sum of time needed for performing the shift back and forth. In many cases the gain in fitness is high enough to easily compensate all the possible energetic costs of being plastic that are not considered explicitly in the model.

Plasticity in the breadth of adaptation is only weakly selected for if the change is irreversible. But for reversible plastic responses, a large breadth of adaptation becomes advantageous if the expected fitness reduction for inflexible genotypes is large. If t_r is not small compared to t_s, then selection for plasticity in the breadth of adaptation can be even stronger than selection for shifting the mode of the tolerance curve.

The evolution of irreversible inducible defenses is advantageous if the temporal between-generation variance or the spatial variance of the environment is large compared to the within-generation temporal variance. Such conditions are fulfilled, e.g., if a predator appears with low probability but for a long period. Reversible plastic inducible defenses are expected to evolve under much less restrictive conditions, e.g., anytime the defense is advantageous for a short time span compared with the entire life span. Reversible plastic modifications are favored over constitutive defenses or irreversible plastic modifications particularly when the fitness loss without the defense is large.

Plasticity and Environmental Tolerance

The evolution and maintenance of inducible defenses can be viewed as a model system of the evolution of phenotypic plasticity. A review of the current discussion on how to model the genetic basis of plasticity is given in Via et al. (1995). Independent and in many aspects complementary to the models described there is the approach by Gabriel and Lynch (1992) which is based on the concept of environmental tolerance (Lynch and Gabriel 1987). By treating mode and variance of the tolerance curve as quantitative genetic traits, predictions are made on the breadth of adaptation depending on the variance of the environment. Lynch and Gabriel (1987) quantified how the interaction of spatial, temporal within-generation, and temporal between-generation variance selects for generalists or specialists. Usually, increased environmental variance implies stronger selection for generalists. But, at first glance counterintuitively, if the between-generation temporal variance is high, then an increased spatial variance selects for specialists.

Gabriel and Lynch (1992) studied the selective advantage of irreversible plastic responses triggered by a reliable environmental cue. A phenotypical modification during development was considered that results either in an adjusted mode of the tolerance curve or in a modified breadth of adaptation. The model assumes the reasonable trade-off that an increased breadth of adaptation is possible only at a cost of reduced maximal fitness at the mode. So, the selection for a modified breadth of adaptation is very weak compared to a shift of the mode. The qualitative results for the selection of specialists and generalists (Lynch and Gabriel 1987) were not changed by the additional plasticity. The evolution of such plasticity, which can be used for a phenotypic modification during development, is favored especially if the sum of the between-generation variance and of the spatial variance is large compared to the within-generation temporal variance of the environment. Therefore, plasticity for induced irreversible modifications (e.g., developmental switches or modulations) is expected to evolve preferably in organisms that experience different environments in successive generations, such as plankton species that are short lived relative to annual cycles.

This model does not specify the kind of character considered, but the calculations were done under the assumption that the phenotypic modification is irreversible. Like other available models (see, e.g. N. Moran 1992), it is not applicable to the large variety of reversible plastic responses. Inducible defenses are often reversible. The degree of reversibility and the time span necessary to perform the phenotypic modifications depend drastically on the character considered and whether it is primarily determined by life history, behavior, physiology, or morphology.

The reversibility of life history traits seems to be very limited. For example, the age of first reproduction can only be delayed or accelerated prior to first reproduction; body length can hardly be reduced if large size becomes suddenly disadvantageous. But flexibility to adjust life history at any time in order to maximize the reproductive value might be highly favorable despite the irreversibility of decisions of the past. The model of Gabriel and Lynch (1992) is applicable to the evolution of plasticity in life history traits if one considers each life history decision separately. The kind and strength of selection can be calculated by life history theory. Such predictions can even be quantitatively very precise. A typical example is the life history model of Taylor and Gabriel (1992, 1993) for the investment of energy in growth and reproduction in *Daphnia*. The predicted life history shifts under different size-selective predation regimes have been confirmed not only qualitatively but also quantitatively by measurements of Tollrian (1995b) and Stibor and Lüning (1994).

Behavioral responses are often very fast and reversible. In many cases, a simple fitness analysis might be sufficient to predict behavioral changes. Diel vertical migration of zooplankton can be interpreted as an induced defense for predator avoidance triggered by light intensity changes in the presence of fish (Lampert 1989). In many species or clones this behavior is reversible, i.e., vertical migration stops if the predation risk disappears. An ESS approach (Gabriel and Thomas 1988) predicts the behavioral shifts by quantifying the frequency-dependent relative fitness. Such qualitative and quantitative predictions have been confirmed by experiments (DeMeester et al. 1995; Loose and Dawidowicz 1994).

Physiology can be as reversible as behavior but it might take a longer time to adjust to a new physiological state. Not much is known about the reversibility of physiological changes in the context of inducible defenses, but it might be a promising research field, especially in plant ecology. Induced morphological changes can be reversible, such as helmets or neckteeth in *Daphnia* (Tollrian, in prep.). There is no theory available to analyze reversible plastic responses like reversible inducible defenses. In the following I formulate a simplifying theory to give some insight at least into expected qualitative dependencies.

Modeling Reversible Plastic Changes: A First Attempt

The following analysis is restricted to a single trait and neglects all complications, such as those from correlated response of the whole character set of an individual to changes of the environment. I use the concept of environmental tolerance introduced by Lynch and Gabriel (1987) where fitness, w, is determined by the phenotypic values of the mode z_1 and the variance z_2 of the tolerance curve measured on an environmental scale φ. The mode gives the value for which the environmental state of an individual's tolerance curve has its maximum:

$$w = (2\pi z_2)^{-1/2} \exp\left(-\frac{(z_1 - \varphi)^2}{2 z_2}\right).$$

There is a trade-off between maximal fitness at the mode and breadth of adaptation $\sqrt{z_2}$: increasing the breadth of adaptation decreases the fitness at the mode. The fitness integrated over the whole range of environmental states, however, remains constant if the breadth of adaptation changes.

The genotypic values g_1 and g_2 underlying the phenotypic characters z_1 and z_2 are assumed to be shaped by long-term evolutionary forces and to be genetically correlated with other traits. Phenotypic plasticity, however, might allow modifications of the phenotype in response to environmental cues.

Figure 16.1a illustrates the effect of plastic responses in the mode z_1 of the tolerance curve. The breadth of adaptation remains unchanged but the tolerance curve undertakes a shift of the mode from $z_1 = 0$ to $z_1 = 2$. If this shift is the optimal response to a change of the environmental state φ from 0 to 2, then the length of the arrow indicates the gain in fitness compared to a non-plastic genotype. For the same change of the environment, figure 16.1b demonstrates how plasticity in breadth of adaptation $\sqrt{z_2}$ enhances fitness while the mode is kept constant.

In reality, such plastic responses might be energetically costly, time consuming, and performed in a suboptimal manner. Whether the plastic modifications are reversible or irreversible might depend on trait, genotype, species, environment, and cue. Important is how well the cue can be perceived, how reliable it is, and what its frequency and probability are.

The case of irreversible plastic modifications during development is analyzed in detail by Gabriel and Lynch (1992). In the following discussion, this model is expanded to cope with the evolution of reversible plastic responses that are performed by an external trigger like chemical cues and are rebuilt if the trigger is no longer present. One favored application is the evolution of inducible defenses including changes in behavioral traits related to colony forming and swarming. The problem is reduced to a very simple case that is, however, sufficient to gain insight into the minimal conditions for the evolution of such plasticity.

In order to simplify the analysis I assume that without the external cue the expected values of an individual's phenotypic traits z_1 and z_2 are equal to its genotypic values g_1 and g_2. Furthermore, I do not treat explicitly (developmental) noise ($= V_e$) and assume that V_e is uncorrelated with z_1 and z_2; it

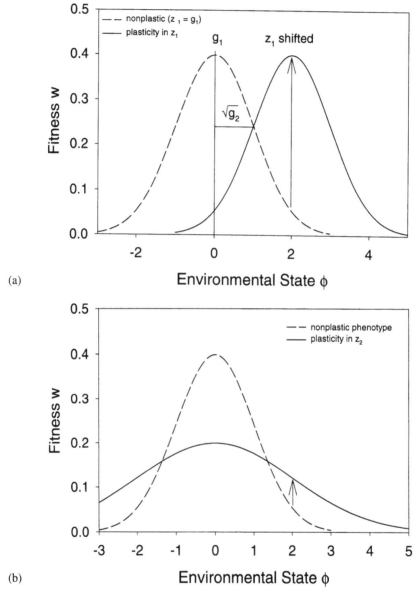

Fig. 16.1. (a) Shifting the mode of a tolerance curve by plasticity in z_1. The environmental state has changed from 0 to 2. If a plastic genotype (full line) shifts its mode also from $z_1 = 0$ to $z_1 = 2$, then the change of the environmental state does not affect fitness, while a nonplastic genotype (broken line) loses fitness. The length of the arrow indicates the gain of fitness by plasticity in z (b) Modifying the breadth of adaptation by plasticity in z_2. As in (a), the environmental state φ has changed from 0 to 2. The modification of the tolerance curve (full line) optimizes fitness if the mode $z_1 = 0$ is kept constant. The length of the arrow indicates the gain of fitness by plasticity in z

REVERSIBLE PLASTIC RESPONSES

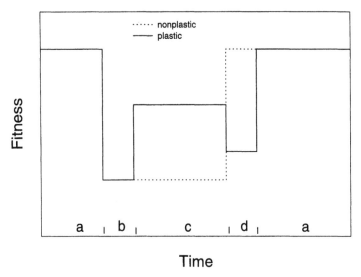

Fig. 16.2. Schematic time course of fitness of nonplastic and reversible plastic genotypes. A predator and its cue are present during the periods b and c. The interval b is needed to perform the shift in z_1 or the modification of z_2. During the interval d these plastic changes are reversed. For a formal description, see the appendix.

may slow down the evolutionary response but it does not affect the direction of selection. It is unknown how energetically costly it is to have the machinery for plastic responses available. These costs might actually be quite low if compared to direct and indirect fitness costs. Looking for minimal conditions I put this purely energetic cost to zero in this analysis and concentrate on direct and indirect fitness costs resulting from maladapted phenotypic values. These can be caused by suboptimal changes in the phenotype or by the time needed to perform the changes (see also Padilla and Adolph 1996). For example, fitness is reduced during periods when a change in phenotype would be advantageous but has not yet occurred, or during periods where the change becomes disadvantageous before rebuilding.

A schematic time course of fitness is given in figure 16.2. Let us assume a predator is absent during the periods denoted by a and d but present during b and c. The fitness of a nonplastic genotype will be reduced during the periods b and c as shown by the dotted line. The phenotypically plastic genotype (full line) needs the period b to perform its modifications. If adjusted, its fitness during period c will then be higher than for a nonplastic genotype. If the predator disappears, then the plastic genotype is maladapted and needs the time span d to rebuild the modifications.

The fitness in the four states a, b, c, and d depends on the phenotypic values with and without plastic modifications and on the actual environmental state. The formal description and the derivation of the corresponding equations are given in the appendix at the end of this chapter.

The following variables are used in the model:

z_1 Phenotypic value of the mode of the tolerance curve. It is a continuous character for which inducible shifts are assumed to be possible. It is measured on an environmental scale φ (see below). For simplification of the analysis, it is assumed that without induced phenotypic shifts phenotypic and genotypic values do not differ so that $z_1 = g_1$, i.e., developmental noise etc. are neglected.

z_2 Phenotypic value of the variance of the tolerance curve. $\sqrt{z_2}$ is the breadth of adaptation.

g_1 Genotypic value of z_1. Without loss of generality, it is assumed that $g_1 = 0$.

g_2 Genotypic value of z_2.

φ Actual environmental state corresponding to z_1. The individual has maximal fitness if $\varphi = z_1$. The individual is assumed to be adapted to the situation without predator, which means $g_1 = \varphi = 0$. In the presence of predators, φ gives the amount of shift necessary to gain maximal fitness. Perfect plasticity in z_1 would imply maximal fitness by keeping $z_1 = \varphi$. For a nonplastic genotype, φ is correlated with the fitness reduction caused, e.g., by the presence of a predator.

t_s "Signal" time span for which a shifted tolerance curve would be optimal, e.g., periods during which predators are present. Time is measured in parts of total lifetime so that t_s also gives the proportion of the lifetime during which an induced defense is optimal.

t_r "Reaction" time span. Time needed to perform phenotypic shifts both for the induced defense and to rebuild the defense after becoming suboptimal. (Because t_s and t_r are normalized to 1, they also can be interpreted as frequencies or long-term probabilities.) Building a defense and removing it are not usually symmetric processes and the time needed might differ. As long as the asymmetry is not large, it is not necessary to treat the corresponding time spans separately (but see the appendix for an exact treatment).

w Fitness.

For the analysis of plasticity in the mode, it is assumed that fitness values during b and d are equal and that the shift is almost optimal and not costly. This implies that after the shift, during the presence of the predator, relative fitness is the same as during times before the predator appears. Therefore, relative fitness returns to 1 during c. This simplifying assumption is biased in the same direction as other simplifications: the fitness gain for plastically shifting z_1 is favored compared to a plastic increase in z_2. Therefore, the conditions derived for selection on reversible modifications of the breadth of adaptation are conservative estimates.

If there is plasticity only in the breadth of adaptation, then the fitness during c will always be lower than during a. As shown in the appendix, modifying z_2 is advantageous only if $\varphi^2 > z_2$. If the predator disappears, the organism immediately has its optimal value z_1 but the increased breadth of adaptation causes a fitness loss. Fitness during d might be higher than during b and will usually be higher than fitness of a genotype that is plastic only for the mode z_1.

Figure 16.2 shows only a single induction period. During a lifetime, several or many such periods might appear and have different consequences for fitness. As a first approximation, let us analyze the simplified case of equal fitness schemes for all induction periods. The different periods a, b, c, and d represent then the sum over the corresponding time intervals. The variables t_r and t_s represent the sum of reactive and signal (cue) time spans.

It is easy to think in terms of shifting a character z_1 but it might be difficult to imagine a modification of the breadth of adaptation $\sqrt{z_2}$ as an inducible defense. To illustrate these two different plastic responses, let me give an example that is, however, highly speculative. It might turn out to be wrong after careful investigations. Nevertheless, it can help to illustrate the problem. Let's compare two inducible defenses of cladocera: neckteeth (or helmet) formation and enhanced alertness. The morphological change is assumed to be energetically almost neutral but the enhanced alertness is an inducible behavioral change that is associated with increased energy consumption caused by physiology. Let's assume that the constraints of physiology do not allow a shift of a mode but only a modification of the shape of a tolerance curve. Let's further assume that neckteeth (or helmets) are morphological modifications that can be described by a shift in a character z_1. Helmets or neckteeth would then guarantee a high fitness in the presence of a specific, e.g., invertebrate, predator; ideally, the fitness would be equal compared to an individual without helmets in the absence of this predator. If the invertebrate predator is absent and its absence is correlated with an increased probability of the presence of another but optical-orientated predator, e.g., a fish, then helmets or neckteeth become disadvantageous because of the enhanced visibility. Let's now assume that enhanced alertness is associated with a lowered vulnerability to the invertebrate predator but it does not reduce the predation risk completely. During the presence of the invertebrate predator, this implies a reduced fitness, though it is a fitness gain compared to individuals without enhanced alertness. If the invertebrate predator disappears and the fish predator appears, then enhanced alertness is not disadvantageous like helmets or neckteeth—and might even help to reduce predation risk by fish compared to nonenhanced alertness because of faster initiation of escapes. Therefore, during the presence of fish, individuals with enhanced alertness would have a higher fitness than individuals with helmets. Alertness might, however, need a not-negligible amount of energy. In the absence of predators, therefore, the fitness of individuals with enhanced alertness will be reduced compared to individuals without this behavioral defense and also compared to individuals with helmets because of the increased metabolic costs and wasted energy by erroneous escape reactions. In this scenario, enhancing alertness fulfills all conditions to be treated as a modification of the breadth of adaptation. Such a defense can be described by an inducible increase of z_2. (Depending on the physiology involved, however, it is feasible that increased alertness can alternatively be described by a change in the mode z_1.)

In the following, I show the general conditions under which temporary plastic responses are advantageous. If phenotypic modifications are performed only once in life (developmental shift), i.e., if they are irreversible, then the evolution of inducible modifications in the breadth of adaptation is unlikely compared to shifts in the mode (see Gabriel and Lynch 1992). I will determine whether this is still valid for reversible modifications.

Selective Advantage of Reversible Plasticity

As shown in the appendix, the relative fitness of a genotype with reversible plasticity in the mode z_1 compared to a nonplastic genotype is

$$w_{\text{rel}} = \exp\left\{\frac{1}{2}(t_s - t_r)\frac{\varphi^2}{z_2}\right\},$$

where φ^2/z_2 determines the fitness reduction caused by the predator. It measures the square of the deviation of the environmental state from the genotypic mode if expressed in units of breadth of adaptation. The relative fitness advantage is proportional to the difference $t_s - t_r$ (and only the difference enters); t_s denotes the sum over all periods at which the predator is present; t_r is the sum over periods during which the cue is present but the shift has not yet been performed, or the cue is already absent but the performed shift has not yet been taken back. Shifting the mode can be selected for only if $t_s > t_r$, i.e., if the total reaction time is shorter than the time span of the predation risk. This is shown analytically in the appendix but seems also intuitively obvious if one compares the fitness of inflexible genotypes with genotypes capable of phenotypically shifting the mode z_1. In terms of the model, this statement holds independently of φ, i.e., of the amount of fitness reduction caused by the predator. The strength of selection for an inducible defense, however, depends on φ. How fast the fitness is reduced by an increasing φ depends on the breadth of adaptation. Therefore, the selection pressure to evolve an inducible defense is much higher in specialists than in more broadly adapted genotypes. Independent of the breadth of adaptation there might be selection for shortening t_r unless t_s is large compared to t_r. Because of the exponential increase of relative fitness, there is substantial selection for evolving plastic responses in z_1, even if the probability of the cue is low and if the energetic costs of plasticity are not negligible.

Shifts in z_1 that do not optimize fitness can still be advantageous for a large range of values. As shown in the appendix, the theoretically optimal shift in z_1 is not φ but

$$z^*_{1,c} = \varphi \left(1 - \frac{t_r}{2t_s}\right).$$

It is very unlikely that an individual can perform the optimal shift because it would have to know the time of the presence of the predator in advance. But $z^*_{1,c}$ equals φ for small t_r/t_s. Figures 16.3a and b demonstrate the gain in fitness compared to a nonplastic genotype for two different values. As it can be seen from the figures, any shift in z_1 (into the correct direction) is advantageous if smaller than $2\,z^*_{1,c}$.

Selection for modification of the breadth of adaptation $\sqrt{z_2}$ is not possible if $\varphi^2 < 1$. But if $t_s > t_r$ and $\varphi^2 > 1$, then modifications of z_2 are advantageous compared to nonplastic genotypes. Figure 16.4 compares the fitness of nonplastic genotypes (dotted line), reversible plastic modifications in z_2, and reversible plastic shifts in z_1. As φ becomes large, the fitness of the genotype with modifications in z_2 has the highest fitness of all three genotypes. This might seem counterintuitive at first glance because during phase c a genotype that shifts z_1 has a much higher fitness than a genotype that modifies z_2. This is even more pronounced if φ is large. The advantage of modifying z_2 results from phase d. During rebuilding the modifications, the genotype with plasticity in z_1 has a very low fitness because of its maladapted mode, while the genotype with plasticity in z_2 still has the correct mode and suffers only from too large a breadth of adaptation.

Figure 16.5 gives a more detailed analysis for the case that adapting z_2 is more favorable than shifting z_1. The lines give equal fitness of both plastic genotypes for different g_2 values. Above the lines, modifications of the breadth of adaptation are more advantageous than shifts of the mode of the tolerance curve. The preferred conditions for selection of reversible plastic response in the breadth of adaptation are, therefore, large φ values and a ratio t_r/t_s that is not too small. The relative advantage of modifications of z_2 over shifting z_1 becomes smaller with increasing genotypic breadth of adaptation. For larger g_2 values the adaptation in z_2 becomes advantageous at larger φ values, and the increase of the selective advantage is slower. The more specialized an organism is and the higher the predation risk, the more efficient the selection for reversible plasticity to increase breadth of adaptation compared to shifting the mode.

Independent of whether modifications in z_1 or z_2 are more advantageous, the evolution of reversible plastic responses is expected to be faster in spe-

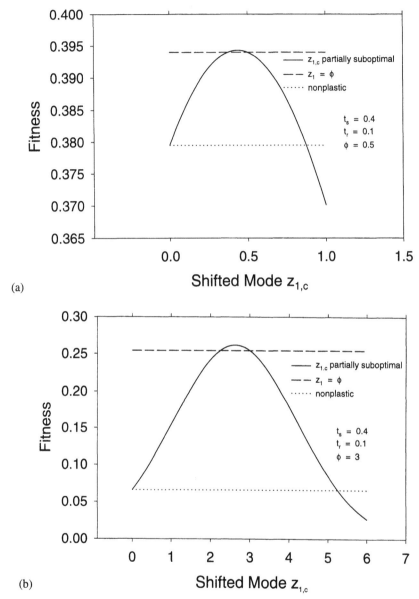

Fig. 16.3. (a) Fitness advantage for reversible plasticity in z_1. The maximum fitness is achieved at $z_{1,c} = z^*_{1,c} = 2\varphi(1 - t_r/(2t_s))$. There is always a fitness advantage to shifting if $z_{1,c} < 2z^*_{1,c}$. The fitness for $z_{1,c} = \varphi$ is given by the dashed line, the fitness of nonplastic genotypes by the dotted line ($t_r = 0.1$, $t_s = 0.4$, $\varphi = 0.5$, $g_2 = 1$). (b) The same as (a) but with $\varphi = 3.0$.

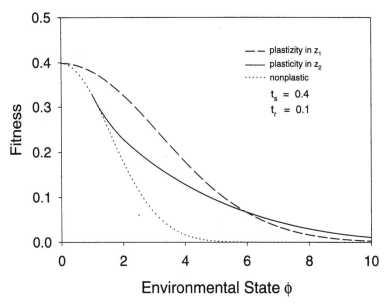

Fig. 16.4. Comparison of fitness of nonplastic genotypes (dotted line) with reversible plastic genotypes: modifications of the breadth of adaptation are given as a full line, shifts of the mode of the tolerance curve as a dashed line ($t_r = 0.1$, $t_s = 0.4$, $g_2 = 1$).

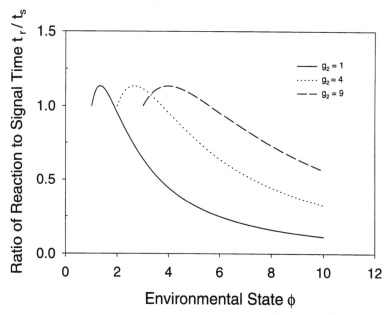

Fig. 16.5. Areas of preferred selection for reversible modifications of the breadth of adaptation $\sqrt{z_2}$. The lines give equal fitness for reversible plasticity in z_1 and z_2. Above the lines selection for modifications of z_2 is stronger ($\sqrt{g_2} = 1, 2, 3$).

cialists than in generalists because of the higher strength of selection in specialists. Therefore, inducible defenses are most likely for organisms with a narrow tolerance curve. Modifications in the breadth of adaptation are unlikely for a behavioral trait because of its short reaction time. A pure modification of the breadth of adaptation can be imagined for physiological traits or heavily constrained morphological characters.

Outlook

The model presented here demonstrates that a further analysis of inducible defenses in the context of environmental tolerance will be fruitful. Just a first attempt has been made to develop some theory on reversible plastic responses. Because of the special emphasis on inducible defenses, only multiplicative fitness contributions during a lifetime have been considered. Many details that might be important for the analysis of real and therefore special cases are neglected because they are—according to preliminary investigations—believed to have only marginal effects on the general qualitative results.

The analysis was performed under the assumption that the genotypic values g_1 and g_2, which determine the tolerance curve, are shaped by long-term evolution and constrained by other traits. Therefore, I did not look for alternative evolutionary responses like evolving an increased genotypic breadth of adaptation without plasticity and a slightly shifted but inflexible mode. Such an analysis could be performed but is outside the scope of this paper. I analyzed the potential selection pressure for plasticity under the assumption of already evolved genotypic values. I have not considered the implications of reversible plastic responses on the genotypic value itself. This is not an easy task but should be attempted in the future, e.g., by the study of the evolution of the genotypic breadth of adaptation depending on temporal and spatial variation of the environment. Additional insight can be expected from a thorough stochastic analysis (e.g., using the approach of Padilla and Adolph 1996), especially considering the probability of occurrence of a predator, the reliability of the cue, and the potential inaccuracy of the plastic responses.

Because I was mainly interested in the selective forces for reversible plasticity, I did not analyze a simultaneous response in z_1 and z_2. Response in just one of the two characters is plausible if the other trait is heavily constrained by selection on other phenotypically or genotypically correlated characters of the individual. For example, for physiological traits it seems plausible that the mode remains fixed and that flexibility is left only for modifications of the breadth of the tolerance curve.

The analysis was restricted to a single life of a single individual. The calculated selective advantages are very high. This is the rationa for stating that there is potentially strong selection for reversible inducible defenses. The actual selection pressure, however, can be estimated only if missing components are added to the model: stochasticity of events, reliability of the cue, energetic costs, density-dependent modulations of response, dependence of the level of response on the level of stimulus beyond an initial threshold, learning and memory effects, etc. But even if such components lower the fitness advantage of plasticity drastically, and even if the probability of the cue is low, there will still be many cases with high enough selection pressure to support the evolution of reversible inducible defenses and—more generally—evolution of reversible plastic responses.

According to Gabriel and Lynch (1992) (irreversible) plastic responses are selected for if the between-generation variance of the environment is large, such as in short-lived organisms or organismal states that experience different environments in successive generations. This might partly explain the increased frequency of (some kinds of) inducible defenses in invertebrate and clonal organisms because they often match these requirements. We do not know enough about the evolution of reversibility. Depending on how reversibility evolves, the conditions favorable for the evolution of irreversible plastic modifications might be advantageous also for the evolution of reversible plasticity—but probably only if further specific conditions are met, e.g., if the periods with predators present are too short to make induced irreversible modifications beneficial.

I would guess, however, that in many cases irreversible plastic responses evolved secondarily after reversible plastic responses had been established. This can be imagined because reversible plastic responses can be selected for under much less restrictive conditions than irreversible plasticity. If the fitness loss for a nonplastic genotype is very large for events that occur with low probability, at low frequency, or during periods short compared to a lifetime, then evolution of reversible plastic responses is expected to be favored over constitutive adaptations and irreversible plastic modifications. It is unknown how costly it is to evolve and to maintain reversible plasticity, but the expected gain in fitness will compensate for high costs. If the energetic costs are low, the advantage of reversible plastic responses is obvious even for events with high probability or high frequency. Reversible inducible defenses are not restricted to any organismal group, as illustrated by the evolved plasticity of behavior in all animal groups.

Appendix

The effect of predation is assumed to be multiplicative over time. The fitness is therefore the product of its components during the four time intervals t_a,

t_b, t_c, and t_d. The values of z_1, z_2, and φ are denoted corresponding to the actual time interval, e.g., $z_{1,b}$. A measure of (relative) fitness is then given by

$$w = \left\{(2\pi z_{2,a})^{-1/2} \exp\left(-\frac{(z_{1,a} - \varphi_a)^2}{2 z_{2,a}}\right)\right\}^{t_a}$$

$$\left\{(2\pi z_{2,b})^{-1/2} \exp\left(-\frac{(z_{1,b} - \varphi_b)^2}{2 z_{2,b}}\right)\right\}^{t_b}$$

$$\left\{(2\pi z_{2,c})^{-1/2} \exp\left(-\frac{(z_{1,c} - \varphi_c)^2}{2 z_{2,c}}\right)\right\}^{t_c}$$

$$\left\{(2\pi z_{2,d})^{-1/2} \exp\left(-\frac{(z_{1,d} - \varphi_d)^2}{2 z_{2,d}}\right)\right\}^{t_d}$$

Time will be normalized so that

$$t_a + t_b + t_c + t_d = 1.$$

Further, the following definitions and relations are obvious:

$$t_s := t_b + t_c$$
$$t_r := t_b + t_d$$

If $t_b = t_d$ then

$$t_b = t_d = \frac{t_r}{2}$$

$$t_a = 1 - t_s - \frac{t_r}{2}$$

$$\varphi_b = \varphi_c := \varphi$$

$$\varphi_a = \varphi_d := \varphi_0.$$

Without loss of generality we can put $\varphi_0 = 0$. A nonplastic genotype keeps mode and breadth of adaptation fixed and has the fitness

$$w_{\text{nonplastic}} = \left\{(2\pi z_2)^{-1/2} \exp\left(-\frac{(z_1 - \varphi_0)^2}{2 z_2}\right)\right\}^{1-t_s}$$

$$\left\{(2\pi z_2)^{-1/2} \exp\left(-\frac{(z_1 - \varphi)^2}{2 z_2}\right)\right\}^{t_s}.$$

REVERSIBLE PLASTIC RESPONSES

With $z_1 = \varphi_0 = 0$, we get

$$w_{\text{nonplastic}} = (2\pi\, z_2)^{-1/2} \exp\left(-\frac{t_s\, \varphi^2}{2\, z_2}\right).$$

If a genotype is totally plastic in the mode, it could shift z_1 so that it equals φ. With $z_{1,a} = z_{1,b} = 0$ and $z_{1,c} = z_{1,d} = \varphi$, we have

$$w_{\text{shift } z_1} = \left\{(2\pi\, z_2)^{-1/2} \exp\left(-\frac{(0-0)^2}{2\, z_2}\right)\right\}^{t_a}$$

$$\left\{(2\pi\, z_2)^{-1/2} \exp\left(-\frac{(0-\varphi)^2}{2\, z_2}\right)\right\}^{t_b}$$

$$\left\{(2\pi\, z_2)^{-1/2} \exp\left(-\frac{(\varphi-\varphi)^2}{2\, z_2}\right)\right\}^{t_c}$$

$$\left\{(2\pi\, z_2)^{-1/2} \exp\left(-\frac{(\varphi-0)^2}{2\, z_2}\right)\right\}^{t_d},$$

and therewith

$$w_{\text{shift } z_1} = (2\pi\, z_2)^{-1/2} \exp\left(-\frac{t_r\, \varphi^2}{2\, z_2}\right).$$

Therefore, the fitness advantage of shifting the mode is

$$\frac{w_{\text{shift } z_1}}{w_{\text{nonplastic}}} = \exp\left\{\frac{(t_s - t_r)\, \varphi^2}{2\, z_2}\right\}.$$

From this follows

$$w_{\text{shift } z_1} > w_{\text{nonplastic}} \Leftrightarrow t_s > t_r.$$

For many cases it might be more realistic to assume that $|z_{1,c}| = |z_{1,d}| < |\varphi|$. As a more general formula of fitness under shifting the mode z_1, we get

$$w_{\text{shift } z_1 \text{ gen}} = \left\{(2\pi\, z_2)^{-1/2} \exp\left(-\frac{(0-0)^2}{2\, z_2}\right)\right\}^{t_a}$$

$$\left\{(2\pi\, z_2)^{-1/2} \exp\left(-\frac{(0-\varphi)^2}{2\, z_2}\right)\right\}^{t_b}$$

$$\left\{(2\pi z_2)^{-1/2} \exp\left(-\frac{(z_{1,c} - \varphi)^2}{2 z_2}\right)\right\}^{t_c}$$

$$\left\{(2\pi z_2)^{-1/2} \exp\left(-\frac{(z_{1,c} - 0)^2}{2 z_2}\right)\right\}^{t_d}.$$

With $t_b = t_d = t_r/2$ we get

$$w_{\text{shift } z_1 \text{ gen}} = (2\pi z_2)^{-1/2} \left\{\exp\left(-\frac{\varphi^2}{2 z_2}\right)\right\}^{\frac{t_r}{2}}$$

$$\left\{\exp\left(-\frac{(z_{1,c} - \varphi)^2}{2 z_2}\right)\right\}^{t_s - \frac{t_r}{2}}$$

$$\left\{\exp\left(-\frac{(z_{1,c})^2}{2 z_2}\right)\right\}^{\frac{t_r}{2}}$$

$$= (2\pi z_2)^{-1/2} \exp\left(-\frac{1}{2 z_2}\left((z_{1,c} - \varphi)^2 t_s + z_{1,c}\, \varphi\, t_r\right)\right).$$

It can easily be seen that the optimal $z_{1,c}$ is not φ but

$$z_{1,c}^* = \varphi\left(1 - \frac{t_r}{2 t_s}\right).$$

The maximal fitness would then be

$$\max(w_{\text{shift } z_1 \text{ gen}}) = (2\pi z_2)^{-1/2} \exp\left(-\frac{1}{2 z_2} \varphi^2\, t_r \left(1 - \frac{t_r}{4 t_s}\right)\right).$$

It can easily be proven that

$$t_r\left(1 - \frac{t_r}{4 t_s}\right) \leq t_s.$$

Therefore max ($w_{\text{shift } z_1 \text{ gen}}$) is always greater or equal than $w_{\text{nonplastic}}$ which implies that selection for shifting is always favored with this modified fitness function. The individual, however, can reach this maximum possible fitness only if it can predict the time span t_s and adjusts its shift accordingly. This would be possible only under very special conditions. Therefore, in the following, only formula $w_{\text{shift } z_1}$ as derived earlier will be considered.

By comparing the fitness of a general shift in z_1 with the nonplastic case, it can easily be calculated that a shift in z_1 into the correct direction is advantageous even if it is smaller or larger than optimal shifts. For $\varphi > 0$ a shift is advantageous if

$$z_1 < z_{1,c} < 2 z^*_{1,c} = 2 \varphi \left(1 - \frac{t_r}{2t_s}\right).$$

Analogous to shifting z_1, the fitness for adapting (increasing) z_2 can be calculated. There are, however, more components to consider. In the absence of the predator, fitness is the same as for the inflexible genotype. Fitness is reduced during the period before the shift is performed. With changing the breadth of adaptation, however, the fitness loss can never be fully compensated like shifting the mode. But there is an optimal breadth of adaptation that maximizes the fitness during this period. During rebuilding, the fitness is not the same as during induction. For some of the following calculations, let us assume that t_r splits into equal halves for building and rebuilding. There is no gain in fitness by adjusting z_2 as long as $\varphi^2 < z_2$ (an analogous analysis is made in Gabriel and Lynch 1992). Assuming that z_1 remains at 0 during all time intervals, we get

$$w_{\text{adapt } z_2} = \left\{(2\pi z_2)^{-1/2} \exp\left(-\frac{(0-0)^2}{2 z_2}\right)\right\}^{t_a}$$
$$\left\{(2\pi z_2)^{-1/2} \exp\left(-\frac{(0-\varphi)^2}{2 z_2}\right)\right\}^{t_b}$$
$$\left\{(2\pi (z_2 + z_{2,\text{add}}))^{-1/2} \exp\left(-\frac{(0-\varphi)^2}{2(z_2 + z_{2,\text{add}})}\right)\right\}^{t_c}$$
$$\left\{(2\pi (z_2 + z_{2,\text{add}}))^{-1/2} \exp\left(-\frac{(0-0)^2}{2(z_2 + z_{2,\text{add}})}\right)\right\}^{t_d}.$$

The optimal value for $z_{2,\text{add}}$ that maximizes the fitness gain can easily be calculated (see Gabriel and Lynch 1992) as

$$z_2 + z_{2,\text{add}} = (z_1 - \varphi)^2.$$

With $z_1 = 0$ we obtain

$$z_{2,\text{add}} = \varphi^2 - z_2.$$

Assuming optimal increase in z_2 and $t_b = t_d = t_r/2$, we get

$$w_{\text{adapt } z2} = \left\{(2\pi z_2)^{-1/2}\right\}^{1 - t_s - \frac{t_r}{2}} \left\{(2\pi z_2)^{-1/2} \exp\left(-\frac{\varphi^2}{2 z_2}\right)\right\}^{\frac{t_r}{2}}$$

$$\left\{(2\pi\varphi 2)^{-1/2} \exp\left(-\frac{1}{2}\right)\right\}^{t_s - \frac{t_r}{2}} \left\{(2\pi\varphi^2)^{-1/2}\right\}^{\frac{t_r}{2}}$$

$$= \left\{(2\pi z_2)\left(\frac{\varphi^2}{z_2}\right)^{t_s}\right\}^{-1/2} \exp\left\{-\frac{1}{2}\left(t_s + \frac{t_r}{2}\left(\frac{\varphi^2}{z_2} - 1\right)\right)\right\}.$$

Let us now ask for the conditions that shifting in the breadth of adaptation is advantageous if compared to an unshifted phenotype.

$$\frac{w_{\text{adapt } z2}}{w_{\text{noplastic}}} = \sqrt{\frac{2\pi z_2 \exp\left\{\frac{\varphi^2 t_s}{z_2}\right\}}{2\pi z_2 \left(\frac{\varphi^2}{z_2}\right)^{t_s} \exp\left\{t_s + \frac{t_r}{2}\left(\frac{\varphi^2}{z_2} - 1\right)\right\}}}$$

$$= \sqrt{\left(\frac{z_2}{\varphi^2}\right)^{t_s} \exp\left\{\left(t_s - \frac{t_r}{2}\right)\left(\frac{\varphi^2}{z_2} - 1\right)\right\}}.$$

A fitness advantage for increasing the breadth of adaptation is possible only if this expression becomes larger than 1. This is equivalent to

$$\left(\frac{z_2}{\varphi^2}\right)^{t_s} \exp\left\{\left(t_s - \frac{t_r}{2}\right)\left(\frac{\varphi^2}{z_2} - 1\right)\right\} > 1.$$

If $t_s > t_r/2$, then a necessary condition is that $\varphi^2 > z_2$. The inequality can be easily fulfilled for large enough values of φ^2, and for $t_s > t_r$ it becomes always valid if $\varphi^2 \gtrsim z_2$. The inequality is equivalent to

$$\left(t_s - \frac{t_r}{2}\right)\left(\frac{\varphi^2}{z_2} - 1\right) > t_s \ln\left(\frac{\varphi^2}{z_2}\right)$$

or

$$\frac{t_r}{2 t_s} < 1 - \frac{\ln\left(\dfrac{\varphi^2}{z_2}\right)}{\dfrac{\varphi^2}{z_2} - 1}.$$

This implies that selection for inducible reversible adaptations in the breadth of adaptation is possible. But are there regions where adapting z_2 is more advantageous than shifting the mode z_1? The condition for $w_{\text{shift } z1} < w_{\text{adapt } z2}$ is equivalent to

$$\frac{1}{\sqrt{2\pi z_2} \exp\left(\dfrac{t_r \varphi^2}{z}\right)} < \frac{1}{\sqrt{2\pi z_2}\left(\dfrac{\varphi^2}{z_2}\right)^{t_s} \exp\left\{t_s + \dfrac{t_r}{2}\left(\dfrac{\varphi^2}{z^2} - 1\right)\right\}}$$

or

$$\frac{t_r}{2t_s} > \frac{\ln\dfrac{\varphi^2}{z_2} + 1}{\dfrac{\varphi^2}{z_2} + 1}$$

or

$$t_s < \frac{t_r}{2} \frac{\left(\dfrac{\varphi^2}{z_2} + 1\right)}{\left(\ln\dfrac{\varphi^2}{z_2} + 1\right)}.$$

For large enough φ we will always find $w_{\text{shift } z1} < w_{\text{adapt } z2}$.

17

The Evolution of Inducible Defenses: Current Ideas

RALPH TOLLRIAN AND C. DREW HARVELL

Abstract

An organism's ability to cope with the biotic environment—i.e., with attacks by predators, herbivores, competitors, parasites or pathogens—determines in good part its evolutionary success. In this book we have detailed inducible defensive reactions spanning a wide range, from behavior to morphology, life history, physiology, and biochemistry. Perhaps even more remarkable than the range of inducible responses is the organismal breadth, from protozoans and plants through fish and amphibians. Accounts of biotically induced changes in form and function include cell-shape changes in protozoans, changes in spines of rotifers and bryozoans, pronounced allometric changes in barnacles and fish, changes from solitary to colonial form in freshwater algae, changes in secondary chemistry of plants, changes in vertical migration behavior in freshwater zooplankton, and, in *Daphnia*, simultaneous changes in morphology, life history, and behavior.

The previous chapters have highlighted specific aspects of several inducible defense systems, and in this chapter we will summarize our views about the ideas raised in this book and, in particular, the insights generated from a comprehensive examination of the study of inducible defenses across taxa and habitats. In the introduction, we suggested four prerequisites for inducible defenses, and we return now to evaluate our conclusions about them here and in the following sections:

1. The selective agent has to have a variable impact, which is sometimes low. The higher the baseline of attack oscillations and the more uniform the selective impact, the more likely will be the evolution of constitutive defenses. The variability selecting for inducible defenses can be a combination of predictable and unpredictable components. For example, many inducing herbivores or predators are predictably seasonal, but the precise onset and magnitude of attack are variable. The frequency of attacks may influence the evolution of defenses. Overall, there is insufficient data to evaluate what frequencies and amplitudes of variability select for inducible defenses in some groups and constitutive defenses in others. In fact, we are only just beginning to amass enough examples of inducible defenses to develop the study systems to address this issue or to examine macroevolutionary or phylogenetic patterns in inducible defenses.

2. Cues have to indicate a potential danger. A possible weak link in the entire success of an inducible response is the timely availability of reliable cues.

3. The defense has to have a benefit. In most systems, effectiveness of defense is the major benefit. It might not be possible to develop effective inducible defenses against every attacker. The organism's ability to evolve in response to attacks is limited by constraints that are a product of the environment and the species' evolutionary history. Besides effectiveness of defense, other benefits (e.g., slowing herbivore coevolution) can favor inducible defenses.

4. Most defenses should incur direct allocation costs or other trade-offs. Inducibility will then reduce the effects of these costs or trade-offs. However, in this volume we have emphasized that the type of cost needs to be defined, and that cost saving is not necessarily the most important agent selecting for inducible defenses.

Discussions in various chapters of the book have emphasized the importance of considering other benefits of inducibility in addition to allocation cost savings. The most discussed of these as an overriding function of inducible defenses is the flexibility in response that allows organisms to cope with complex biotic environments, to which several noncorrelated defensive responses may be required. For example, in multipredator environments, alternative prey phenotypes are produced, depending on the specific predator. Thus a possible fifth prerequisite favoring the evolution of inducible defenses is an environment populated with multiple biotic agents requiring a diversity of phenotypic responses from a single prey.

1. Variability in Selective Agents

All the studies in this book confirm the hypothesis that inducible defenses are favored when the selective pressure is variable. Järemo et al. (chapter 2) conclude from their model that induced defenses in plants are likely to evolve when the risk of herbivory is relatively low and the initial damage is not lethal. The importance of variable nonlethal attacks is certainly also true for the vertebrate immune system (Frost, chapter 6) and for colonial animals (Harvell, chapter 13). Although the attacked zooid will die, the colony usually will not be consumed totally and can recover again. In contrast, in most animals attacks by predators are lethal. Thus, a nonfatal attack is clearly not a prerequisite for inducibility but nevertheless provides a very reliable cue and thus favors the evolution of inducible defense systems. If attacks are lethal, more complex less reliable cues may indicate the proximity of danger.

It was earlier suggested that inducible defenses are disproportionately favored in clonal organisms (Harvell 1990a), with the rationale being that in clonal animals, the entire genet is not killed even if parts of a clone die. Thus the attack is nonlethal for the genotype. Indeed, several clonal organisms form inducible defenses: rotifers (Gilbert, chapter 7), ciliates (Kuhlmann et al., chapter 8), cladocerans (De Meester et al., chapter 9; Tollrian and Dodson, chapter 10), bryozoans (Harvell, chapter 13), and of course unicellular algae (Van Donk et al., chapter 5) and clonal plants. However, we now find many examples of inducible defenses in nonclonal organisms: copepods (De Meester et al., chapter 9), crucian carp (Brönmark et al., chapter 11), amphibians (Anholt and Werner, chapter 12), barnacles (Lively, chapter 14), and snails (Appleton and Palmer 1988; Crowl and Covich 1990). Also in plants we find inducible defenses in both clonal and sexual types. Thus, there is no clear support for the clonal hypothesis, and our view is that some clonal organisms may be predisposed to inducible defenses because of the reliability of nonfatal cues to the presence of an attacker. It may also be possible that alarm substances evolve more rapidly in genetically closely related organisms. For example, Alarm substances are reported, to induce chemical defenses in "talking trees" (Baldwin and Schultz 1983), but also to induce morphological defenses in crucian carp (Stabell and Lwin 1997). Nevertheless, Järemo et al. (chapter 2) emphasize that interplant communication may be favored only if there is a synergistic benefit for the attacked and alarmed plant. At least there has to be a net benefit for the "talking" plant. This benefit possibly can be found in a more effective suppression of the herbivore population by a group of defended organisms.

The specificity of cues inducing defenses varies widely. General defenses act against a large variety of enemies and can be induced nonspecifically by many enemy species or even by damage-associated cues. General defenses may result from diffuse coevolution, e.g., of a plant and its feeding enemies (Iwao and Rausher 1997). General defenses are reported from plants (chapters 1–5), rotifers (chapter 7), some ciliated protozoans (chapter 8), *Daphnia* behavioral reactions to fish (chapter 9), and *Daphnia* life history reactions to fish (chapter 10). Many escape reactions are general defenses.

Because general defenses could be induced by rare herbivores (or even just by damage), that are actually not a danger for the plant, the notion that an attacker has to have a selective impact is not always confirmed. For example, Berenbaum and Zangerl (chapter 1) list several species that induce furanocoumarin production in wild parsnip though these are not its natural enemies. Similar other defenses that are induced by general cues can be induced by organisms that have no impact. For example, even piscivorous fish induce vertical migration as a predator avoidance mechanism in zooplankton (De Meester et al., chapter 9). For many inducible defenses, we do not yet know how specific the cues are.

Specific defenses are activated by particular consumers or competitors. They may result from pairwise coevolution. In our book, clear specific reactions have been reported from several plants (chapter 1), ciliated protozoa (chapter 7), *Daphnia* species (chapter 10), barnacles (chapter 14), and bryozoans (chapter 13), but most strikingly from the vertebrate immune system (chapter 6). The vertebrate immune system emerges as the most complex example of a diversity of chemical responses operating against specific selective agents, often in concert. The immune system is composed of both a fast-acting nonspecific component and a more slowly produced specific immunity. Although most work on the immune system is more mechanistic than evolutionary, in chapter 6 Frost evaluates the functional significance of inducible components in the immune response and examines how the invertebrate defense system succeeds despite being less specific than the vertebrate system.

Frequency of attack may also influence the evolution of defenses. Gabriel (chapter 16) points out that an unspoken assumption in previous models of plastic responses is the irreversibility of the change. And yet most inducible defenses are reversible; after some relaxation time, the undefended phenotype is resumed. Gabriel finds that it is the degree of within-generation variability in attacks that determines whether permanent or reversible defenses should evolve. In situations where defenses are selected for, low within-generation variability should favor either reversible or irreversible inducible defenses. But only organisms with permanent or reversible inducible defenses can survive high levels of within-generation variability in attacks.

2. Cues

In all inducible defense systems, cues indicate the potential danger and the availability of timely, reliable cues may be the "weak link" in determining whether an inducible defense can evolve. Defenses can be induced by mechanical, tactile, visual, or chemical cues. Relevant cues depend on both the selective agent's "detectability" and the threatened organism's sensory capacities and capabilities. In many cases, coevolution might have shaped the existing "information systems." The usual first step in the coevolution of cues and sensory systems will be that prey evolve sensitivity to an existing cue. All organisms leave "traces" that can be used by other organisms for their own purpose. The second step in specific coevolution could be counterselection against the release of the "inducing" cue. Adler and Grünbaum (chapter 15) develop a theory appropriate for considering the issues involved in coevolutionary "cue races," and emphasize the types of constraints that might face an attacker in evading the inducible responses of its prey. This previously uninvestigated area of inducible defenses promises fruitful insights into the evolutionary dynamics of predator and prey.

If no reliable cue is available or sensitivity to a cue has not yet evolved, "moving target" strategies might evolve where prey change their phenotypes randomly (Adler and Grünbaum, chapter 15), or environmental "cues" might be used as proximate factors. For example, occurrence of predators in ponds, lakes, and streams is often related to water temperature and season. Thus, temperature and photoperiod might be good indicators of a danger and could be used as cues (e.g., temperature-induced diapause as a potential fish avoidance mechanism in copepods; Hairston et al. 1990).

Cues are likely a crucial determinant of which type of defense will evolve, and vice versa. Escape reactions require immediate—e.g., visual or mechanical (including acoustical)—cues. Avoidance mechanisms require a good indication of a potential dangerous area by longer-lasting—e.g., chemical—cues. Morphological defenses that deter feeding often require a latent period between the detection of a cue and the time when the defense is formed (Padilla and Adolph 1996). Here, chemical cues will be good signals. In general, chemical cues can accumulate and thus are perfect if defenses require a threshold to avoid errors in defense formation, if defenses require information about predator distribution, or if the defenses require cues to be transferred over spatial or temporal scales. In contrast, visual or mechanical cues can rarely accumulate (given that each mechanical damage is a single signal) and thus are ideal for inducing rapid reactions, e.g., escape reactions. Nevertheless, mechanical cues might induce delayed defenses in systems where an attack is not immediately fatal, e.g., in plant-herbivore interactions.

Chemical cues might be more specific and reliable than mechanical cues in the identification of biotic enemies. For example, specific plant responses to pathogens are induced by elicitors, and spit factors of herbivores induce specific defenses in plants (Berenbaum and Zangerl, chapter 1), or signal production in tritrophic systems (Dicke, chapter 4). Also, ciliates (Kuhlmann et al., chapter 8), rotifers (Gilbert, chapter 7), daphnids (Tollrian and Dodson, chapter 10), bryozoans (Harvell, chapter 13), and, most of all, the vertebrate immune system (Frost, chapter 6) respond to specific chemical cues.

Environmental conditions influence which cues can be used. Copepods in the marine environment, which is generally characterized by much higher transparency compared to freshwater habitats, respond to the visual or mechanical cues of their predators (Bollens and Frost 1989). In contrast, copepods in freshwater environments, which often live in dark, turbid habitats, respond to chemicals released by their predators (Neill 1992).

In some systems, two signals are required (Dicke, chapter 4). The information about an herbivore attack is perceived by the plant and passed on to the next trophic level. The volatile cue emitted by the plant can attract specific parasitoids, which attack the herbivore.

However, defenses are only as good as the cues, and cues are not always perfect. Lively (chapter 14) reports for barnacles that high-risk patches are not detectable because the predators are not present in summer when many barnacle larvae settle. Nevertheless, inducibility will be favored over bet-hedging in the presence of reliable cues (Lively, chapter 14).

3. Benefits and Constraints

Direct protective benefits have been demonstrated for most defenses discussed in this volume, and yet many so-called defenses are only small phenotypic modifications. Why aren't they greater? Are these responses effective enough or are there substantial developmental, environmental, phylogenetic, or cost constraints in the evolution of inducible defenses? Inducible defenses may be especially favored in systems where small changes can be protective. Large changes are more likely to be constrained and to have high costs.

Synthesis and modification of secondary metabolites for toxin production offer plants very good options to cope with herbivores. Herbivores might coevolve, develop detoxification systems, and specialize on a food source. Plants can respond with new toxins and with mixed toxins. Small modification of already existing toxins often leads to synthesis of new toxins, as described by Berenbaum and Zangerl (chapter 1) for furanocoumarins.

There are some remarkably effective defenses against herbivores that nevertheless have a limited taxonomic distribution. For example, some plants even synthesize insect hormones (e.g., ecdysone, juvenile hormones) that seriously interfere with the herbivore's physiology. Insects appear unable to coevolve against this defense. Hendrix (1980) reported that ferns and gymnosperms that produce phytoecdysones are nearly free of herbivores, possibly as an effect of the hormones. In contrast, nearly all other deterrent-producing plants are attacked by some specialist herbivores. Insect hormones are general defenses acting against a large variety of insect herbivores. Why did more plants not evolve this defense? Are other groups of plants restricted by constraints or by costs?

The types of defense that can be used depend on the evolutionary history and the ecological environment of the organism.

Developmental constraints can prevent the use of several types of inducible defenses or of available cues. Hairston et al. (1990) showed that copepods use temperature cues to time the onset of diapause and thereby avoid times of high fish predation, while Slusarczyk (1995) showed that resting egg production could be induced by fish kairomones in cladocerans. Hairston et al. (1990) speculated that fish cues cannot be used by copepods because their generation time is too long, while in cladocerans generation times are short enough to allow formation of an effective defense.

Similarly, environmental constraints can prevent the use of several types of inducible defenses or cues. Predator avoidance can be a very effective defense if the environment allows for spatial or temporal segregation. For example, zooplankton in deep lakes can avoid fish predation by diel vertical migration, while such an avoidance will be less effective in shallow ponds, where predator encounters are still possible.

Phylogenetic constraints have been implicated in the evolution of many behavioral and morphological traits; understanding the evolutionary history of inducible defenses is new and fertile ground. Since at least some of the responses are morphological and are found in organisms with a fossil record, paleontological studies could provide insight into which came first, constitutive or inducible responses. Related questions are as follows: Did inducibility arise multiple independent times or only in certain lineages? Do some lineages have a propensity for inducibility? Harvell (1994) suggested that some colonial marine invertebrates, already having a high incidence of discrete morphological novelties, called polymorphisms (avicularia of bryozoans, spines of bryozoans, or sweeper tentacles of corals), might also have a high propensity for inducibility, often in these same characters. A phylogenetic analysis in this group would allow estimation of rates of origination of novelties, and for some groups this could be coupled with an analysis of the origins of inducibility in characters. Most evolutionary models of induced resistance treat it as a derived state of constitutive resistance, the presumed ancestral state (reviewed in Karban and Baldwin 1997). Thaler and Karban (1997) investigated this question with a phylogenetic analysis of the current distribution of inducibility in the commercial cotton genus *Gossypium*. They were able to map spider mite resistance onto a previously known phylogeny. They concluded that both induced and constitutive resistance were derived and not ancestral traits. Induced resistance arose multiple times and was not correlated with constitutive resistance. Similar phylogenetic reconstructions of inducible and constitutive defenses in other groups are crucial in understanding the evolution of inducibility and the relationship between inducible and constitutive resistance.

Besides protection, other benefits can favor the evolution of inducible defenses:

- Coevolution on the attacker side can be slowed down by limiting the defenses to periods where they are necessary and by creating variability (e.g., Agrawal and Karban, chapter 3).
- Inducibility allows more options. Organisms in environments with attackers that have different selectivities can defend against each predator (e.g., Tollrian and Dodson, chapter 10) or competitor (Harvell, chapter 13). This is especially important when it is highly disadvantageous to choose the wrong strategy.

4. Costs

Rhoades (1979) developed his optimal defense theory based on observations of chemical defenses in plants and influenced by optimal foraging theory of animals (e.g., MacArthur and Pianka 1966; Stephens and Krebs 1986). The optimal defense theory has two main components:

1. Organisms evolve and allocate defenses in a way that maximizes individual inclusive fitness.
2. Defenses are costly in terms of fitness to organisms.

From number 2, Rhoades concluded that less well defended organisms have a higher fitness in the absence of enemies, that organisms evolve defenses in direct proportion to their risk from enemies and in inverse proportion to the cost of defense, and that internal allocation of defenses is in direct proportion to the risk and value of a specific tissue and indirectly related to the cost of defense.

Although in several systems some costs have been identified, it remains to be proven that other types of "costs" are not also at work. For plants in particular, several studies have failed to detect costs of defense (reviewed in Schultz 1988; Simms 1992). It is becoming increasingly clear that the old idea, that all defenses incur costs and that cost saving is a major reason for inducibility, needs revision. First, it is possible that some defenses might not incur *any* cost. For example, a defense might be combined with a beneficial behavior or result from the modification of a metabolite. It is predicted that these "cheap" defenses will become permanent, even if the other factors that favor inducibility are present. A mutation that renders an organism already defended before an attack can happen should be selected for. Second, independent of whether there are costs or not, there could be an advantage to inducibility per se. For example, Agrawal and Karban (chapter 3) suggest that inducibility could help to prevent coevolution of the attackers. Finally, the general term "costs" should be categorized to separate the different types of costs that might differently affect the evolution of the trait. For example, Harvell (1990a) classified costs as opportunity, construction, maintenance, plasticity, and genetic. In table 17.1, we suggest a further classification of costs, intended to focus study on measurable, ecologically and evolutionarily relevant units.

1. *Plasticity costs*, costs resulting from an organism's ability to be plastic. Plasticity costs are decrements for providing and maintaining the facilities to detect cues and to respond (e.g., chemosensors, genetic mechanisms, hormones). Thus, they represent permanent costs for an inducible organism. Plasticity costs per se cannot explain why defenses are inducible. They act exclusively on organisms with inducible defenses and not on organisms with permanent defenses. Thus, they can act as a factor influencing evolutionary stability between constitutive and inducible defenses (Lively, chapter 14). Plasticity costs include genetic costs, which can be a result of negative pleiotropic effects (Berenbaum and Zangerl, chapter 1) and infrastructure costs, which result from providing the mechanisms to react to cues and to be

TABLE 17.1
Classification of Possible Costs of Inducible Defenses

Type of Cost	Subtype	Origin
Plasticity costs	Genetic costs	Genetic correlations (e.g., negative pleiotropy)
	Infrastructure costs	Costs of providing mechanisms to respond, (e. g., receptors, hormones)
Allocation costs	Construction costs	Resources and energy for building the defense (e.g., synthesis of feeding deterrents)
	Maintenance costs	Resources and energy for maintaining the defense
	Operation costs	Resources and energy for using defenses (e.g., energy for escape)
Self-damage costs		Costs caused by self-damage or for protection against self-damage (e.g., autotoxicity)
Opportunity costs		Long-term consequences of allocation or developmental constraints
Environmental costs		Interaction of the defense with the environment. Often in variable environments (e.g., misplaced strategies)

plastic (e.g., chemosensors, hormones). Plasticity costs can be estimated indirectly by comparing the fitness of inducible and undefended or constitutively defended organisms of the same species. To detect the potential for cost to constrain the evolution of an inducible defense, this is the most useful cost to measure, but for many organisms the appropriate mating design may not be possible. However, in some taxa (Rotifers, Gilbert, pers. comm.; Bryozoans, chapter 13), both constitutive and inducible genotypes occur within the same species. For plants, genetic correlations have been examined (Simms and Rausher 1987; Simms 1992).

2. *Allocation costs*, costs resulting from changes in energy and materials associated with forming and maintaining inducible defenses. These are the "classical" costs, which have also been termed "internal" costs. Allocation costs may, for example, result from higher energy expenditures for escape movements, from the energy and material required for the synthesis of defensive chemicals, or from higher material allocation to morphological defenses. These costs act during the period when defenses are formed and maintained. Allocation costs could be subdivided into construction, mainte-

nance and, operation costs, depending on whether the costs are incurred for construction, whether they also persist after construction when a defense is maintained, or if are incurred during operation when the defense is used. The use of most escape movements will cause an operation cost, e.g., due to higher energy requirements. In most systems it might not be possible to differentiate between all types of allocation costs. In ciliates the fitness loss due to the initial construction is high compared to the costs for the maintenance of defense. Initially, ciliates double their generation time by skipping one cell division while the larger cells continue to divide in only slightly longer intervals compared to the undefended morph (Kuhlmann et al., chapter 8). Allocation (operation) costs are found to be very low in migrating zooplankton (De Meester et al., chapter 9). Allocation (construction) costs in plants are found by Berenbaum and Zangerl (chapter 1). They review evidence that induced furanocoumarin production stops regular cell growth in cell cultures. In wild parsnip they could measure an increase in respiration due to toxin synthesis; this is the first such physiological measurement of a cost of chemical defense in plants.

3. *Self-damage costs*, such as autotoxicity or other forms of self-damage. Producing, storing, or using defenses could harm the defended organism. For example, in plants or marine invertebrates, the production of toxins could lead to autotoxicity (Orians and Janzen 1974; Simms 1992; Harvell et al. 1996). Allelopathic compounds produced by organisms that suppress competitors, and many toxins that deter feeding enemies, should bear the potential to inhibit the producing organism as well (see McKey 1974). Existence of such a cost would imply that no effective self-protection has evolved. Evolution is often predicted to lead to an association between toxin production and self-tolerance (see Baldwin and Callahan 1993), but mechanisms to prevent self-damage, such as encapsulation of toxins, might be costly as well. Another alternative is that autotoxicity might produce some real constraints on the use of toxic compounds.

4. *Opportunity costs*, incurred after the initial defense has been formed. Gulmon and Mooney (1986) initially described opportunity costs in plants as the lost growth (and competitive) potential incurred as a result of early allocation to defense. Thus, opportunity costs can be seen as long-term allocation costs. However, long-term consequences of developmental constraints that might not carry allocation costs also fit into this category. Barnacles pay their defense later with a reduced fecundity, because the bent morph has a smaller brood cavity. Furthermore, growth is lower because it is restricted to one side (Lively, chapter 14). Exact opportunity costs are hard to measure because they require lifetime observation and possibly could act in later generations as well. They might play an important role in many systems.

5. *Environmental costs*, in contrast to other types of costs, are only occasional costs that, depending on the environment, may or may not occur. They

result from the interaction of a defense with often varying environments and might be a major cost beside allocation costs. They have also been termed "external" costs. In some systems, environmental costs might even be the only detectable costs. Environmental costs can include the following. In zooplankton, morphological changes can alter swimming and feeding hydrodynamics and probably result in higher energy requirements, especially in colder water (Tollrian and Dodson, chapter 10). Diel vertical migration, as a predator avoidance, forces zooplankton to move to strata with low temperature and often low food, thus reducing growth rate (De Meester et al., chapter 9). Van Donk et al. (chapter 5) could not measure costs of induced colony formation in algal growth rates. They suggest that grazer-induced colony formation in plankton algae has the cost of higher sinking rates. In bryozoans, at least part of the slowed growth associated with inducible spines is caused by a decreased feeding efficiency of spined colonies in some flow environments (Harvell, chapter 13; Grünbaum, 1996). Furthermore, spine formation in bryozoans could cause fouling during periods of high sedimentation. Anholt and Werner (chapter 12) describe the dilemma of tadpoles under predation: reduced activity decreases encounter rates with predators but negatively interferes with searching for food and therefore decreases food uptake in low food environments. In crucian carp, environmental costs are detectable. Brönmark et al. (chapter 11) report evidence that cost results from higher drag of the changed morphology. Some environmental costs result from the constraint that adaptations to one predator can prevent adaptation to changing environments. Fish-induced production of smaller offspring in cladocerans could lead to higher starvation if food becomes depleted. Tollrian and Dodson (chapter 10) describe that reactions to one predator can be disadvantageous in the presence of other predators. Predation regimes in lakes frequently change, and fixed defenses or induced defenses against single predator types could be highly disadvantageous in the long run. Thus, phenotypic plasticity can create environmental costs. On the other hand, phenotypic plasticity can help to minimize environmental costs by allowing a switch between adaptations for various situations. Clearly, inducible defended organisms have more options than constitutive organisms. These options even increase if inducible reactions are reversible.

Most cost analyses so far have been designed to measure allocation costs. Measurement of environmental costs requires complex experimental setups that allow the testing of different environmental situations. Unsuccessful attempts to find costs may be a consequence of the experimental design rather than an indication for the lack of trade-offs. On the other hand, if costs are found, they will not necessarily result from a single cause.

In animals and plants, inducible defenses allow for testing the cost hypothesis with genetically identical or similar organisms from single clones/populations. So far, plant studies have been hampered by the problem that induction requires mechanical damage, which in turn influences photosynthesis rates. Berenbaum and Zangerl (chapter 1) found genotype-specific differences in furanocoumarin production and therefore could solve that problem by detecting a negative correlation between photosynthesis rate and furanocoumarin production. Progress in cost analysis in plants might be possible with new approaches. Baldwin (pers. comm.) reported that alkaloid synthesis can be induced by hormone application (e.g., jasmonate addition to the roots). Thus, induction did not require simulated herbivory. However, it has to be shown whether these hormone-induced effects are comparable to herbivory-induced defenses.

It will be interesting to see where the costs are in tritrophic systems (reviewed in Dicke, chapter 4). The plants produce and emit a volatile signal. They leave the "dirty work" to the parasitoids of their attackers. Will this be a mechanism to reduce costs? Clearly, the benefit depends on the population dynamics of the parasitoids, and this type of defense will be unreliable if parasitoids are rare.

Other Factors Influencing the Evolution of Defenses

Variability—Reversibility

Many inducible defenses can be reversible: e.g., chemical defense in plants (Berenbaum and Zangerl, chapter 1; Järemo et al., chapter 2), tritrophic interactions (Dicke, chapter 4), immune reactions (Frost, chapter 6), morphological reactions in ciliates (Kuhlmann et al., chapter 8), morphological reactions in cladocerans (Tollrian and Dodson, chapter 10), and of course most behavioral defenses (chapters 8, 9, 10, and 12). Some reactions can be considered irreversible after they are formed: e.g., morphological defenses in rotifers (Gilbert, chapter 7), life history changes in cladocerans (Tollrian and Dodson, chapter 10), morphological defenses in crucian carp (Brönmark et al., chapter 11), spines in bryozoans (Harvell, chapter 13), morphology in barnacles (Lively, chapter 14), and shell thickness in gastropods (Appleton and Palmer 1988; Crowl and Covich 1990). Between reversible and irreversible reactions there is a continuous scale. A crucial difference is whether defenses are induced by one developmental switch or whether they require induction during several developmental stages. In the first case, the reaction will be fixed after the sensitive period, while in the latter case the reaction or the degree of the reaction can be changed during ontogeny.

Reversibility of defenses may allow reduction of both allocation and environmental costs (see Gabriel, chapter 16). Reversibility also creates variability and thus could slow down coevolution on the side of the attacker. Agrawal and Karban (chapter 3) conclude that variability may hinder herbivore performance in ecological time as well as the evolutionary ability of herbivores to adapt to host plants. Constant levels of deterrents would allow for the evolution of resistance to defenses. Amplification mechanisms found in many defense systems (e.g., Kuhlmann et al., chapter 8; Tollrian and Dodson, chapter 10; Harvell, chapter 13) rapidly create high levels of defense, which may additionally prevent coevolution.

Multiple Defenses—Multiple Predators

Various species have evolved multiple inducible defenses or combinations of inducible and constitutive defenses. Single plants synthesize structurally different furanocoumarins (Berenbaum and Zangerl, chapter 1) and have alternative structural defenses; ciliates form morphological and behavioral defenses (Kuhlmann et al., chapter 8); cladocerans (even single clones) form morphological, life history, and behavioral defenses (De Meester et al., chapter 9; Tollrian and Dodson, chapter 10); colonial marine invertebrates have chemical and morphological defenses (Harvell, chapter 13); and amphibians form morphological and behavioral defenses (Anholt and Werner, chapter 12) as well as chemical antiparasite responses. We can only begin to speculate about why different types of defenses evolve and are favored.

First, multiple defenses might allow the switch in defenses and could prevent adaptation of attackers by creating variability (Agrawal and Karban, chapter 3). Second, multiple defenses might allow adaptation to changes in attacker densities. Third, multiple defenses allow for adaptations to different predators or competitors (e.g., Berenbaum and Zangerl, chapter 1; Kuhlmann et al., chapter 8; Tollrian and Dodson, chapter 10; Harvell, chapter 13). Fourth, multiple inducible defenses will be highly advantageous if predators with contrasting selectivities occur (Tollrian and Dodson, chapter 10).

Multiple defenses, as well as polymorphisms in inducibility, occur in bryozoans (Harvell, chapter 13). Within single populations, colonies occur that make either constitutive spines, inducible spines, or no defense at all. Within the same population, many of the colonies are capable of an inducible response against competitors as well. The polymorphism in spine production could be maintained by trade-offs due to costs, as well as frequency- and density-dependent effects of predation and competition. Lively (chapter 14) found evidence for the presence of both constitutive undefended and inducible individuals in barnacles and calculated strategy models to explain the evolution of polymorphism. He found that genetic polymorphism can be evolutionarily stable. Similar, De Meester et al. (chapter 9) report that multiple defenses against single attacker species can help to maintain genetic variation and genetic polymorphism for predator-induced defenses within populations, because multiple optima might exist.

Timescales

Plant defenses can be rapid or delayed by entire seasons. Rapid defenses within hours or days will be necessary if the initial damage is relatively high (Järemo et al., chapter 2). Delayed induced defenses might evolve if an attack in one year indicates a likelihood of attack in the following year, e.g., if the offspring of attackers from one year will become a danger in the following year, as is the case in some insects (Schultz 1988).

Timing of defense also varies in animals, although long lags of entire seasons have not been reported (Tollrian and Dodson, chapter 10). Defenses that require a longer latency period before they are effective can evolve in addition to rapid defenses if the threat is actually present, or if a cue indicates a delayed danger. For example, pike will colonize ponds frequently as juveniles. This might give a crucian carp population enough time to change their morphology (Brönmark et al., chapter 11).

Spatial Scales

Plant defenses and defenses of animal colonies can be localized or systemic. Localized defenses will be favored if the initial damage is not high and/or if the attack is local (e.g., pathogen infections). Systemic defenses will be favored if attackers are mobile and are likely to spread out and/or if the initial damage is high (Järemo et al., chapter 2).

Population Biology

Information is still limited on the population-dynamic consequences of inducible defenses. Defenses will influence population dynamics of the defended organism, and possibly of the attacking organism (e.g., Myers 1988). In many cases, competitive relationships will be also influenced (Anholt and Werner, chapter 12).

It is becoming increasingly clear that defenses have to be evaluated in their environmental context to allow estimations of consequences. Resource availability will influence the costs of defenses (Gilbert, chapter 7; De Meester et al., chapter 9; Tollrian and Dodson, chapter 10; Brönmark et al., chapter 11) and thus the population dynamics of the prey.

"Defenses" can even be directed against competitors, as it is described for allelopathic interactions in plants (although these are not yet known to be inducible). Also, sessile marine organisms form inducible, "aggressive" defenses against competitors (Harvell, chapter 13). Defenses can be induced by interference competitors also in rotifers (Gilbert, chapter 7); here however, the larger competitors can mechanically damage rotifers, which has a consequence similar to predation.

Clearly, individual performance of herbivores or predators can be reduced as a consequence of induced defenses of their prey. Empirical observations suggest that delayed induced defenses can cause herbivore population cycles (Haukioja 1980), and theoretical models indicate that inducible defenses can regulate herbivore populations, and, in conjunction with other factors, they can even significantly depress herbivore populations (Edelstein-Keshet and Rausher 1989; Lundberg et al. 1994). Besides temporal dynamics, spatial dynamics could be also influenced. Morris and Dwyer (1997) calculated that for mobile herbivores that are sensitive to food quality, inducible defenses may have no effect on spatial spread even though they reduce the rate of population growth, while constitutive defenses, in contrast, can accelerate herbivore spatial spread. Despite all these theoretical considerations, proof is still lacking that inducible defenses influence the population dynamics of attackers in the field (reviewed in Schultz 1988; Baldwin 1996).

Final Perspectives

In choosing the chapters for this book, we placed a value on taking a comprehensive, multiorganismal and multihabitat approach to the problem of why and how inducible defenses evolve. We have found the most useful insights generated from this approach to be the following.

1. Inducible defenses are almost ubiquitous and are found in diverse organisms, from protozoans to plants to fish; and in all habitats, from terrestrial to freshwater to marine. We can expect inducible defenses in systems where the factors, which we pointed out in this volume, are fulfilled. Although we cannot yet apply a formula to predict where inducible defenses will arise, we have more confidence in predicting when inducible defenses will not be favored, i.e., when costs cannot be saved, time lags in response are substantial, and selective agents are not variable. For many groups, sufficient comparative data on the incidence of inducibility are now being tallied to permit phylogenetic analyses of inducible characters, and thus will address the major open questions about how often and in what circumstances inducibility arises.

2. Evaluating the role of costs as a constraint in the evolution of inducible defenses continues to be a complex issue. Costs studies need to carefully specify different types of costs, and rarely can all possible costs be considered.

3. By necessity, most researchers have oversimplified the nature of the biotic environment in their assessments of inducible defenses. Many of us have focused on a single prey and predator in a simplified interaction to evaluate some of the issues in the evolution of these defenses. This volume

sends the clear message that prey must deal simultaneously with (a) multiple predators, (b) predators and competitors, (c) predators and pathogens, and (d) even wider trophic influences. The most important selective agent for an inducible defense may not be the predator it is directed against or even the cost of defense, but rather an alternative biotic agent or the biotic environment as a whole.

4. A big gap in understanding the dynamics of inducible defenses, whether of neckteeth in cladocerans or macrophages in the vertebrate immune system, is in assessing the precise population consequences to the consumer of the defended organism. Under what conditions are inducible defenses a controlling force in the population dynamics of consumer and prey?

References

References are arranged alphabetically by a single author's last name. Two or more authors are arranged, first, alphabetically by the first author's last name, then chronologically by date.

Abrams, P. A. 1982. Functional responses of optimal foragers. Am. Nat. 120: 382–390.
Abrams, P. A. 1984. Foraging time optimization and interactions in food webs. Am. Nat. 24: 80–96.
Abrams, P. A. 1990. The effects of adaptive behavior on the type-2 functional response. Ecology 71: 877–885.
Abrams, P. A. 1991a. Strengths of indirect effects generated by optimal foraging. Oikos 62: 167–176.
Abrams, P. A. 1991b. Life history and the relationship between food availability and foraging effort. Ecology 72: 1242–1252.
Abrams, P. A. 1992a. Predators that benefit prey and prey that harm predators: Unusual effects of interacting foraging mechanisms. Am. Nat. 140: 573–560.
Abrams, P. A. 1992b. Why don't predators have positive effects on prey populations? Evol. Ecol. 6: 56–72.
Abrams, P. A. 1993a. Why predation rates should not be proportional to predator density. Ecology 74: 726–733.
Abrams, P. A. 1993b. Indirect effects arising from optimal foraging. Pages 255–279 in H. Kawanabe, J. E. Cohen, and K. Iwasaka, eds., Mutualism and Community Organisation: Behavioural, Theoretical, and Food Web Approaches. Oxford University Press, Oxford.
Abrams, P. A. 1995. Implications of dynamically variable traits for identifying, classifying and measuring direct and indirect effects in ecological communities. Am. Nat. 146: 112–134.
Adler, F. R., and C. D. Harvell. 1990. Inducible defenses, phenotypic plasticity and biotic environments. Trends Ecol. Evol. 5: 407–410.
Adler, F. R., and R. Karban. 1994. Defended fortresses or moving targets? Another model of inducible defenses inspired by military metaphors. Am. Nat. 144: 813–832.
Adler, M. W., ed. 1993. ABC of AIDS. BMJ Publishing Group, London.
Afek, U., N. Aharoni, and S. Carmeli. 1995. The involvement of marmesin in celery resistance to pathogens during storage and the effect of temperature on its concentration. Am. Phytopath. Soc. 85: 1033–1036.
Agelopoulos, N. G., and M. A. Keller. 1994a. Plant-natural enemy association in the tritrophic system, *Cotesia rubecula-Pieris rapae*-Brassicaceae (Crucifera): I. Sources of infochemicals. J. Chem. Ecol. 20: 1725–1734.
Agelopoulos, N. G., and M. A. Keller. 1994b. Plant-natural enemy association in the tritrophic system, *Cotesia rubecula-Pieris rapae*-Brassicaceae (Crucifera): III. Col-

lection and identification of plant and frass volatiles. J. Chem. Ecol. 20: 1955–1967.

Agrawal, A. A., and R. Karban. 1997. Domatia mediate plant-arthropod mutualism. Nature 387: 562–563.

Ahmed, R., and D. Gray. 1996. Immunological memory and protective immunity—understanding their relation. Science 272: 54–60.

Ajlan, A. M., and D. A. Potter. 1991. Does immunization of cucumber against anthracnose by *Colletotrichum lagenarium* affect host suitability for arthropods? Entomol. Exp. Appl. 58: 83–92.

Ajlan, A. M., and D. A. Potter. 1992. Lack of effect of tobacco mosaic virus-induced systemic acquired resistance on arthropod herbivores in tobacco. Phytopathology 82: 647–651.

Alados, C. L., F. G. Barroso, A. Aguirre, and J. Escos. 1996. Effects of early season defoliation of *Anthyllis cytisoides* (a Mediterranean browse species) on further herbivore attack. J. Arid Environ. 34: 455–463.

Alarcon, J. J., and M. Malone. 1995. The influence of plant age on wound induction of proteinase inhibitors in tomato. Physiologia Plantarum 95: 423–427.

Alcaraz, M., and J. R. Strickler. 1988. Locomotion in copepods: Pattern of movements and energetics of *Cyclops*. Hydrobiologia 167: 409–414.

Allen, J. E., and R. M. Maizels. 1997. Th1-Th2: Reliable paradigm or dangerous dogma? Immunol. Today 18: 387–392.

Anderson, R. M., and R. M. May. 1978. Regulation and stability of host-parasite population interactions. I. Regulatory processes. J. Anim. Ecol. 47: 219–247.

Anholt, B. R. 1990. An experimental separation of exploitative and interference competition in a larval damselfly. Ecology 71: 1483–1493.

Anholt, B. R. 1997. How should we test for the role of behavior in population dynamics? Evol. Ecol. 11: 663–640.

Anholt B. R., and E. E. Werner. 1995. Interaction between food availability and predation mortality mediated by adaptive behavior. Ecology: 76: 2230–2234.

Anholt B. R., D. K. Skelly, and E. E. Werner. 1996. Factors modifying antipredator behavior in larval toads. Herpetologica 52: 301–313.

Anholt, B. R., and E. E. Werner. In press. Predictable changes in predation mortality as a consequence of changes in food availability and predation risk. Evol. Ecol.

Antia, R., J. C. Koella, and V. Perrot. 1996. Models of the within-host dynamics of persistent mycobacterial infections. Proc. Roy. Soc. Lond. B 263: 257–263.

Antonovics, J., and P. H. Thrall. 1994. The cost of resistance and the maintenance of genetic polymorphism in host-pathogen systems. Proc. Roy. Soc. Lond. B 257: 105–110.

Appert, C., E. Longemann, K. Hahlbrock, J. Schmid, and N. Armhein. 1994. Structural and catalytic properties of the four phenylalanine ammonia-lyase isoenzymes from parsley (*Petroselinum crispum* Nym.). Eur. J. Biochem. 225: 491–499.

Appleton R. D., and A. R. Palmer. 1988. Water-borne stimuli released by predatory crabs and damaged prey induce more predator-resistant shells in a marine gastropod. Proc. Natl. Acad. Sci. USA 85: 4387–4391.

Apriyanto, D., and D. A. Potter. 1990. Pathogen-activated induced resistance of cucumber: Response of arthropod herbivores to systemically protected leaves. Oecologia 85: 25–31.

REFERENCES

Arts, M. T., E. J. Maly, and M. Pasitschniak. 1981. The influence of *Acilius* (Dytiscidae) predation on *Daphnia* in a small pond. Limnol. Oceanogr. 26: 1172–1175.
Åström, M., and P. Lundberg. 1994. Plant defence and stochastic risk of herbivory. Evol. Ecol. 8: 288–298.
Augner, M. 1994. Should a plant always signal its defence against herbivores? Oikos 70: 322–332.
Azevedo-Ramos, C., M. Van Sluys, J.-M. Hero, and W. E. Magnusson. 1992. Influence of tadpole movement on predation by odonate naiads. J. Herpetol. 26: 335–338.
Baalen, M. van, and M. W. Sabelis. 1993. Coevolution of patch selection strategies of predator and prey and the consequences for ecological stability. Am. Nat. 142: 646–670.
Baker, H. G., and I. Baker. 1975. Studies of nectar-constitution and pollinator-plant coevolution. Pages 100–140 in L. E. Gilbert and P. H. Raven, eds., Coevolution of Animals and Plants. University of Texas Press, Austin.
Baldwin, I. T. 1988. Short-term damage-induced increases in tobacco alkaloids protect plants. Oecologia 75: 367–370.
Baldwin, I. T. 1991. Damage-induced alkaloids in wild tobacco. Pages 47–69 in D. W. Tallamy and M. J. Raupp, eds., Phytochemical Induction by Herbivores. John Wiley, New York.
Baldwin, I. T. 1994. Chemical changes rapidly induced by folivory. Pp. 1–23 in E. Bernays, ed., Insect Plant Interactions, vol. 5. CRC Press, Baton Rouge, Louisiana.
Baldwin I. T. 1996. Inducible defenses and population biology. Trends Ecol. Evol. 11: 104–105.
Baldwin I. T., and J. C. Schultz. 1983. Rapid changes in tree leaf chemistry induced by damage: Evidence for communication between plants. Science 221: 277–279.
Baldwin, I. T., C. L. Sims, and S. E. Kean. 1990. The reproductive consequences associated with inducible alkaloidal responses in wild tobacco. Ecology 71: 252–262.
Baldwin, I. T., and P. Callahan. 1993. Autotoxicity and chemical defense: Nicotine accumulation and carbon gain in solanaceous plants. Oecologia 94: 534–541.
Baldwin, I. T., and M. J. Karb. 1995. Plasticity in allocations of nicotine to reproductive parts in *Nicotiana attenuata*. J. Chem. Ecol. 21: 897–909.
Baldwin, I. T., and E. A. Schmelz. 1996. Immunological "memory" in the induced accumulation of nicotine in wild tobacco. Ecology 77: 236–246.
Baldwin, I. T., Z. Zhang, N. Diab, T. E. Ohnmeiss, E. S. McCloud, G. Y. Lynds, and E. A. Schmelz. 1997. Quantification, correlations, and manipulations of wound-induced changes in jasmonic acid and nicotine in *Nicotiana sylvestris*. Planta 201: 397–404.
Barbosa, P., P. Gross, and J. Kemper. 1991a. Influence of plant allelochemicals on the tobacco hornworm and its parasitoid, *Cotesia congregata*. Ecology 72: 1567–1575.
Barbosa, P., V. A. Krischick, and C. G. Jones, eds. 1991b. Microbial Mediation of Plant-Herbivore Interactions. John Wiley, New York.
Barbour, J. D., R. R. Farrar, Jr., and G. G. Kennedy. 1993. Interaction of *Manduca sexta* resistance in tomato with insect predators of *Helicoverpa zea*. Entomol. Exp. Appl. 68: 143–155.
Barry, M. J. 1994. The cost of crest induction in *Daphnia carinata*. Oecologia 97: 278–288.

Baskin, J. M., and C. M. Baskin. 1979. Studies on the autecology and population biology of the weedy monocarpic perennial, *Pastinaca sativa*. J. Ecol. 67: 601–610.

Baur, R., V. Kostal, and E. Städler. 1996. Root damage by conspecific larvae induces preference to oviposition in cabbage root flies. Entomol. Exp. Appl. 80: 224–227.

Bayly, I.A.E. 1986. Aspects of diel vertical migration in zooplankton, and its enigma variations. Pages 349–368 in P. De Deckker and W. D. Williams, eds., Limnology in Australia. Junk, Dordrecht, The Netherlands.

Bayly, I.A.E. and W. D. Williams. 1973. Inland waters and their ecology. Longman, Melbourne.

Bazzaz, F. A., N. R. Chiariello, P. D. Coley, and L. F. Pitelka. 1987. Allocating resources to reproduction and defence. BioScience 37: 58–67.

Beaton, M. J., and P.D.N. Hebert. 1997. The cellular basis of divergent head morphologies in *Daphnia*. Limnol. Oceanogr. 42: 346–356.

Beauchamp, P. de. 1952a. Un facteur de la variabilité chez les rotifères du genre *Brachionus*. Comptes rendus des séances de l'Académie des Sciences 234: 573–575.

Beauchamp, P. de. 1952b. Variation chez les rotifères du genre *Brachionus*. Comptes rendus des séances de l'Académie des Sciences 235: 1355–1356.

Beauchamp, P. de. 1965. Classe des rotifères. Pages 1225–1379 in P.-P. Grassé, ed., Traité de zoologie. Vol. 4(3). Masson & Cie, Paris.

Beddington, J. R. 1975. Mutual interference between parasites and predators and its effect on searching efficiency. J. Anim. Ecol. 44: 331–340.

Beier, R. C., and E. H. Oertli. 1983. Psoralen and other linear furocoumarins as phytoalexins in celery. Phytochem. 22: 2595–2597.

Bell, G. 1982. The Masterpiece of Nature: The Evolution and Genetics of Sexuality. University of California Press, Berkeley.

Bendelac, A., O. Lantz, M. E. Quimby, J. W. Yewdell, J. R. Bennick, and R. R. Brutkiewicz. 1995. CD1 recognition by mouse NK1.1+ T lymphocytes. Science 268: 863–865.

Bender, E. A., T. J. Case, and M .E. Gilpin. 1984. Perturbation experiments in community ecology: Theory and practice. Ecology 65: 1–13.

Benhamou, N. 1996. Elicitor-induced plant defence pathways. Trends Plant Sci. 1: 233–240.

Benndorf, J., and M. Henning. 1989. *Daphnia* and toxic blooms of *Microcystis aeruginosa* in Bautzen Reservoir (GDR). Int. Rev. Ges. Hydrobiol. 74: 233–248.

Bennett, R. N., and R. M. Wallsgrove. 1994. Tansley review no. 72. Secondary metabolites in plant defence mechanisms. New Phytol. 127: 617–633.

Berenbaum, M. 1981. Patterns of furanocoumarin distribution and insect herbivory in the Umbelliferae: Plant chemistry and community structure. Ecology 62:1254–1266.

Berenbaum, M. R. 1983. Coumarins and caterpillars: A case for coevolution. Evolution 37: 163–179.

Berenbaum, M. R. 1991. Coumarins. Pages 221–249 in G. A. Rosenthal and M. R. Berenbaum, eds., Herbivores: Their Interactions with Secondary Plant Metabolites. Vol. 1. 2d ed. Academic Press, New York.

Berenbaum, M. R. 1995a. Metabolic detoxification of plant prooxidants. Pages 181–

209 in S. Ahmad, ed., Oxidative Stress and Antioxidant Defense in Biology. Chapman and Hall, New York.

Berenbaum, M. R. 1995b. Phototoxicity of plant secondary metabolites: Insect and mammalian perspectives. Arch. Insect Biochem. Physiol. 29: 119–134.

Berenbaum, M. R., A. R. Zangerl, and J. K. Nitao. 1986. Constraints on chemical coevolution: Wild parsnips and the parsnip webworm. Evolution 40: 1215–1228.

Berenbaum, M. R., and A. R. Zangerl. 1988. Stalements in the coevolutionary arms race: Synergism, syntheses, and sundry other sins. Pages 113–132 in K. Spencer, ed., Chemical Coevolution. Academic Press, New York.

Berenbaum, M. R., A. R. Zangerl, and K. Lee. 1989. Chemical barriers to adaptation by a specialist herbivore. Oecologia 89: 501–506.

Berenbaum, M. R., and A. R. Zangerl. 1992. Genetics of physiological and behavioral resistance to host furanocoumarins in the parsnip webworm. Evolution 46: 1373–1384.

Berenbaum, M. R., and A. R. Zangerl. 1993. Furanocoumarin metabolism in *Papilio polyxenes*: Genetic variability, biochemistry, and ecological significance. Oecologia 94: 370–375.

Berenbaum, M. R., and A. R. Zangerl. 1994. Costs of inducible defense: Protein limitation, growth, and detoxification in parsnip webworms. Ecology 75: 2311–2317.

Berenbaum, M. R., and A. R. Zangerl. 1996. Phytochemical diversity: Adaptive or random variation? Rec. Adv. Phytochem. 30: 1–24.

Bergelson, J., S. Fowler, and S. Hartley. 1986. The effects of foliage damage on casebearing moth larvae, *Coleophora serratella*, feeding on birch. Ecol. Entomol. 11: 241–250.

Bergelson, J. M., and J. H. Lawton. 1988. Does foliage damage influence predation on the insect herbivores of birch? Ecology 69: 434–445.

Bergelson, J., and C. B. Purrington. 1996. Surveying patterns in the cost of resistance in plants. Am. Nat. 148: 536–558.

Bernays, E., and M. Graham. 1988. On the evolution of host specificity in phytophagous arthropods. Ecology 69: 886–892.

Bernstein, C. 1984. Prey and predator emigration responses in the acarine system *Tetranychus urticae-Phytoseiulus persimilis*. Oecologia 61: 134–142.

Berryman, A. A. 1988. Towards a unified theory of plant defense. Pages 39–55 in W. J. Mattson, J. Levieux, and C. Bernard-Dagan, eds., Mechanisms of Woody Plant Defenses against Insects: Search for Pattern. Springer-Verlag, New York.

Birch, A.N.E., D. W. Griffiths, R. J. Hopkins, and W.H.M. Smith. 1996. A time-course study of chemical and physiological responses in Brassicas induced by turnip root fly (*Delia floralis*) larval feeding. Entomol. Exp. Appl. 80: 221–223.

Blaakmeer, A., J.B.F. Geervliet, J.J.A. van Loon, M. A. Posthumus, T. A. van Beek, and A. E. de Groot. 1994a. Comparative headspace analysis of cabbage plants damaged by two species of *Pieris* caterpillars: Consequences for in-flight host location by *Cotesia* parasitoids. Entomol. Exp. Appl. 73: 175–182.

Blaakmeer, A., D. Hagenbeek, T. A. van Beek, A. E. de Groot, L. M. Schoonhoven, and J.J.A. van Loon. 1994b. Plant response to eggs vs. host marking pheromone as factors inhibiting oviposition by *Pieris brassicae*. J. Chem. Ecol. 20: 1657–1665.

Black, A. R. 1993. Predator-induced phenotypic plasticity in *Daphnia pulex*: Life-

history and morphological responses to *Notonecta* and *Chaoborus*. Limnol. Oceanogr. 38: 986–996.
Black, A. R., and S. I. Dodson. 1990. Demographic costs of *Chaoborus*-induced phenotypic plasticity in *Daphnia pulex*. Oecologia 83: 117–122.
Black, R., and L. Slobodkin. 1987. What is cyclomorphosis? Freshw. Biol. 18: 373–378.
Blau, P. A., P. Feeny, L. Contardo, and D. S. Robson. 1978. Allylglucosinolate and herbivorous caterpillars: A contrast in toxicity and tolerance. Science 200: 1296–1298.
Bluestone, J. A. 1995. New perspectives of CD28-B7-mediated T cell costimulation. Immunity 2: 555–559.
Bodnaryk, R. P. 1992. Effects of wounding on glucosinolates in the cotyledons of oilseed rape and mustard. Phytochemistry 31: 2671–2677.
Boethel, D. J., and R. D. Eikenbary. 1986. Interactions of plant resistance and parasitoids and predators of insects. John Wiley, New York.
Bogahawatte, C.N.L., and H. F. van Emden. 1996. The influence of the host plant of diamond-back moth (*Plutella xylostella*) on the plant preferences of its parasitoid *Cotesia plutellae* in Sri Lanka. Physiol. Entomol. 21: 93–96.
Bogdan, K. G., and J. J. Gilbert. 1982. The effects of posterolateral spine length and body length on feeding rate in the rotifer, *Brachionus calyciflorus*. Hydrobiologia 89: 263–268.
Bohlmann, J., E. Gibraltarskaya, and U. Eilert. 1995. Elicitor induction of furanocoumarin biosynthetic pathway in cell cultures of *Ruta graveolens*. Plant Cell, Tissue and Organ Culture 43: 155–161.
Bollens, S. M., and B. W. Frost. 1989. Predator-induced diel vertical migration in a planktonic copepod. J. Plankton Res. 11: 1047–1065.
Bollens, S. M., and B. W. Frost. 1991. Ovigerity, selective predation, and variable diel vertical migration in *Euchaeta elongata* (Copepoda: Calanoida). Oecologia 87: 155–161.
Bollens, S. M., B. W. Frost, D. S. Thoreson, and S. J. Watts. 1992. Diel vertical migration in zooplankton: Field evidence in support of the predator avoidance hypothesis. Hydrobiologia 234: 33–39.
Bollens, S. M., B. W. Frost, and J. R. Cordell. 1994. Chemical, mechanical and visual cues in the vertical migration behavior of the marine planktonic copepod *Acartia hudsonica*. J. Plankton Res. 16: 555–564.
Bolter, C. J., M. Dicke, J.J.A. van Loon, J. H. Visser, and M. A. Posthumus. 1997. Role of volatiles from herbivore-damaged potato plants in attraction of Colorado potato beetle during herbivory and after its termination. J. Chem. Ecol. 23: 1003–1023.
Bosma, M. J., and A. M. Carroll. 1993. The SCID mouse mutant: Definition, characterisation, and potential uses. Ann. Rev. Immunol. 9: 323–350.
Brattsten, L. B., J. H. Samuelian, K. Y. Long, S. A. Kincaid, and C. K. Evans. 1983. Cyanide as a feeding stimulant for the southern armyworm, *Spodoptera eridania*. Ecol. Entomol. 8: 125–132.
Brewer, M. C., and J. N. Coughlin. 1995. Virtual plankton: A novel approach to the investigation of aquatic predator-prey interactions. Mar. Freshw. Behav. Physiol. 26: 91–100.

Brönmark, C., and J. G. Miner. 1992. Predator-induced phenotypical change in body morphology in crucian carp. Science 258: 1348–1350.

Brönmark, C., and L. B. Pettersson. 1994. Chemical cues from piscivores induce a change in morphology in crucian carp. Oikos 70: 396–402.

Brönmark, C., C. A. Paszkowski, W. M. Tonn, and A. Hargeby. 1995. Predation as a determinant of size structure in populations of crucian carp (*Carassius carassius*) and tench (*Tinca tinca*). Ecol. Freshw. Fish 4: 85–92.

Bronner, R., E. Westphal, and F. Dreger. 1991a. Enhanced peroxidase activity associated with the hypersensitive response of *Solanum dulcamara* to the gall mite *Aceria cladophthirus* (Acari: Eriophyoidea). Can. J. Bot. 69: 2192–2196.

Bronner, R., E. Westphal, and F. Dreger. 1991b. Pathogenesis-related proteins in *Solanum dulcamara* L. resistant to the gall mite *Aceria cladophthirus* (Nalepa) (synonym *Eriophyes cladophthirus* Nal.). Physiol. Mol. Plant Pathol. 38: 93–104.

Brooks, J. L. 1947. Turbulence as an environmental determinant of relative growth in *Daphnia*. Proc. Natl. Acad. Sci. USA 33: 141–148.

Brooks, J. L. 1957. The systematics of North American *Daphnia*. Mem. Conn. Acad. Arts and Sci., vol. 13.

Brooks, J. L., and S. I. Dodson. 1965. Predation, body size, and composition of plankton. Science 150: 28–35.

Brown, D. G. 1988. The cost of plant defense: An experimental analysis with inducible proteinase inhibitors in tomato. Oecologia 76: 467–470.

Brown, W. L. Jr., T. Eisner and R. H. Whittaker. 1970. Allomones and kairomones: Transspecific chemical messengers. Bio. Sci. 20: 21–22.

Bruin, J., M. Dicke, and M. W. Sabelis. 1992. Plants are better protected against spider mites after exposure to volatiles from infected conspecifics. Experientia 48: 525–529.

Bruin, J., M. W. Sabelis, and M. Dicke. 1995. Do plants tap SOS signals from their infested neighbours? Trends Ecol. Evol. 10: 167–170.

Brunt, A. A., K. Crabtree, M. J. Dallwitz, A. J. Gibbs, L. Watson, and E. J. Zurcher, eds. 1996. Plant Viruses online: Descriptions and lists from the VIDE database. Version: 20 August 1996. URL http://biology.anu.edu.au/Groups/MES/vide/.

Brusca, R. C. 1980. Common intertidal invertebrates of the Gulf of California. University of Arizona Press, Tucson.

Bruton, O. C. 1952. Agammaglobulinemia. Pediatrics 9: 722–728.

Bryant, B. P., and J. Atema. 1987. Diet manipulation affects social behaviour of catfish: Importance of body odour. J. Chem. Ecol. 13: 645–659.

Bryant, J. P., I. Heitkonig, P. Kuropat, and N. Owen-Smith. 1991a. Effects of severe defoliation on the long-term resistance of insect attack and on leaf chemistry in six woody species of the southern African savanna. Am. Nat. 137: 50–63.

Bryant, J. P., K. Danell, F. Provenza, P. B. Reichardt, T. A. Clausen, and R. A. Werner. 1991b. Effects of mammal browsing on the chemistry of deciduous woody plants. Pages 135–154 in D. W. Tallamy and M. J. Raupp, eds., Phytochemical Induction by Herbivores. John Wiley, New York.

Bryant, J. P., P. B. Reichardt, T. A. Clausen, and R. A. Werner. 1993. Effects of mineral nutrition on delayed inducible resistance in Alaska paper birch. Ecology 74: 2072–2084.

Bull, J. J. 1987. Evolution of phenotypic variance. Evolution 41: 303–315.

Bulmer, M. 1994. Theoretical Evolutionary Ecology. Sinauer Associates, Sunderland, Mass.

Burns, C. W. 1968. The relationship between body size of filter-feeding Cladocera and the maximum size of particles ingested. Limnol. Oceanogr. 13: 675–678.

Burns, C. W., and J. J. Gilbert. 1986a. Effects of daphnid size and density on interference between *Daphnia* and *Keratella cochlearis*. Limnol. Oceanogr. 31: 848–858.

Burns, C. W., and J. J. Gilbert 1986b. Direct observations of the mechanisms of interference between *Daphnia* and *Keratella cochlearis*. Limnol. Oceanogr. 31: 859–866.

Buss, L. W., J.B.C. and Jackson. 1979. Competitive networks: Nontransitive competitive relationships in cryptic coral reef environments. Am. Nat. 113: 223–234.

Campbell, B. C., and S. S. Duffey. 1979. Tomatine and parasitic wasps: Potential incompatibility of plant-antibiosis with biological control. Science 205: 700–702.

Cappuccino, N., and P. W. Price. 1995. Population Dynamics: New Approaches and Hypotheses. Academic Press, New York.

Carroll, C. R., and C. A. Hoffman. 1980. Chemical feeding deterrent mobilized in response to insect herbivory and counteradaptation by *Epilachna tredecimnotata*. Science 209: 414–416.

Cerkauskas, R. F. 1986. Susceptibility of parsnip cultivars to canker caused by *Phoma complanata*. Can. J. Plant Pathol. 8: 455–458.

Cerkauskas, R. F., and M. Chiba. 1990. Association of phoma canker with photocarcinogenic furocoumarins in parsnip cultivars. Can. J. Plant Pathol. 12: 349–357.

Cerkauskas, R. F., and M. Chiba. 1991. Soil densities of *Fusarium oxysporum* f. sp. *apii* race 2 in Ontario, and the association between celery cultivar resistance and photocarcinogenic furocoumarins. Can. J. Plant Path. 13: 305–314.

Ceska, O., S. Chaudhary, P. Warrington, G. Poulton, and M. Ashwood-Smith. 1986. Naturally occurring crystals of photocarcinogenic furocoumarins on the surface of parsnip roots sold as food. Experientia 42: 1302–1304.

Chambliss, O. L., and C. M. Jones. 1966. Cucurbitacins: Specific insect attractants in Cucurbitaceae. Science 153: 1392–1393.

Channon, A. G. 1965. Studies on parsnip canker. IV. *Centrospora acerina* (Hartig) Newhall—a further cause of black canker. Ann. Appl. Biol. 56: 119–128.

Channon, A. G., and M. C. Thompson. 1981. Parsnip canker caused by *Cylindrocarpon destructions*. Plant Path. 30: 181.

Charlesworth, B. 1980. Evolution in age-structured populations. Cambridge University Press, New York.

Charnov, E. L. 1976. Optimal foraging: The marginal value theorem. Theor. Pop. Biol. 9: 129–136.

Charnov E. L., G. H. Orians, and K. Hyatt. 1976. Ecological implications of resource depression. Am. Nat. 110: 247–259.

Cheah, C. A., and T. H. Coaker. 1992. Host finding and discrimination in *Diglyphus isaea*, a parasitoid of the chrysanthemum leaf miner, *Chromatomyia syngenesiae*. Biocontrol Sci. Technol. 2: 109–118.

Cheng, T. C. 1991. Is parasitism symbiosis? A definition of terms and the evolution of concepts. Pages 15–36 in C. A. Toft, A. Aeschlimann, and L. Bolis, eds., Parasite-host associations: Coexistence or conflict? Oxford University Press, Oxford.

Chew, F. S. 1980. Foodplant preferences of *Pieris* caterpillars. Oecologia 46: 347–353.

Chew, F. S., and J. E. Rodman. 1979. Plant resources for chemical defense. Pages 271–307 in G. A. Rosenthal and D. H. Janzen, eds., Herbivores: Their Interaction with Secondary Plant Metabolites. Academic Press, New York.

Chew, F. S., and J.A.A. Renwick. 1995. Host plant choice in *Pieris* butterflies. Pages 214–238 in W. J. Bell and R. T. Cardé, eds., Chemical Ecology of Insects, vol. 2. Chapman and Hall, New York.

Chivers, D. P., and R.J.F. Smith. 1995. Free-living fathead minnows rapidly learn to recognize pike as predators. J. Fish Biol. 46: 949–954.

Chornesky, E. A. 1983. Induced development of sweeper tentacles on the reef coral *Agaricia agaricites*: A response to direct competition. Biol. Bull. 165: 569–581.

Chornesky, E. A., and S. L. Williams. 1983. Distribution of sweeper tentacles on *Montastrea cavernosa*. Pages 61–68 in M. L. Reaka, ed., The Ecology of Deep and Shallow Reefs. Symp. for Undersea Res. NURC., Rockville, Md.

Christiansen, F. B. 1991. On conditions for evolutionary stability for a continuously varying character. Am. Nat. 138: 37–50.

Clark, C. W. 1990. Mathematical Bioeconomics: The Optimal Management of Renewable Resources. John Wiley, New York.

Clark, C. W., and C. D. Harvell. 1992. Inducible defenses and the allocation of resources: A minimal model. Am. Nat. 139: 521–539.

Clark, W. C. 1978. Metabolite mediated density-dependent sex determination in a free-living nematode, *Diplenteron potohikus*. J. Zool. Lond. 184: 245–254.

Coley, P. D., J. P. Bryant, and F. S. Chapin III. 1985. Resource availability and plant anitherbivore defence. Science 230: 895–899.

Coll, M. 1996. Feeding and ovipositing on plants by an omnivorous insect predator. Oecologia 105: 214–220.

Collins, J. P., and J. E. Cheek. 1983. Effect of food and density on development of typical and cannibalistic salamander larvae in *Ambystoma tigrinum nebulosum*. Am. Zool. 23: 77–84.

Colten, H. R., and F. S. Rosen. 1992 Complement deficiencies. Annual Review of Immunology 10: 809–834.

Confer, J. L., G. Applegate, and C. A. Evanik. 1980. Selective predation by zooplankton and the response of cladoceran eyes to light. Pages 604–608 in W. C. Kerfoot, ed., Evolution and Ecology of Zooplankton Communities. University Press of New England, Hanover, N.H.

Connell, J. H. 1961. The influence of interspecific competition and other factors on the distribution of the barnacle *Chthamalus stellatus*. Ecology 42: 710–723.

Craig, J. F. 1996. Population dynamics, predation and role in the community. Pages 201–218 in J. F. Craig, ed., Pike: Biology and Exploitation. Chapman and Hall, London.

Crawley, M. J. 1983. Herbivory: The Dynamics of Animal-Plant Interactions. Blackwell Scientific, Oxford.

Crespi, B. J. 1988. Adaptation, compromise, and constraint: The development, morphometrics, and behavioral basis of a fighter-flier polymorphism in male *Hoplothrips karnyi* (Insecta: Thysanoptera). Behav. Ecol. Sociobiol. 23: 93–104.

Croft, K. P., F. Juttner, and A. J. Slusarenko. 1993. Volatile products of the lipoxygenase pathway evolved from *Phaseolus vulgaris* (L.) leaves inoculated with *Pseudomonas syringae* pv *phaseolicola*. Plant Physiol. 101: 13–24.

Cronin, G., and M. E. Hay. 1996. Amphipod grazing and induction of seaweed chemical defenses. Ecology 77: 2287–2301.

Crowl, T. A., and A. P. Covich. 1990. Predator-induced life-history shifts in a freshwater snail. Science 247: 949–951.

Curio, E. 1976. The Ethology of Predation. Springer-Verlag, Berlin.

Cushing, D. H. 1951. The vertical migration of planctonic Crustacea. Biol. Rev. 26: 158–192.

Da Costa, C. P., and C. M. Jones. 1971. Cucumber beetle resistance and mite susceptibility controlled by the bitter gene in *Cucumis sativus* L. Science 172: 1145–1146.

Danell, K., and K. Huss-Danell. 1985. Feeding by insects and hares on birches earlier affected by moose browsing. Oikos 4: 75–81.

David, W.A.L., and B.O.C. Gardiner. 1966. Mustard oil glucosides as feeding stimulants for *Pieris brassicae* larvae in a semi-synthetic diet. Entomol. Exp. Appl. 9: 247–255.

Dawidowicz, P. 1993. Diel vertical migration in *Chaoborus flavicans*: population patterns vs. individual tracks. Arch. Hydrobiol., Beih. Ergebn. Limnol. 39: 19–28.

Dawidowicz, P., J. Pijanowska, and K. Ciechomski. 1990. Vertical migration of *Chaoborus* larvae is induced by the presence of fish. Limnol. Oceanogr. 35: 1631–1637.

Dawidowicz, P., and C. J. Loose. 1992a. Cost of swimming by *Daphnia* during diel vertical migration. Limnol. Oceanogr. 37: 665–669.

Dawidowicz, P., and C. J. Loose. 1992b. Metabolic costs during predator-induced diel vertical migration of *Daphnia*. Limnol. Oceanogr. 37: 1589–1595.

De Boer, R. J., and A. S. Perelson. 1993. How diverse should the immune system be? Proc. Roy. Soc. Lond. B. 252: 171–175.

De Meester, L. 1990. Evidence for intra-population genetic variability for phototactic behaviour in *Daphnia magna* Straus, 1820. Biol. Jb. Dodonaea 58: 84–93.

De Meester, L. 1991. An analysis of the phototactic behaviour of *Daphnia magna* clones and their sexual descendants. Hydrobiologia 225: 217–227.

De Meester, L. 1993a. Genotype, fish-mediated chemicals and phototaxis in *Daphnia*. Ecology 74: 1467–1474.

De Meester, L. 1993b. The vertical distribution of *Daphnia magna* genotypes selected for different phototactic behaviour: Outdoor experiments. Arch. Hydrobiol., Beih. Ergebn. Limnol. 39: 137–155.

De Meester, L. 1994a. Life histories and habitat selection in *Daphnia*: Divergent life histories of *D. magna* clones differing in phototactic behaviour. Oecologia 97: 333–341.

De Meester, L. 1994b. Habitat partitioning in *Daphnia*: Coexistence of *Daphnia magna* clones differing in phototactic behaviour. Pages 323–335 in A. R. Beaumont, ed., Genetics and Evolution of Aquatic Organisms. Chapman and Hall, London.

De Meester, L. 1995. Life history characteristics of *Daphnia magna* clones differing in phototactic behaviour. Hydrobiologia 307: 167–175.

De Meester, L. 1996a. Evolutionary potential and local genetic differentiation in a

phenotypically plastic trait of a cyclical parthenogen, *Daphnia magna*. Evolution 50: 1293–1298.
De Meester, L. 1996b. Local genetic differentiation and adaptation in freshwater zooplankton populations: Patterns and processes. Ecoscience 3: 385–399.
De Meester, L., and N. Beenaerts. 1993. Heritable variation in carotenoid content in *Daphnia magna*. Limnol. Oceanogr. 38: 1193–1200.
De Meester, L., J. Vandenberghe, K. Desender, and H. J. Dumont. 1994. Genotype-dependent daytime vertical distribution of *Daphnia magna* in a shallow pond. Belg. J. Zool. 124: 3–9.
De Meester, L., L. J. Weider, and R. Tollrian. 1995. Alternative antipredator defences and genetic polymorphism in a pelagic predator-prey system. Nature 378: 483–485.
De Meester, L., and J. Pijanowska. 1996. On the trait-specificity of the response of *Daphnia* genotypes to the chemical presence of a predator. Pages 407–417 in P. H. Lenz, D. K. Hartline, J. E. Purcell, and D. L. Macmillan, eds., Zooplankton: Sensory Ecology and Physiology. Gordon and Breach, Amsterdam.
De Meester, L., and C. Cousyn. 1997. The change in phototactic behaviour of a *Daphnia magna* clone in the presence of fish kairomones: The effect of exposure time. Hydrobiologia. 360: 169–175.
DeMott, W. R. 1989. The role of competition in zooplankton succession. Pages 195–252 in U. Sommer, ed., Plankton Ecology. Springer-Verlag, Berlin.
DeMott, W. R. 1995. The influence of prey hardness on *Daphnia*'s selectivity for large prey. Hydrobiologia 307: 127–138.
den Hartog, J. C. 1977. The marginal tentacles of *Rhodactis sanctithomae* (Corallimorpharia) and the sweeper tentacles of *Montastrea cavernosa* (Scleractinia), their cnidom and possible function. Pages 463–469 in D. L. Taylor, ed., Proc. Third Intl. Coral Reef Symp. University of Miami Press, Miami.
Denno, R. F., and M. S. McClure, eds. 1983. Variable Plants and Herbivores in Natural and Managed Systems. Academic Press, New York.
Dercks, W., J. Trumble, and C. Winter. 1990. Impact of atmospheric pollution on linear furanocoumarin content in celery. J. Chem. Ecol. 16: 443–454.
Desjardins, A. E., R. D. Plattner, and G. F. Spencer. 1988. Inhibition of trichothecene toxin biosynthesis by naturally occurring shikimate aromatics. Phytochem. 27: 767–771.
Desjardins, A. E., G. F. Spencer, and R. D. Plattner. 1989a. Tolerance and metabolism of furanocoumarins by the phytopathogenic fungus *Gibberella pulicaris* (*Fusarium sambucinum*). Phytochem. 28: 2963–2969.
Desjardins, A. E., G. F. Spencer, R. D. Plattner, and M. N. Beremand. 1989b. Furanocoumarin phytoalexins, trichothecene toxins, and infection of *Pastinaca sativa* by *Fusarium sporotrichioides*. Phytopath. 79: 170–175.
Dicke, M. 1986. Volatile spider-mite pheromone and host-plant kairomone, involved in spaced-out gregariousness in the spider mite *Tetranychus urticae*. Physiol. Entomol. 11: 251–262.
Dicke, M. 1988. Prey preference of the phytoseiid mite *Typhlodromus pyri*: 1. Response to volatile kairomones. Exp. Appl. Acarol. 4: 1–13.
Dicke, M. 1994. Local and systemic production of volatile herbivore-induced terpenoids: Their role in plant-carnivore mutualism. J. Plant Physiol. 143: 465–472.

Dicke, M. 1995. Why do plants "talk"? Chemoecology 5/6: 159–165.
Dicke, M., and A. Groeneveld. 1986. Hierarchical structure in kairomone preference of the predatory mite *Amblyseius potentillae*: Dietary component indispensable for diapause induction affects prey location behaviour. Ecol. Entomol. 11: 131–138.
Dicke, M., M. W. Sabelis, and A. Groeneveld. 1986. Vitamin A deficiency modifies response of predatory mite *Amblyseius potentillae* to volatile kairomone of two-spotted spider mite, *Tetranychus urticae*. J. Chem. Ecol. 12: 1389–1396.
Dicke, M., and M. W. Sabelis. 1988a. Infochemical terminology: Based on cost-benefit analysis rather than origin of compounds? Funct. Ecol. 2: 131–139.
Dicke, M., and M. W. Sabelis. 1988b. How plants obtain predatory mites as bodyguards. Neth. J. Zool. 38: 148–165.
Dicke, M., M. W. Sabelis, and M. de Jong. 1988. Analysis of prey preference of phytoseiid mites as determined with an olfactometer, predation models and electrophoresis. Exp. Appl. Acarol. 5: 225–241.
Dicke, M., and M. W. Sabelis. 1989. Does it pay to advertise for bodyguards? Towards a cost-benefit analysis of induced synomone production. Pages 341–353 in H. Lambers, M. L. Cambridge, H. Konings and T. L. Pons, eds., Causes and Consequences of Variation in Growth Rate and Productivity of Higher Plants. SPB Academic Publishing, The Hague.
Dicke, M., M. de Jong, M.P.T. Alers, F.C.T. Stelder, R. Wunderink, and J. Post. 1989. Quality control of mass-reared arthropods: Nutritional effects on performance of predatory mites. J. Appl. Entomol. 108: 462–475.
Dicke, M., T. A. van Beek, M. A. Posthumus, N. Ben Dom, H. van Bokhoven, and A. E. de Groot. 1990a. Isolation and identification of volatile kairomone that affects acarine predator-prey interactions: Involvement of host plant in its production. J. Chem. Ecol. 16: 381–396.
Dicke, M., M. W. Sabelis, J. Takabayashi, J. Bruin, and M. A. Posthumus. 1990b. Plant strategies of manipulating predator-prey interactions through allelochemicals: Prospects for application in pest control. J. Chem. Ecol. 16: 3091–3118.
Dicke, M., and O.P.J.M. Minkenberg. 1991. Role of volatile infochemicals in foraging behavior of the leafminer parasitoid *Dacnusa sibirica* (Hymenoptera: Braconidae). J. Insect Behav. 4: 489–500.
Dicke, M., and H. Dijkman. 1992. Induced defence in detached uninfested plant leaves: Effects on behaviour of herbivores and their predators. Oecologia 91: 554–560.
Dicke, M., P. van Baarlen, R. Wessels, and H. Dijkman. 1993. Herbivory induces systemic production of plant volatiles that attract predators of the herbivore: Extraction of endogenous elicitor. J. Chem. Ecol. 19: 581–599.
Dicke, M., and L.E.M. Vet. 1998. Plant-carnivore interactions: Evolutionary and ecological consequences for plant, herbivore and carnivore. In H. Olff, V. K. Brown, and R. H. Drent, eds., Herbivores, Plants and Predators. Proc. 38th Symp. British Ecological Society. In press.
Dicke, M., J. Takabayashi, M. A. Posthumus, C. Schütte, and O. E. Krips. 1998. Plant-phytoseiid interactions mediated by prey-induced plant volatiles: Variation in production of cues and variation in responses of predatory mites. Exp. Appl. Acarol. 22: 311–333.
Doares, S. H., J. Narvaez-Vasquez, A. Conconi, and C. A. Ryan. 1995a. Salicylic acid

inhibits synthesis of proteinase inhibitors in tomato leaves induced by systemin and jasmonic acid. Plant Physiol. 108: 1741–1746.
Doares, S. H., T. Syrovets, E. W. Weiler, and C. A. Ryan. 1995b. Oligogalacturonides and chitosan activate plant defensive genes through the octadecanoid pathway. Proc. Natl. Acad. Sci. USA 92: 4095–4098.
Dodson, S. I. 1972. Mortality in a population of *Daphnia rosea*. Ecology 53: 1011–1023.
Dodson, S. I. 1974. Adaptive change in plankton morphology in response to size-selective predation: A new hypothesis of cyclomorphosis. Limnol. Oceanogr. 19: 721–729.
Dodson, S. I. 1988a. Cyclomorphosis in *Daphnia galeata mendotae* Birge and *Daphnia retrocurva* Forbes as a predator-induced response. Freshw. Biol. 19: 109–114.
Dodson, S. I. 1988b. The ecological role of chemical stimuli for zooplankton: Predator-avoidance behavior in *Daphnia*. Limnol. Oceanogr. 33: 1431–1439.
Dodson, S. I. 1989a. The ecological role of chemical stimuli for the zooplankton: Predator-induced morphology in *Daphnia*. Oecologia 78: 361–367.
Dodson, S. 1989b. Predator-induced reaction norms. Bioscience 39: 447–453.
Dodson, S. I. 1990. Predicting diel vertical migration of zooplankton. Limnol. Oceanogr. 35: 1195–1200.
Dodson, S. I., and J. E. Havel. 1988. Indirect prey effects: Some morphological and life history responses of *Daphnia pulex* exposed to *Notonecta undulata*. Limnol. Oceanogr. 33: 1274–1285.
Dodson, S. I., and C. W. Ramcharan. 1991. Size-specific swimming behavior of *Daphnia pulex*. J. Plankton Res. 13: 1365–1379.
Dodson, S. I., and D. G. Frey. 1991. Cladocera and other Branchiopoda. Pages 723–786 in J. H. Thorp and A. P. Covich, eds., Ecology and Classification of North American Freshwater Invertebrates. Academic Press, New York.
Dodson, S. I., T. A. Crowl, B. L. Peckarsky, L. B. Kats, A. P. Kovich, and J. M. Culp. 1994. Non-visual communication in freshwater benthos: An overview. J. N. Am. Benthol. Soc. 13: 268–282.
Doherty, H. M., R. R. Selvendran, and D. J. Bowles. 1988. The wound response of tomato plants can be inhibited by aspirin and related hydroxybenzoic acids. Physiol. Mol. Plant Pathol. 33: 377–384.
Doughty, K. J., G. A. Kiddle, B. J. Pye, R. M. Wallsgrove, and J. A. Pickett. 1995. Selective induction of glucosinolates in oilseed rape leaves by methyl jasmonate. Phytochemistry 38: 347–350.
Douglas, C., H. Hoffman, W. Schulz, and K. Hahlbrock. 1987. Structure and elicitor or UV light-stimulated expression of two 4-coumarate: CoA ligase genes in parsley. EMBO J. 6: 1189–1195.
Downum, K. R. 1992. Tansley review no. 43. Light-activated plant defence. New Phytol. 122: 401–420.
Drukker, B., P. Scutareanu, and M. W. Sabelis. 1995. Do anthocorid predators respond to synomones from *Psylla*-infested pear trees under field conditions? Entomol. Exp. Appl. 77: 193–203.
Du, Y.-J., G. M. Poppy, and W. Powell. 1996. The relative importance of semiochemicals from the first and second trophic level in the host foraging behavior of *Aphidius ervi*. J. Chem. Ecol. 22: 1591–1605.

Duffey, S. S., and G. W. Felton. 1989. Plant enzymes in resistance to insects. Pages 289–313 in J. R. Whitaker and P. E. Sonnet, eds., Biocatalysis in Agricultural Biotechnology. American Chemical Society, Toronto.

Duffey, S. S., K. Hoover, B. Bonning, and B. D. Hammock. 1995. The impact of host plant on the efficacy of baculoviruses. Pages 137–275 in R. M. Roe and R. J. Kuhr, eds., Review of Pesticides and Toxicology, vol. 3. Toxicology Commincations, Raleigh, N.C.

Duhr, B. 1955. Über Bewegung, Orientierung und Beutefang der Corethralarve (*Chaoborus crystallinus* de Geer). Zool. Jahrb. 65: 387–429.

Dumont, H. J., I. Miron, V. D'Allasta, W. Decraemer, C. Claus, and D. Somers. 1973. Limnological aspects of some Moroccon Atlas lakes, with reference to some physical and chemical variables, the nature and distribution of the phyto- and zooplankton, including a note on possibilities for the development of an inland fishery. Int. Rev. Ges. Hydrobiol. 58: 33–60.

Dungan, M. L. 1985. Competition and the morphology, ecology and evolution of acorn barnacles: An experimental test. Paleobiology 11: 165–173.

Dungan, M. L. 1987. Indirect mutualism: Complementary effects of grazing and predation on a rocky intertidal community. Pages 188–200 in W. C. Kerfoot and A. Sih, eds., Predation: Direct and Indirect Impacts on Aquatic Communities. University Press of New England, Hanover, N.H.

Dunn, J. A., and J. Kirkley. 1966. Studies on the aphid *Cavariella aegopodii*. Scop. II. On secondary hosts other than carrot. Ann. Appl. Biol. 58: 213–217.

Du Pasquier, L. 1982. Antibody diversity in lower vertebrates—why is it so restricted? Nature 296: 311–313.

Ebel, J. 1986. Phytoalexin synthesis: The biochemical analysis of the induction process. Ann. Rev. Phytopathol. 24: 235–264.

Eberhard, W. 1982. Beetle horn dimorphism: Making the best of a bad lot. Am. Nat. 119: 420–426.

Edelstein-Keshet, L., and M. D. Rausher. 1989. The effects of inducible plant defenses on herbivore populations: 1. Mobile herbivores in continuous time. Am. Nat. 133: 787–810.

Edwards, P. J., and S. D. Wratten. 1983. Wound-induced defenses in plants and their consequences for patterns of insect grazing. Oecologia 59: 88–93.

Edwards, P. J., S. D. Wratten, and R. M. Gibberd. 1991. The impact of inducible phtyochemicals on food selection by insect herbivores and its consequences for the distribution of grazing damage. Pages 205–221 in D. W. Tallamy and M. J. Raupp, eds., Phytochemical Induction by Herbivores. John Wiley, New York.

Edwards, P. J., S. D. Wratten, and E. A. Parker. 1992. The ecological significance of rapid wound-induced changes in plants: Insect grazing and plant competition. Oecologia 91: 266–272.

Egan, P. F., and F. R. Trainor. 1989. The role of unicells in the polymorphic *Scenedesmus armatus* (Chlorophyceae). J. Phycol. 25: 65–70.

Ekström, C. U. 1838. Iakttagelser öfver formförändringen hos rudan (*Cypr. carassius* Lin.). Kungliga Vetenskapsakademins Handlingar 1838: 213–225.

Ellner, S., and N. G. Hairston, Jr. 1994. Role of overlapping generations in maintaining genetic variation in a fluctuating environment. Am. Nat. 143: 403–417.

Elnagar, S., and A. F. Murant. 1976. The role of the helper virus, anthriscus yellows,

in the transmission of parsnip yellow fleck virus by the aphid *Cavariella aegopodii*. Ann. Appl. Biol. 84: 169–181.
Erman, L. A. 1962. Cyclomorphosis and feeding of plankton Rotifera [in Russian]. Zoolohichnyi Zhurnal Ukrayiny 41: 998–1003.
Eshel, I. 1983. Evolutionary and continuous stability. J. Theor. Biol. 103: 99–111.
Etter, R. J. 1988. Asymmetrical developmental plasticity in an intertidal snail. Evolution 42: 322–334.
Facey, D. E., and G. D. Grossman. 1990. The metabolic cost of maintaining position for four North American stream fishes: Effects of season and velocity. Physiol. Zool. 63: 757–776.
Faeth, S. H. 1991. Variable induced responses: direct and indirect effects on oak folivores. Pages 293–323 in D. W. Tallamy and M. J. Raupp, eds., Phytochemical Induction by Herbivores. John Wiley, New York.
Fagerström, T. 1989. Anti-herbivore chemical defence in plants: A note on the concept of cost. Am. Nat. 133: 281–287
Fairweather, P. G. 1988. Predation creates haloes of bare space among prey on rocky seashore in New South Wales. Austral. J. Ecol. 13: 401–410.
Farmer, E. E., and C. A. Ryan. 1990. Interplant communication: Airborne methyl jasmonate induces synthesis of proteinase inhibitors in plant leaves. Proc. Natl. Acad. Sci. USA 87: 7713–7716.
Fearon, D. T., and R. M. Locksley. 1996. The instructive role of innate immunity in the acquired immune response. Science 272: 50–53.
Feeny, P. 1976. Plant apparency and chemical defense. Rec. Adv. Phytochem. 10: 1–40.
Feeny, P. 1977. Defensive ecology of the Cruciferae. Ann. Miss. Bot. Gar. 64: 221–234.
Feminella, J. W., and C. P. Hawkins. 1994. Tailed frog tadpoles differentially alter their feeding behavior in response to non-visual cues from four predators. JNABS 13: 310–320.
Fernandes, G. W. 1990. Hypersensitivity: A neglected plant resistance mechanism against insect herbivores. Environ. Entomol. 19: 1173–1182.
Finidori-Logli, V., A. G. Bagneres, and J. L. Clement. 1996. Role of plant volatiles in the search for a host by parasitoid *Diglyphus isaea* (Hymenoptera: Eulophidae). J. Chem. Ecol. 22: 541–558.
Flecker, A. S. 1992. Fish predation and the evolution of invertebrate drift periodicity: Evidence from neotropical streams. Ecology 73: 438–448.
Flick, B.J.G., and J. Ringelberg. 1993. Influence of food availability on the initiation of diel vertical migration (DVM) in lake Maarsseveen. Arch. Hydrobiol. Beih. 39: 57–65.
Folt, C., P. C. Schulze, and K. Baumgartner. 1993. Characterizing a zooplankton neighborhood: Small-scale patterns of association and abundance. Freshw. Biol. 30: 289–300.
Forward, R. B., Jr. 1988. Diel vertical migration: Zooplankton photobiology and behaviour. Oceanogr. Mar. Biol. Ann. Rev. 26: 361–393.
Forward, R. B., Jr. 1993. Photoresponses during diel vertical migration of brine shrimp larvae: Effect of predator exposure. Arch. Hydrobiol. Beih. Ergebn. Limnol. 39: 37–44.

Forward, R. B., Jr., and W. F. Hettler, Jr. 1992. Effects of feeding and predator exposure on photoresponses during diel vertical migration of brine shrimp larvae. Limnol. Oceanogr. 37: 1261–1270.

Fowden, L., and P. J. Lea. 1979. Mechanism of plant avoidance of autotoxicity by secondary metabolites, especially by nonprotein amino acids. Pages 135–160 in G. A. Rosenthal and D. H. Janzen, eds., Herbivores: Their Interaction with Secondary Plant Metabolites. Academic Press, New York.

Fowler, S. V., and J. H. Lawton. 1985. Rapidly induced defenses and talking trees: The devil's advocate position. Am. Nat. 126: 181–195.

Francko, D. A., and R. G. Wetzel. 1982. Production and release of cAMP by *Daphnia pulex*: Implications of grazing activity. J. Freshw. Ecol. 1: 365–371.

Frank, S. A. 1996. Models of parasite virulence. Quart. Rev. Biol. 71: 37–78.

Fry, J. D. 1989. Evolutionary adaptation to host plants in a laboratory population of the phytophagous mite *Tetranychus urticae* Koch. Oecologia 81: 559–565.

Fulton, R. S., and H. W. Paerl. 1987. Effects of colonial morphology on zooplankton utilization of algal resources during blue-green algal (*Microcystis aeruginosa*) blooms. Limnol. Oceanogr. 32: 634–644.

Futuyma, D. J., and M. Slatkin. 1983. Coevolution. Sinauer Associates, Sunderland, Mass.

Fyda, J., and K. Wiąckowski. 1992. Predator-induced morphological defences in the ciliate *Colpidium*. Europ. J. Protistol. 28: 341.

Gabriel, W., and B. Thomas. 1988. Vertical migration of zooplankton as an evolutionary stable strategy. Am. Nat. 132: 199–216.

Gabriel, W., and M. Lynch. 1992. The selective advantage of reaction norms for environmental tolerance. J. Evol. Biol. 5: 41–59.

Garrity, S. D., and S. C. Levings. 1981. A predator-prey interaction between two physically and biologically constrained rocky shore gastropods: Direct, indirect and community effects. Ecol. Monogr. 51: 267–286.

Geervliet, J.B.F., L.E.M. Vet, and M. Dicke. 1994. Volatiles from damaged plants as major cues in long-range host-searching by the specialist parasitoid *Cotesia rubecula*. Entomol. Exp. Appl. 73: 289–297.

Geervliet, J.B.F., L.E.M. Vet, and M. Dicke. 1996. Innate responses of the parasitoids *Cotesia glomerata* and *C. rubecula* (Hymenoptera: Braconidae) to volatiles from different plant-herbivore complexes. J. Insect Behav. 9: 525–538.

Geervliet, J.B.F., A. I. Vreugdenhil, M. Dicke, and L.E.M. Vet. 1998. Learning to discriminate between infochemicals from different plant-host complexes by the parasitoids *Cotesia glomerata* and *C. rubecula*. Entomol. Exp. Appl. 86: 241–252.

Geervliet, J.B.F., M. A. Posthumus, L.E.M. Vet, and M. Dicke. 1997. Comparative analysis of headspace from different caterpillar-infested or uninfested food plants of *Pieris* caterpillars. J. Chem. Ecol. 23: 2935–2954.

George, D. G. 1983. Interrelation between the vertical distribution of *Daphnia* and chlorophyll-a in two large limnetic enclosures. J. Plankton Res. 5: 457–475.

Georgiadis, N. J., and S. J. McNaughton. 1988. Interactions between grazers and a cyanogenic grass, *Cynodon plectostachys*. Oikos 51: 343–350.

Gerritsen, J., and J. R. Strickler. 1977. Encounter probabilities and community structure in zooplankton: A mathematical model. J. Fish. Res. Board Can. 34: 73–82.

Giamoustaris, A., and R. Mithen. 1995. The effect of modifying the glucosinolate content of leaves of oilseed rape (*Brassica napus* ssp. oleifera) on its interaction with specialist and generalist pests. Ann. Appl. Biol. 126: 347–363.

Giamoustaris, A., and R. Mithen. 1996. The effect of flower colour and glucosinolates on the interaction between oilseed rape and pollen beetles. Entomol. Exp. Appl. 80: 206–208.

Giamoustaris, A., and R. Mithen. 1997. Glucosinolate and disease resistance in oilseed rape (*Brassica napus* ssp. oleifera). Plant Pathol. 46: 271–275.

Gibbons, R. J. 1992. How microorganisms cause disease. Pages 7–10, in S. L. Gorbach, J. G. Bartlett, and N. R. Blacklow, eds., Infectious Diseases. W. B. Saunders, London.

Gilbert, J. J. 1966. Rotifer ecology and embryological induction. Science 151: 1234–1237.

Gilbert, J. J. 1967. *Asplanchna* and posterolateral spine induction in *Brachionus calyciflorus*. Arch. Hydrobiol. 64: 1–62.

Gilbert, J. J. 1980. Further observations on developmental polymorphism and its evolution in the rotifer *Brachionus calyciflorus*. Freshw. Biol. 10: 281–294.

Gilbert, J. J. 1985. Escape response of the rotifer *Polyarthra*: A high-speed cinematographic analysis. Oecologia 66: 322–331.

Gilbert, J. J. 1987. The *Polyarthra* escape response: Defense against interference from *Daphnia*. Hydrobiologia 147: 235–238.

Gilbert, J. J. 1988a. Susceptibilities of ten rotifer species to interference from *Daphnia pulex*. Ecology 69: 1826–1838.

Gilbert, J. J. 1988b. Suppression of rotifer populations by *Daphnia*: A review of the evidence, the mechanisms, and the effects on zooplankton community structure. Limnol. Oceanogr. 33: 1286–1303.

Gilbert, J. J. 1992. Rotifera. Pages 115–136 in K. G. Adiyodi and R. G. Adiyodi, eds., Reproductive Biology of Invertebrates, vol. 5: Sexual Differentiation and Behaviour. Oxford and IBH publishing Company, New Delhi.

Gilbert, J. J. 1994. Susceptibility of planktonic rotifers to a toxic strain of *Anabaena flos*-aquae. Limnol. Oceanogr. 39: 1286–1297.

Gilbert, J. J. 1996. Effect of temperature on the response of planktonic rotifers to a toxic cyanobacterium. Ecology 77: 1174–1180.

Gilbert, J. J., and J. K. Waage. 1967. *Asplanchna*, *Asplanchna*-substance and posterolateral spine length variation of the rotifer *Brachionus calyciflorus* in a natural environment. Ecology 48: 1027–1031.

Gilbert, J. J., and R. S. Stemberger. 1984. *Asplanchna*-induced polymorphism in the rotifer *Keratella slacki*. Limnol. Oceanogr. 29: 1309–1316.

Gilbert, J. J., and R. S. Stemberger. 1985. Control of *Keratella* populations by interference competition from *Daphnia*. Limnol. Oceanogr. 30: 180–188.

Gilbert, J. J., and K. L. Kirk. 1988. Escape response of the rotifer *Keratella*: Description, stimulation, fluid dynamics, and ecological significance. Limnol. Oceanogr. 33: 1440–1450.

Gilbert, J. J., and H. J. MacIsaac. 1989. The susceptibility of *Keratella cochlearis* to interference from small cladocerans. Freshw. Biol. 22: 333–339.

Gillen, A. L., R. A. Stein, and R. F. Carline. 1981. Predation by pellet-reared tiger muskellunge on minnows and bluegills in experimental systems. Trans. Am. Fish. Soc. 110: 197–209.

Givnish, T. J. 1986. Economics of biotic interactions. Pages 667–679 in T. J. Givnish, ed., On the Economy of Plant Form and Function. Cambridge University Press, Cambridge, U.K.

Gliwicz, M. Z. 1986. Predation and the evolution of vertical migration in zooplankton. Nature 320: 746–748.

Gliwicz, Z. M. 1990. Food thresholds and body size in cladocerans. Nature 343: 638–640.

Gliwicz, Z. M., and J. Pijanowska. 1989. The role of predation in zooplankton succession. Pages 253–296 in U. Sommer, ed., Plankton Ecology: Succession in Plankton Communities. Springer-Verlag, Berlin.

Godfray, H.C.J. 1995. Communication between the first and third trophic levels: An analysis using biological signaling theory. Oikos 72: 367–374.

Goldberg, W. M., K. R. Grange, G. T. Taylor, and A. L. Zuniga. 1990. The structure of sweeper tentacles in the black coral *Antipathes fiordensis*. Biol. Bull. 179: 96–104.

Gotthard, K., and N. Sören. 1995. Adaptive plasticity and plasticity as an adaptation: A selective review of plasticity in animal morphology and life history. Oikos 74: 3–17.

Gould, F. 1979. Rapid host range evolution in a population of the phytophagous mite *Tetranychus urticae* Koch. Evolution 33: 791–802.

Gould, F. 1986a. Simulation models for predicting durability of insect-resistant germ plasm: A deterministic diploid, two-locus model. Environ. Entomol. 15: 1–10.

Gould, F. 1986b. Simulation models for predicting durability of insect-resistant germ plasm: Hessian fly (Diptera: Cecidomyiidae) resistant winter wheat. Environ. Entomol. 15: 11–23.

Gould, F. 1995. The evolutionary potential of crop pests. Pages 190–201 in M. Slatkin, ed., Exploring Evolutionary Biology: Readings from *American Scientist*. Sinauer Associates, Sunderland, Mass.

Gould, F., and A. Anderson. 1991. Effects of *Bacillus thuringiensis* and HD-73 delta-endotoxin on growth, behavior, and fitness of susceptible and toxin-adapted strains of *Heliothis virescens* (Lepidoptera, Noctuidae). Environ. Entomol. 20: 30–38.

Gould, F., G. G. Kennedy, and M. T. Johnson. 1991. Effects of natural enemies on the rate of herbivore adaptation to resistant host plants. Entomol. Exp. Appl. 58: 1–14.

Grant, J.W.G., and I.A.E. Bayly. 1981. Predator induction of crests in morphs of the *Daphnia carinata* King complex. Limnol. Oceanogr. 26: 201–218.

Grasswitz, T. R., and T. D. Paine. 1993. Effect of experience on in-flight orientation to host-associated cues in the generalist parasitoid *Lysiphloebus testaceipes*. Entomol. Exp. Appl. 68: 219–229.

Green, J., and O. B. Lan. 1974. *Asplanchna* and the spines of *Brachionus calyciflorus* in two Javanese sewage ponds. Freshw. Biol. 4: 223–226.

Green, T. R., and C. A. Ryan. 1972. Wound-induced proteinase-inhibitors in plant leaves: A possible defense mechanism against insects. Science 175: 776–777.

Grieve, M. 1971. A Modern Herbal, vol. 2. Dover Publications, New York.

Griffiths, D. W., A.N.E. Birch, and W. H. Macfarlane-Smith. 1994. Induced changes in the indole glucosinolate content of oilseed and forage rape (*Brassica napus*) plants in response to either turnip root fly (*Delia floralis*) larval feeding or artificial root damage. J. Sci. Food Agric. 65: 171–178.

Grimm, M. P., and Klinge, M. 1996. Pike and some aspects of its dependence on vegetation. Pages 125–156 in J. F. Craig, ed., Pike: Biology and Exploitation. Chapman and Hall, London.

Grünbaum, D. 1997. Hydrodynamical mechanisms of colony organization and costs of defense in an encrusting bryozoan. Limnol. Oceanogr. 42: 741–752.

Guerrieri, E., F. Pennacchio, and E. Tremblay. 1993. Flight behaviour of the aphid parasitoid *Aphidius ervi* (Hymenoptera: Braconidae) in response to plant and host volatiles. Eur. J. Entomol. 90: 415–421.

Guilford, T. 1995. Animal signals: All honesty and light? Trends Ecol. Evol. 10: 100–101.

Guisande, C., A. Duncan and W. Lampert. 1991. Trade-offs in *Daphnia* vertical migration strategies. Oecologia 87: 357–359.

Gulland, F.M.D. 1995. Impact of infectious diseases on wild animal populations: A review. Pages 20–51 in B. T. Grenfell and A. P. Dobson, eds., Ecology of Infectious Diseases in Natural Populations. Cambridge University Press, Cambridge, U.K.

Gulland, F.M.D., S. D. Albon, J. M. Pemberton, P. R. Moorcroft, and T. H. Clutton-Brock. 1993. Parasite-associated polymorphism in a cyclic ungulate population. Proc. Roy. Soc. Lond. B 254: 7–13.

Gulmon, S. L., and H. A. Mooney. 1986. Costs of defense and their effects on plant productivity. Pages 681–699 in T. J. Givnish and R. Robichaux, eds., On the Economy of Plant Form and Function. Cambridge University Press, Cambridge, U.K.

Guzmán, H. M., and Cortés, J. 1984. Mortandad de *Gorgonia flabellum* Linnaeus (Octocorallia: Gorgoniidae) en la Costa Caribe de Costa Rica. Rev. Biol. Trop. 32: 305–308.

Hahlbrock, K., and D. Scheel. 1989. Physiology and molecular biology of phenylpropanoid metabolism. Ann. Rev. Plant Physiol. Plant Mol. Biol. 40: 347–369.

Hairston, N. G., Jr. 1979. The adaptive significance of color polymorphism in two species of *Diaptomus* (Copepoda). Limnol. Oceanogr. 24: 15–37.

Hairston, N. G., Jr. 1980. The vertical distribution of diaptomid copepods in relation to body pigmentation. Pages 98–110 in W. C. Kerfoot, ed., Evolution and Ecology of Zooplankton Communities. University Press of New England, Hanover, N.H.

Hairston, N. G., Jr. 1987. Diapause as a predator avoidance. Pages 281–290 in W. C. Kerfoot and A. Sih, eds., Predation: Direct and Indirect Impacts on Aquatic Communities. University Press of New England, Hanover, N.H.

Hairston, N. G., F. E. Smith, and L. B. Slobodkin. 1960. Community structure, population control, and competition. Am. Nat. 94: 421–425.

Hairston, N. G., Jr., and B. T. DeStasio. 1988. Rate of evolution slowed by a dormant propagule pool. Nature 36: 239–242.

Hairston, N. G., Jr., T. A. Dillon, and B. T. DeStasio. 1990. A field test for the cues of diapause in a freshwater copepod. Ecology 71: 2218–2223.

Halbach, U. 1969a. Räuber und ihre Beute: Anpassungswert von Dornen bei Rädertieren. Naturwissenschaften 56: 142–143.

Halbach, U. 1969b. Das Zusammenwirken von Konkurrenz und Räuber-Beute Beziehungen bei Rädertieren. Zoologischer Anzeiger (Suppl.) 33: 72–79.

Halbach, U. 1970. Die Ursachen der Temporalvariation von *Brachionus calyciflorus* Pallas (Rotatoria). Oecologia 4: 262–318.

Halbach, U. 1971a. Zum Adaptivwert der zyklomorphen Dornenbildung von *Brachionus calyciflorus* Pallas (Rotatoria): I. Räuber-Beute-Beziehung in Kurzzeit-Versuchen. Oecologia 6: 267–288.

Halbach, U. 1971b. Das Rädertiere *Asplanchna*—ein ideales Untersuchungsobjekt: III. Ein Räuber liefert seiner Beute die Abwehrwaffen. Mikrokosmos 60: 360–365.

Halbach, U., and J. Jacobs. 1971. Seasonal selection as a factor in rotifer cyclomorphosis. Naturwissenschaften 57: 326.

Hall, D. J., S. T. Threlkeld, C. W. Burns, and P. H. Crowley. 1976. The size-efficiency hypothesis and the size structure of zooplankton communities. Ann. Rev. Ecol. Syst. 7: 177–208.

Hambright, K. D. 1991. Experimental analysis of prey selection by largemouth bass: Role of predator mouth width and prey body depth. Trans. Am. Fish. Soc. 120: 500–508.

Hambright, K. D., R. W. Drenner, S. R. McComas, and N. G. Hairston, Jr. 1991. Gape-limited piscivores, planktivore size refuges, and the trophic cascade hypothesis. Arch. Hydrobiol. 121: 389–404.

Hamerski, D., R. C. Beier, R. E. Kneusel, U. Matern, and K. Himmelspach. 1990. Accumulation of coumarins in elicitor-treated cell suspension cultures of *Ammi majus*. Phytochemistry 29: 1137–1142.

Hamilton, W. D. 1982. Pathogens as causes of genetic diversity in their host populations. Pages 269–296 in R. M. Anderson and R. M. May, eds., Population Biology of Infectious Diseases. Springer-Verlag, New York.

Hamm, A. 1964. Untersuchungen über die Ökologie und Variabilität von *Aspidisca costata* (Hypotricha) im Belebtschlamm. Arch. Hydrobiol. 60: 286–339.

Hanazato, T. 1991. Induction of development of high helmets by a *Chaoborus*-released chemical in *Daphnia galeata*. Arch. Hydrobiol. 122: 167–175.

Hanazato, T., and S. I. Dodson. 1993. Morphological responses of four species of cyclomorphic *Daphnia* to a short term exposure to the insecticide carbaryl. J. Plankton Res. 15: 1087–1095.

Haney, J. F. 1993. Environmental control of diel vertical migration behaviour. Arch. Hydrobiol., Beih. Ergebn. Limnol. 39: 1–17.

Haney, J. F., J. J. Sasner, and M. Ikawa. 1994. Effects of products released by *Aphanizomenon flos-aquae* and purified saxitoxin in the movements of *Daphnia carinata* feeding appendages. Limnol. Oceanogr. 40: 263–272.

Hara, T. J., ed. 1992. Fish Chemoreception. Chapman and Hall, London.

Hara, T. J. 1993. Role of olfaction in fish behaviour. In T. J. Pitcher, ed., Behaviour of Teleost Fishes. 2d ed. Chapman and Hall, London.

Harari, A. R., D. Ben-Yakir, and D. Rosen. 1994. Mechanism of aggregation behavior in *Maladera matrida* Argaman (Coleoptera: Scarabeidae). J. Chem. Ecol. 20: 361–371.

Hare, J. D. 1992. Effects of plant variation on herbivore-natural enemy interactions. Pages 278–300 in R. S. Fritz and E. L. Simms, eds. Plant Resistance to Herbivores and Pathogens: Ecology, Evolution, and Genetics. University of Chicago Press, Chicago.

Hart, P.J.B., and B. Connellan. 1984. Cost of prey capture, growth rate and ration in size in pike, *Esox lucius* L., as functions of prey weight. J. Fish Biol. 25: 279–292.

Hart, P., and S. F. Hamrin. 1988. Pike as a selective predator. Effects of prey size, availability, cover and pike jaw dimensions. Oikos 51: 220–226.

Hartley, S. E. 1988. The inhibition of phenolic biosynthesis in damaged and undamaged birch foliage and its effect on insect herbivores. Oecologia 76: 65–70.
Hartwig, W. 1897. Zur Verbreitung der niederen Crustaceen in der Provinz Brandenburg. Plöner Berichte 5: 115–149.
Harvell, C. D. 1984a. Predator-induced defenses in a marine bryozoan. Science 224: 1357–1359.
Harvell, C. D. 1984b. Why nudibranchs are partial predators: Intracolonial variation in bryozoan palatability. Ecology 63: 716–724.
Harvell, C. D. 1986. The ecology and evolution of inducible defenses in a marine bryozoan: Cues, costs, and consequences. Am. Nat. 128: 810–823.
Harvell, C. D. 1990a. The ecology and evolution of inducible defenses. Quart. Rev. Biol. 65: 323–340.
Harvell, C. D. 1990b. The evolution of inducible defence. Parasitology 100: 53–61.
Harvell, C. D. 1991. Coloniality and inducible polymorphism. Am. Nat. 138: 1–14.
Harvell, C. D. 1992. Inducible defenses and allocation shifts in a marine bryozoan. Ecology 73: 1567–1576.
Harvell, C. D. 1994. The evolution of polymorhism in colonial invertebrates and in social insects. Q. Rev. Biol. 69: 155–185.
Harvell, C. D. 1998. Genetic variation and polymorphism in the inducible spines of a marine bryozoan. Evolution 52: 36–42.
Harvell, C. D., and R. K. Grosberg. 1988. On timing reproduction in clonal animals. Ecology 69: 1855–1864.
Harvell, C. D., W. Fenical, and C. H. Greene. 1988. Chemical and structural defenses of Caribbean gorgonians (*Pseudopterogorgia* spp.): I. Development of an in situ feeding assay. Mar. Ecol. Prog. Ser. 49: 287–294.
Harvell, C. D., and D. K. Padilla. 1990. Inducible morphology, heterochrony, and size hierarchies in a colonial monoculture. Proc. Natl. Acad. Sci. USA 87: 508–512.
Harvell, C. D., H. Caswell, and P. Simpson. 1990. Density effects in a bryozoan monoculture: Experimental sudies with a marine bryozoan (*Membranipora membranacea*). Oecologia 82: 227–237.
Harvell, C. D., and R. H. Helling. 1993. Experimental induction of localized reproduction in a marine bryozoan. Biol. Bull. 184: 286–295.
Harvell, C. D., W. Fenical, V. Roussis, J. L. Ruesink, C. C. Griggs, and C. H. Greene. 1993. Local and geographic variation in the defensive chemistry of a West Indian gorgonian coral (*Briareum asbestinum*). Mar. Ecol. Prog. Ser. 93: 165–173.
Harvell, C. D., J. M. West, and C. Griggs. 1996. Chemical defense of embryos and larvae of a West Indian gorgonian coral, *Briareum asbestinum*. Invertebr. Reprod. and Devel. 30: 239–246.
Harvell, C.D., K. Kim, R. Taylor, E. Rodriguez, S. Merkel. 1997. Mechanisms of sea fan resistance to a fungal pathogen. II. Chemical Resistance. Am. Zool. 37: 14A.
Hassell, M. P. 1978. Dynamics of Arthropod Predator-Prey Systems. Princeton University Press, Princeton, N.J.
Hatcher, P. E. 1995. Three-way interactions between plant pathogenic fungi, herbivorous insects and their host plants. Biol. Rev. 70: 639–694.
Haudenschild, C., and M.-A. Hartmann. 1995. Inhibition of sterol biosynthesis during elicitor-induced accumulation of furanocoumarins in parsley cell suspension cultures. Phytochemistry 40: 1117–1124.

Hauffe, K. D., K. Hahlbrock, and D. Scheel. 1985. Elicitor-stimulated furanocoumarin biosynthesis in cultured parsley cells: S-adenosylmethionine: bergaptol and S-adenosylmethionine: xanthotoxol O-methyltransferases. Z. Naturforsch. 41c: 228–239.

Haukioja, E. 1977. The mechanism of *Oporinia autumnata* cycles. Proc. Circumpolar Conf. on Northern Ecology 1: 235–242.

Haukioja, E. 1980. On the role of plant defences in the fluctuations of herbivore populations. Oikos 35: 202–213.

Haukioja, E., and S. Neuvonen. 1985. Induced long-term resistance of birch foliage against defoliators: Defensive or incidental? Ecology 66: 1303–1308.

Haukioja, E., K. Ruohomaki, J. Senn, J. Suomela, and M. Walls. 1990. Consequences of herbivory in the mountain birch *(Betula pubescens* ssp. *tortuosa*): Importance of the functional organization of the tree. Oecologia 82: 238–247.

Havel, J. E. 1985a. Cyclomorphosis of *Daphnia pulex* spined morphs. Limnol. Oceanogr. 30: 853–861.

Havel, J. E. 1985b. Predation of common invertebrate predators on long- and short-featured *Daphnia retrocurva*. Hydrobiologia. 124: 141–149.

Havel, J. E. 1987. Predator-induced defenses: A review. In Predation: Direct and indirect impacts on aquatic communities. Pages 263–278 in W. C. Kerfoot and A. Sih, eds., Direct and Indirect Impacts on Aquatic Communities. University Press of New England, Hanover, N.H.

Havel, J. E., and S. I. Dodson. 1984. *Chaoborus* predation on typical and spined morphs of *Daphnia pulex*: Behavioral observations. Limnol. Oceanogr. 29: 487–494.

Havel, J. E., and S. I. Dodson. 1987. Reproductive costs of *Chaoborus*-induced polymorphism in *Daphnia pulex*. Hydrobiologia 150: 273–281.

Havel, J. E., W. R. Mabee, and J. R. Jones. 1995. Invasion of the exotic cladoceran *Daphnia lumholtzi* into North American reservoirs. Can. J. Fish. Aquat. Sci. 52: 151–160.

Hay, M. E. 1996. Marine chemical ecology: What's known and what's next? J. Exp. Mar. Biol. Ecol. 200: 103–134.

Hays, G. C., C. A. Proctor, A.W.G. John, and A. J. Warner. 1994. Interspecific differences in the diel vertical migration of marine copepods: The implications of size, color and morphology. Limnol. Oceanogr. 39: 1621–1629.

Hazel, W. N., and D. A. West. 1979. Environmental control of pupal colour in swallowtail butterflies (Lepidoptera: Papiloininae). Ecol. Entomol. 4: 393–400.

Hazel, W. N., R. Smock, and M. D. Johnson. 1990. A polygenic model of the evolution and maintenance of conditional strategies. Proc. Roy. Soc. Lond. B 242: 181–187.

Hazel, W. N., and R. Smock. 1993. Modeling selection on conditional strategies in stochastic environments. Pages 147–154 in J. Yoshimura and C. W. Clark, eds., Adaptation in Stochastic Environments. Springer-Verlag, Berlin.

Heath-Pagliuso, S., S. A. Matlin, N. Fang, R. H. Thompson, and L. Rappaport. 1992. Stimulation of furanocoumarin accumulation in celery and celeriac tissues by *Fusarium oxysporum*. Phytochemistry 31: 2683–2688.

Hebert, P.D.N. 1978. Cyclomorphosis in natural populations of *Daphnia cephalata* King. Freshw. Biol. 8: 79–90.

Hebert, P.D.N. and P. M. Grewe. 1985. *Chaoborus*-induced shifts in the morphology of *Daphnia ambigua*. Limnol. Oceanogr. 30: 1291–1297.

REFERENCES

Heckmann, K. 1995. Räuber-induzierte Feindabwehr bei Protozoen. Naturwissenschaften 82: 107–116.
Hendrix, S. D. 1980. An evolutionary and ecological perspective of the insect fauna of ferns. Am. Nat. 115: 171–196.
Herms, D. A., and W. J. Mattson. 1992. The dilemma of plants: To grow or defend. Quart. Rev. Biol. 67: 283–335
Hessen, D. O., and E. van Donk. 1993. Morphological changes in *Scenedesmus* induced by substances released from *Daphnia*. Arch. Hydrobiol. 127: 129–140.
Hews, D. 1988. Alarm response in larval western toads, *Bufo boreas*: Release of larval chemicals by a natural predator and its effect on predator capture efficiency. Anim. Behav. 36: 125–133.
Hews, D., and A. R. Blaustein. 1985. An investigation of the alarm response in *Bufo boreas* and *Rana cascadae* tadpoles. Behav. Neur. Biol. 43: 47–57.
Hidaka, M., and K. Yamazato. 1984. Sweeper tentacles in *Madracis decactis*. Coral Reefs 3: 77–85.
Hildebrand, D. F., G. C. Brown, D. M. Jackson, and T. R. Hamilton-Kemp. 1993. Effects of some leaf-emitted volatile compounds on aphid population increase. J. Chem. Ecol. 19: 1875–1887.
Hill, A. V., J. Elvin, A. C. Willis, M. Aidoo, C.E.M. Allsopp, F. M. Gotch, X. M. Gao, M. Takiguchi, B. M. Greenwood, A.R.M. Townsend, A. J. McMichael, and H. C. Whittle. 1992. Molecular analysis of the association of B53 and resistance to severe malaria. Nature 360: 434–440.
Hjalten, J., K. Danell, and L. Ericson. 1994. The impact of herbivory and competition on the phenolic concentration and palatability of juvenile birches. Oikos 71: 416–422.
Hoffman, J. A., C. A. Janeway, and S. Natori, eds. 1994. Phylogenetic Perspectives in Immunity: The Insect Host Defense. R. G. Landes Company, Austin, Texas.
Holopainen, I. J., and A. K. Pitkänen. 1985. Population size and structure of crucian carp (*Carassius carassius* (L.)) in two small, natural ponds in Eastern Finland. Annales Zoologici Fennici 22: 397–406.
Horat, P., and R. D. Semlitsch. 1994. Effects of predation risk and hunger on the behaviour of two species of tadpoles. Behav. Ecol. Sociobiol. 34: 393–401.
Houston, A. I., J. M. McNamara, and J.M.C. Hutchinson. 1993. General results concerning the trade-off between gaining energy and avoiding predators. Phil. Trans. Roy. Soc. Lond. 341: 375–397.
Hoyle, J. A., and A. Keast. 1987. Prey handling time in two piscivores, *Esox americanus vermiculatus* and *Micropterus salmoides*, with contrasting mouth morphologies. Can. J. Zool. 66: 540–542.
Hrbácek, J. 1962. Species composition and the amount of zooplankton in relation to fish stock. Rozpr. CSAV, Rad. Mat. Prir. Ved. 72: 1–117.
Huang, X., and J.A.A. Renwick. 1993. Differential selection of host plants by two *Pieris* species: The role of oviposition stimulants and deterrents. Entomol. Exp. Appl. 68: 59–69.
Hudson, P. J., A. P. Dobson, and D. Newborne. 1985. Cyclic and non-cyclic populations of red grouse: A role for parasitism? Pages 77–90 in D. Rollinson and R. M. Anderson, eds., Ecology and Genetics of Host-Parasite Interactions. Academic Press, London.

Hudson, P. J., and A. P. Dobson. 1995. Macroparasites: Observed patterns. Pages 144–176 in B. T. Grenfell and A. P. Dobson, eds., Ecology of Infectious Diseases in Natural Populations. Cambridge University Press, Cambridge, U.K.

Huey, R., and E. C. Pianka. 1981. Ecological consequences of foraging mode. Ecology 62: 991–999.

Huffaker, C. B. 1958. Experimental studies on predation: Dispersion factors and predator-prey oscillations. Hilgardia 27: 343–383.

Hughes, A. L., and M. Nei. 1988. Pattern of nucleotide substitution at major histocompatibility complex class I loci reveals overdominant selection. Nature 335: 167–170.

Hughes, A. L., and M. Nei. 1989. Nucleotide substitution at major histocompatibility complex class II loci: Evidence for overdominant selection. Proc. Natl. Acad. Sci. USA 88: 958–962.

Hunter, M. D., and J. C. Schultz. 1993. Induced plant defenses breached? Phytochemical induction protects an herbivore from disease. Oecologia 94:195–203.

Hunter, M. D., and J. C. Schultz. 1995. Fertilization mitigates chemical induction and herbivore responses within damaged oak trees. Ecology 76: 1226–1232.

Huntley, M., and E. R. Brooks. 1982. Effects of age and food availability on diel vertical migration of *Calanus pacificus*. Mar. Biol. 71: 23–31.

Hutchings, J. A. 1993. Adaptive life histories effected by age-specific survival and growth rate. Ecology 74: 143–157.

Hutchinson, G. E. 1967. A Treatise on Limnology, vol. 2: Introduction to Lake Biology and Limnoplankton. John Wiley, New York.

Ives, A. R., and A. P. Dobson. 1987. Antipredator behavior and the population dynamics of simple predator-prey systems. Am. Nat. 130: 431–447.

Ivker, F. 1972. A hierarchy of histo-incompatibility in Hydractinia echinata. Biol. Bull 143: 162–74.

Iwao, K., and M. D. Rausher. 1997. Evolution of plant resistance to multiple herbivores: Quantifying diffuse coevolution. Am. Nat. 149: 316–335.

Jacobs, J. 1967. Untersuchungen zur Funktion und Evolution der Zyklomorphose bei *Daphnia*, mit besonderer Berücksichtigung der Selektion durch Fische. Arch. Hydrobiol. 62: 467–541.

Jacobs, J. 1980. Environmental control of cladoceran cyclomorphosis via target-specific growth factors in the animal. Pages 429–437 in W. C. Kerfoot, ed., Evolution and Ecology of Zooplankton Communities. University Press of New England, Hanover, N.H.

Janeway, C. A. 1992. The immune system evolved to discriminate infectious nonself from noninfectious self. Immunol. Today 13: 11–16.

Janeway, C. A. 1994. The role of microbial pattern recognition in self–non-self discrimination in innate and adaptive immunity. Pages 115–122 in J. A. Hoffman, C. A. Janeway, and S. Natori, eds., Phylogenetic Perspectives in Immunity: The Insect Host Defense. R. G. Landes Company, Austin, Texas.

Janeway, C. A., and P. Travers. 1994. Immunobiology: The Immune System in Health and Disease. Garland Publishing, New York.

Janssen, A., C. D. Hofker, A. R. Braun, N. Mesa, M. W. Sabelis, and A. C. Bellotti. 1990. Preselecting predatory mites for biological control: The use of an olfactometer. Bull. Entomol. Res. 80: 177–181.

Jensen, P. R., C. D. Harvell, K. Wirtz, and W. Fenical. 1996. The incidence of antimicrobial activity among Caribbean gorgonians. Mar. Biol. 125: 411–421.
Jerka-Dziadosz, M., C. Dosche, H.-W. Kuhlmann, and K. Heckmann. 1987. Signal-induced reorganization of the microtubular cytoskeleton in the ciliated protozoon *Euplotes octocarinatus*. J. Cell Sci. 87: 555–564.
Jermy, T. 1988. Can predation lead to narrow food specialization in phytophagous insects? Ecology 69: 902–904.
Johnsen, G. J., and P. J. Jakobsen. 1987. The effect of food limitation on vertical migration in *Daphnia longispina*. Limnol. Oceanogr. 32: 873–880.
Johnson, C., D. R. Brannon, and J. Kuc. 1973. Xanthotoxin: A phytoalexin of *Pastinaca sativa* root. Phytochemistry 12: 2961–2962.
Jones, C. G., and R. D. Firn. 1991. On the evolution of plant secondary chemical diversity. Phil. Trans. Roy. Soc. Lond. B 333: 273–280.
Jones, C. G., R. F. Hopper, J. S. Coleman, and V. A. Krischik. 1993. Control of systemically induced herbivore resistance by plant vascular architecture. Oecologia 93: 452–456.
Jones, D. A. 1972. Cyanogenic glucosides and their function. Pages 103–124 in J. B. Harborne, ed., Phytochemical Ecology. Academic Press, London.
Jones, D. A., and A. D. Ramnani. 1985. Altruism and movement of plants. Evol. Theory 7: 143–148.
Jones, K. A. 1992. Food search behaviour in fish and the use of chemical lures in commercial and sports fishing. Pages 288–320 in T. J. Hara, ed., Fish Chemoreception. Chapman and Hall, London.
Jones, P. P., A. P. Begovich, F. M. Tacchini-Cottier, and T. H. Vu. 1990. Evolution of class II genes: Role of selection in both maintenance of polymorphism and the retention of non-expresses alleles. Immunol. Res. 9: 200–211.
Jongsma, M. A., P. L. Bakker, J. Peters, D. Bosch, and W. J. Stiekema. 1995. Adaptation of *Spodoptera exigua* larvae to plant proteinase inhibitors by induction of gut proteinase activity insensitive to inhibition. Proc. Natl. Acad. Sci. USA 92: 8041–8045.
Jude, D. J., and J. Pappas. 1992. Fish utilization of Great Lakes coastal wetlands. Great Lakes Res. 18: 651–672.
Kajak, Z., and B. Ranke-Rybicka. 1970. Feeding and production efficiency of *Chaoborus flavicans* Meigen (Diptera, Culicidae) larvae in eutrophic and dystrophic lake. Pol. Arch. Hydrobiol. 17: 225–232.
Karban, R. 1993a. Induced resistance and plant density of a native shrub, *Gossypium thurberi*, affect its herbivores. Ecology 74: 1–8.
Karban, R. 1993b. Costs and benefits of induced resistance and plant density for a native shrub, *Gossypium thurberi*. Ecology 74: 9–19.
Karban, R., and J. R. Carey. 1984. Induced resistance of cotton seedlings to mites. Science 225: 53–54.
Karban, R., and J. H. Myers. 1989. Induced plant response to herbivory. Ann. Rev. Ecol. Syst. 20: 331–348.
Karban, R., A. K. Brody, and W. C. Schnathorst. 1989. Crowding and a plant's ability to defend itself against herbivores and diseases. Am. Nat. 134: 749–760.
Karban, R., and C. Niiho. 1995. Induced resistance and susceptibility to herbivory: Plant memory and altered plant development. Ecology 76: 1220–1225.

Karban, R., and I. T. Baldwin. 1997. Induced Responses to Herbivory. University of Chicago Press, Chicago.

Karban, R., A. A. Agrawal, and M. Mangel. 1997. The benefits of induced defenses against herbivores. Ecology 78: 1351–1355.

Karban, R., and F. R. Adler. 1996. Induced resistance to herbivores and the information content of early season attack. Oecologia 107: 379–385.

Kauffman, W. C., and G. G. Kennedy. 1989. Toxicity of allelochemicals from wild insect-resistant tomato, *Lycopersicon hirsutum f. glabratum* to *Campoletis sonorensis*, a parasitoid of *Heliothis zea*. J. Chem. Ecol. 15: 2051–2060.

Kauss, H. 1994. Systemic signals condition plant cells for increased elicitation of diverse responses. Biochem. Soc. Symp. 60: 95–100

Keefe, M. L. 1992. Chemically mediated avoidance behavior in wild brook trout, *Salvelinus fontinalis*: The response to familiar and unfamiliar predaceous fishes and the influence of fish diet. Can. J. Zool. 70: 288–292.

Keen, M. A. 1971. Sea Shells of Tropical West America: Marine Mollusks from Baja California to Peru. Stanford University Press, Stanford, Calif.

Keller, M. A., and P. A. Horne. 1993. Sources of host-location cues for the parasitic wasp *Orgilus lepidus* (Braconidae). Austral. J. Zool. 41: 335–341.

Kelsoe, G., and D. H. Schulze, eds. 1987. Evolution and Vertebrate Immunity: The antigen-Receptor and MHC Gene Families. University of Texas Press, Austin.

Kerfoot, W. C. 1978. Combat between predatory copepods and their prey: *Cyclops*, *Epischura*, and *Bosmina*. Limnol. Oceanogr. 23: 1089–1102.

Kerfoot, W. C. 1987. Translocation experiments: *Bosmina* responses to copepod predation. Ecology 68: 596–586.

Ketola, M., and I. Vuorinen. 1989. Modification of life history parameters of *Daphnia pulex* Leydig and *Daphnia magna* Straus by the presence of *Chaoborus* sp. Hydrobiology 179: 149–155.

Ketterson, E. D., and V. Nolan, Jr. 1994. Hormones and life history: An integrative approach. Pages 327–353 in L. A. Real, ed., Behavioral Mechanisms in Evolutionary Ecology. University of Chicago Press, Chicago.

Kim, K. C., and B. A. McPheron, eds. 1993. Evolution of Insect Pests: Patterns of Variation. John Wiley, New York.

Kim, K, C.D. Harvell, and G. Smith. 1997. Mechanisms of sea fan resistance to a fungal epidemic. I. Role of sclerites. Am. Zool. 37: 132A.

Kiman, Z. B., and K. V. Yeargan. 1985. Development and reproduction of the predator *Orius insidiosus* (Hemiptera: Anthocoridae) reared on diets of selected plant paterial and arthropod prey. Ann. Entomol. Soc. Am. 78: 464–467.

King, C. E., and M. R. Miracle. 1995. Diel vertical migration by *Daphnia longispina* in a Spanish lake: Genetic sources of distributional variation. Limnol. Oceanogr. 40: 226–231.

Kirk, K. L., and J. J. Gilbert. 1988. The escape behavior of *Polyarthra* in response to artificial flow stimuli. Bull. Mar. Sci. 43: 551–560.

Klasing, K. C., D. E. Laurin, R. K. Peng, and D. M. Fry. 1987. Immunologically mediated growth depression in chicks: Influence of feed intake, corticosterone and interleukin-1. J. Nutr. 117: 1629–1637.

Kluger, M. J. 1979. Fever. Princeton University Press, Princeton, N.J.

Kombrink, E., and I. E. Somssich. 1995. Defense responses of plants to pathogens. Adv. Bot. Res. 21: 1–34.

Koptur, S. 1992. Extrafloral nectary-mediated interactions between insects and plants. Pages 81–129 in E. A. Bernays, ed., Insect-Plant Interactions. Vol 4. CRC Press, Boca Raton, Florida.

Koritsas, V. M., J. A. Lewis, and G. R. Fenwick. 1991. Glucosinolate responses of oilseed rape, mustard and kale to mechanical wounding and infestation by cabbage stem flea beetle (*Psylliodes chrysocephala*). Ann. Appl. Biol. 118: 209–222.

Koveos, D. S., N. A. Kouloussis, and G. D. Broufas. 1995. Olfactory responses of the predatory mite *Amblyseius andersoni* Chant (Acari, Phytoseiidae) to bean plants infested by the spider mite *Tetranychus urticae* Koch (Acari, Tetranychidae). J. Appl. Entomol. 119: 615–619.

Kozlowski, J., and R. G. Wiegert. 1987. Optimal age and size at maturity in annuals and perennials with determinate growth. Evol. Ecol. 1: 231–244.

Krips, O. E., P.E.L. Willems, and M. Dicke. 1996. Suitability of the ornamental crop *Gerbera jamesonii* for spider mites and the attraction of predators in response to spider mite damage. Bulletin IOBC/WPRS 19: 81–87.

Krueger, D. A., and S. I. Dodson. 1981. Embryological induction and predation ecology in *Daphnia pulex*. Limnol. Oceanogr. 26: 219–223.

Kuc, J. 1982. Induced immunity to plant disease. BioScience 32: 854–860.

Kuc, J. 1983. Induced systemic resistance in plants to disease caused by fungi and bacteria. Pages 191–221 in J. A. Bailey and B. J. Deverall, eds., The Dynamics of Host Defense. Academic Press, New York.

Kuc, J. 1987. Plant immunization and its applicability for disease control. Pages 255–273 in I. Chet, ed., Innovative Approaches to Plant Disease Control. John Wiley, New York.

Kuc, J. 1995. Phytoalexins, stress metabolism, and disease resistance in plants. Ann. Rev. Phytopathol. 33: 275–297.

Kuc, J., G. Schockley, and K. K. Kearney. 1975. Protection of cucumber against *Colletotrichum lagenarium* by *Colletotrichum lagenarium*. Physiol. Plant Path. 7: 195–199.

Kuhlmann, H.-W. 1989. Defensive Gestalts- und Verhaltensänderungen des Ciliaten *Euplotes octocarinatus* bei Kontakt mit räuberisch lebenden Turbellarien. Verh. Deutsch. Zool. Ges. 82: 321.

Kuhlmann, H.-W. 1990. Zur Effizienz defensiver Zellveränderungen bei *Euplotes*. Verh. Deutsch. Zool. Ges. 83: 589–590.

Kuhlmann, H.-W. 1993. Giants in *Lembadion bullinum* (Ciliophora, Hymenostomata)—General morphology and inducing conditions. Arch. Protistenkd. 143: 325–336.

Kuhlmann, H.-W. 1994. Escape response of *Euplotes octocarinatus* to turbellarian predators. Arch. Protistenkd. 144: 163–171.

Kuhlmann, H.-W., and K. Heckmann. 1985. Interspecific morphogens regulating prey-predator relationships in protozoa. Science 227: 1347–1349.

Kuhlmann, H.-W., and K. Heckmann. 1994. Predation risk of typical ovoid and "winged" morphs of *Euplotes* (Protozoa, Ciliophora). Hydrobiologia 284: 219–227.

Kuhlmann, H.-W. and H. J. Schmidt. 1994. Extracellular nucleotides are active in evoking the "winged" morph of *Euplotes octocarinatus*. Comp. Biochem. Physiol. 109: 455–461.

Kuhn, D. N., J. Chappell, A. Boudet, and K. Hahlbrock. 1984. Induction of phenylalanine ammonia-lyase and 4-coumarate:CoA ligase mRNAs in cultured plant cells by UV light or fungal elicitor. Proc. Natl. Acad. Sci. USA 81: 1102–1106.

Kunkel, B. N. 1996. A useful weed put to work: Genetic analysis of disease resistance in *Arabidopsis thaliana*. Trends Genet. 12: 63–69.

Kusch, J. 1993a. Induction of defensive morphological changes in ciliates. Oecologia 94: 571–575.

Kusch, J. 1993b. Predator-induced morphological changes in *Euplotes* (Ciliata): Isolation of the inducing substance released from *Stenostomum sphagnetorum* (Turbellaria). J. Exp. Zool. 265: 613–618.

Kusch, J. 1993c. Behavioural and morphological changes in ciliates induced by the predator *Amoeba proteus*. Oecologia 96: 354–359.

Kusch, J. 1994a. Predator-released factors that induce defensive responses in *Euplotes*. Pages 56–58 in K. Hausmann and N. Hülsmann, eds., Progress in Protozoology. Fischer-Verlag, Jena, Stuttgart, New York.

Kusch, J. 1994b. Specificity of predator-induced morphological defence in *Euplotes*. Verh. Deutsch. Zool. Ges. 87: 311.

Kusch, J. 1995. Adaptation of inducible defense in *Euplotes daidaleos* (Ciliophora) to predation risks by various predators. Microb. Ecol. 30: 79–88.

Kusch, J., and K. Heckmann. 1992. Isolation of the *Lembadion*-factor, a morphogenetically active signal, that induces *Euplotes* cells to change from their ovoid form into a larger lateral winged morph. Dev. Genet. 13: 241–246.

Kusch, J., and H.-W. Kuhlmann. 1994. Cost of *Stenostomum*-induced morphological defence in the ciliate *Euplotes octocarinatus*. Arch. Hydrobiol. 130: 257–267.

Kusch, J., and K. Heckmann. 1996. Population structure of *Euplotes* ciliates revealed by RAPD fingerprinting. Ecoscience, 3: 378–384.

Kvam, O. V., and O. T. Kleiven. 1995. Diel horizontal migration and swarm formation in *Daphnia* in response to *Chaoborus*. Hydrobiologia 307: 177–184.

Laing, J. E., and C. B. Huffaker. 1969. Comparative studies of predation by *Phytoseiulus persimilis* Athias-Henriot and *Metaseiulus occidentalis* (Nesbitt) (Acarina: Phytoseiidae) on populations of *Tetranychus urticae* Koch (Acarina: Tetranychidae). Res. Pop. Ecol. 11: 105–126.

Lamb, R. J. 1989. Entomology of oilseed Brassica crops. Ann. Rev. Ent. 34: 211–230.

Lampert, W. 1987a. Feeding and nutrition in *Daphnia*. Mem. Ist. Ital. Idrobiol. 45: 143–192.

Lampert, W. 1987b. Predictability in lake ecosystems: The role of biotic interactions. Ecol. Stud. 61: 323–346.

Lampert, W. 1989. The adaptive significance of diel vertical migration of zooplankton. Funct. Ecol. 3: 21–27.

Lampert, W. 1993a. Phenotypic plasticity of the size at first reproduction in *Daphnia*: The importance of maternal size. Ecology 74: 1455–1466.

Lampert, W. 1993b. Ultimate causes of diel vertical migration of zooplankton: New evidence for the predator-avoidance hypothesis. Arch. Hydrobiol., Beih. Ergebn. Limnol. 39: 79–88.

Lampert, W., and C. J. Loose. 1992. Plankton towers: Bridging the gap between laboratory and field experiments. Arch. Hydrobiol. 126: 53–66.

Lampert, W., K. O. Rothhaupt, and E. von Elert. 1994. Chemical induction of colony formation in a green alga (*Scenedesmus acutus*) by grazers (*Daphnia*). Limnol. Oceanogr. 39: 1543–1550.

Landau, I., H. Muller-Scharer, and P. I. Ward. 1994. Influence of cnicin, a sesquiterpene lactone of *Centaurea maculosa* (Asteraceae), on specialist and generalist insect herbivores. J. Chem. Ecol. 20: 929–942.

Landolt, P. J., and B. Lenczewski. 1993. Lack of evidence for the toxic nectar hypothesis: A plant alkaloid did not deter nectar feeding by Lepidoptera. Florida Entomol. 76: 556–566.

Lang, J. C., and E. A. Chornesky. 1990. Competition between scleractinian corals—a review of mechanisms and effects. Pages 209–252 in Z. Dubinsky, ed., Coral Reef Ecosystems of the World. Elsevier, Amsterdam.

Langenheim, J. H. 1994. Higher plant terpenoids: A phytocentric overview of their ecological roles. J. Chem. Ecol. 20: 1223–1280.

Larsson, P., and S. I. Dodson. 1993. Chemical communication in planktonic animals. Arch. Hydrobiol. 129: 129–155.

Law, R. 1979. Optimal life histories under age-specific predation. Am. Nat. 114: 399–417.

Lawler, S. P. 1989. Behavioural responses to predators and predation risk in four species of larval anuran. Anim. Behav. 38: 1039–1047.

Lawton, J. H., and S. McNeill. 1979. Between the devil and the deep blue sea: On the problem of being a herbivore. Pages 223–244 in R. M. Anderson, B. D. Turner, and L. R. Taylor, eds., Population Dynamics. Blackwell Scientific, Oxford.

Lehtilä, K. 1996. Optimal distribution of herbivory and localized compensatory responses within a plant. Vegetatio 127: 99–109.

Leibold, M. A. 1990. Resources and predators can affect the vertical distributions of zooplankton. Limnol. Oceanogr. 35: 938–944.

Lengwiler, U., T.C.J. Turlings, and S. Dorn. 1994. Chemically mediated host searching behaviour in a parasitoid of *Phyllonorycter blancardella* F. (Lepidoptera: Gracillariidae) on apple. Norw. J. Agric. Sci. (Suppl.) 16: 401.

León, J. A. 1993. Plasticity in fluctuating environments. Pages 105–121 in J. Yoshimura and C. W. Clark, eds., Adaptation in Stochastic Environments. Springer-Verlag, Berlin.

Leshem, Y. Y., and P.J.C. Kuiper. 1996. Is there a GAS (general adaptation syndrome) response to various types of environmental stress? Biol. Plantar. 38: 1–18.

Levene, H. 1953. Genetic equilibrium when more that one ecological niche is available. Am. Nat. 87: 331–333.

Levin, B. R., and J. J. Bull. 1994. Short-sighted evolution and the virulence of pathogenic microorganisms. Trends Microbiol. 2: 76–81.

Levin, S. A., J. E. Levin, and R. T. Paine. 1977. Snowy owl predation on short-eared owls. Condor 79: 395.

Levin, S. A., L. A. Segel, and F. R. Adler. 1990. Diffuse coevolution in plant-herbivore communities. Theor. Pop. Biol. 37: 171–191.

Levins, R. 1962. Theory of fitness in a heterogeneous environment: I. The fitness set and adaptive function. Am. Nat. 96: 361–368.

Levins, R. 1963. Theory of fitness in a heterogeneous environment: II. Developmental flexibility and niche selection. Am. Nat. 97: 75–90.

Levins, R. 1968. Evolution in Changing Environments. Princeton University Press, Princeton, N.J.

Levinton, J. 1988. Genetics, Paleontology, and Macroevolution. Cambridge University Press, Cambridge, U.K.
Li, C. 1993. Furanocoumarin responses of wild and cultivated parsnip roots to abiotic and biotic stresses. Master's thesis, University of Illinois at Urbana-Champaign.
Lilljeborg, W. 1900. Cladocera Sueciae. Akademischen Buchdruckerei, Uppsala.
Lima, S. L., and L. M. Dill. 1990. Behavioral decisions made under the risk of predation: A review and prospectus. Can. J. Zool. 68: 619–640.
Lindroth, R. L. 1991. Differential toxicity of plant allelochemicals to insects: Roles of enzymatic detoxification systems. Pages 1–33 in E. A. Bernays, ed., Insect-Plant Interactions, vol. 3. CRC Press, Baton Rouge, Louisiana.
Link, J. 1996. Capture probabilities of Lake Superior zooplankton by an obligate planktivorous fish—the lake herring. Trans. Am. Fish. Soc. 125: 139–142.
Liu, Y., and B. E. Tabashnik. 1997. Experimental evidence that refuges delay insect adaptation to *Bacillus thuringiensis*. Proc. Roy. Soc. Lond. B 264: 605–610.
Lively, C. M. 1986a. Predator-induced shell dimorphism in the acorn barnacle *Chthamalus anisopoma*. Evolution 40: 232–242.
Lively, C. M. 1986b. Competition, comparative life histories, and maintenance of shell dimorphism in a barnacle. Ecology 67: 858–864.
Lively, C. M. 1986c. Canalization versus developmental conversion in a spatially variable environment. Am. Nat. 128: 561–572
Lively, C. M., and P. T. Raimondi. 1987. Desiccation, predation, and mussel-barnacle interactions in the northern Gulf of California. Oecologia 74: 304–309.
Lively, C. M., P. T. Raimondi, and L. F. Delph. 1993. Intertidal community structure: Space-time interactions in the northern Gulf of California. Ecology 74: 162–173
Lively, C. M., and V. Apanius. 1995. Genetic diversity in host-parasite interactions. Pages 421–449 in B. T. Grenfell, and A. P. Dobson, eds., Ecology of Infectious Diseases in Natural Populations. Cambridge University Press, Cambridge, U.K.
Lloyd, D. G. 1984. Variation strategies of plants in heterogeneous environments. Biol. J. Linn. Soc. 27: 357–385.
Loader, C., and H. Damman. 1991. Nitrogen content of food plants and vulnerability of *Pieris rapae* to natural enemies. Ecology 72: 1586–1590.
Logemann, E., S.-C. Wu, J. Schröder, E. Schmelzer, I. E. Somssich, and K. Hahlbrock. 1995. Gene activation by UV light, fungal elicitor or fungal infection in *Petroselinum crispum* is correlated with repression of cell cycle-related genes. Plant J. 8: 865–876.
Lois, R., A. Dietrich, K. Hahlbrock, and W. Schulz. 1989. A phenylalanine ammonia-lyase gene from parsley: Structure, regulation and identification of elicitor and light responsive cis-acting elements. EMBO J. 8: 1641–1648.
Lois, R., and K. Hahlbrock. 1991. Differential wound activation of members of the phenylalanine ammonia-lyase and 4-coumarate: CoA ligase gene families in various organs of parsley plants. Z. Naturforsch. 47: 90–94.
Loose, C. J. 1992. Experimentelle Untersuchungen zur tagesperiodischen Vertikalwanderung von *Daphnia*. Ph.D. diss., University of Kiel, Germany.
Loose, C. J. 1993. *Daphnia* diel vertical migration behavior: Response to vertebrate predator abundance. Arch. Hydrobiol., Beih. Ergebn. Limnol. 39: 29–36.
Loose, C. J., E. Von Elert, and P. Dawidowicz. 1993. Chemically-induced diel vertical

migration in *Daphnia*: A new bioassay for kairomones exuded by fish. Arch. Hydrobiol. 126: 329–337.

Loose, C. J., and P. Dawidowicz. 1994. Trade-offs in diel vertical migration by zooplankton: The costs of predator avoidance. Ecology 75: 2255–2263.

Lord, K. M., H.A.S. Epton, and R. R. Frost. 1988. Virus infection and furanocoumarins in celery. Plant Pathol. 37: 385–389.

Loughrin, J. H., A. Manukian, R. R. Heath, T.C.J. Turlings, and J. H. Tumlinson. 1994. Diurnal cycle of emission of induced volatile terpenoids in herbivore-injured cotton plants. Proc. Natl. Acad. Sci. USA 91: 11836–11840.

Loughrin, J. H., A. Manukian, R. R. Heath, and J. H. Tumlinson. 1995a. Volatiles emitted by different cotton varieties damaged by feeding beet armyworm larvae. J. Chem. Ecol. 21: 1217–1227.

Loughrin, J. H., D. A. Potter, and T. R. Hamilton-Kemp. 1995b. Volatile compounds induced by herbivory act as aggregation kairomones for the Japanese beetle (*Popilia japonica* Newman). J. Chem. Ecol. 21: 1457–1467.

Lowman, M. D. 1982. Effects of different rates and methods of leaf area removal on rain forest seedlings of coachwood (*Ceratopetalum apletalum*). Austral. J. Bot. 30: 477–483.

Lozoya, E., A. Block, R. Lois, K. Hahlbrock, and D. Scheel. 1991. Transcriptional repression of light-induced flavonoid synthesis by elicitor treatment of cultured parsley cells. Plant J. 1: 227–234.

Lundberg, S., J. Järemo, and P. Nilsson. 1994. Herbivory, inducible defence and population oscillations: A preliminary theoretical analysis. Oikos 71: 537–540

Lüning, J. 1994. Anti-predator defenses in *Daphnia*—are life-history changes always linked to induced neck spines? Oikos 69: 427–436.

Lürling, M., and E. Van Donk. (1996). Zooplankton induced unicell-colony transformation in *Scenedesmus acutus* and the effect on herbivore *Daphnia*. Oecologia 108: 432–437.

Lürling, M., and E. Van Donk. 1997. Morphological changes in the alga *Scenedesmus* induced by an infochemical released in situ from zooplankton grazers. Limnol. Oceanogr. 42: 783–788.

Lynch, M. 1980a. *Aphanizomenon* blooms: Alternate control and cultivation by *Daphnia pulex*. Pages 299–304 in W. C. Kerfoot, ed., Evolution and Ecology of Zooplankton Communities. University Press of New England, Hanover, N.H.

Lynch, M. 1980b. The evolution of cladoceran life histories. Quart. Rev. Biol. 55: 23–42.

Lynch, M. 1989. The life history consequences of resource depression in *Daphnia pulex*. Ecology 70: 246–256.

Lynch, M., and W. Gabriel. 1987. Environmental tolerance. Am. Nat. 129: 283–303.

MacArthur, R. H., and E. R. Pianka. 1966. An optimal use of a patchy environment. Am. Nat. 100: 603–609.

Machácek, J. 1991. Indirect effects of planktivorous fish on the growth and reproduction of *Daphnia galeata*. Hydrobiologia 225: 193–197.

MacIsaac, H. J., and J. J. Gilbert. 1991. Discrimination between exploitative and interference competition between Cladocera and *Keratella cochlearis*. Ecology 72: 924–937.

Mackie, G. 1986. From aggregates to integrates: Physiological aspects of modularity in colonial animals. Philos. Trans. Roy. Soc. Lond. B 313: 175–176.
Maddox, D., and R. Root. 1990. Structure of the encounter between goldenrod (*Solidago altissima*) and its diverse insect fauna. Ecology 71: 2115–2124.
Maitland, P. S., and R. N. Campbell. 1992. Freshwater Fishes of the British Isles. HarperCollins, London.
Malcolm, S. B., and M. P. Zalucki. 1996. Milkweed latex and cardenolide induction may resolve the lethal plant defense paradox. Entomol. Exp. Appl. 80: 193–196.
Malej, A., and G. P. Harris. 1993. Inhibition of copepod grazing by diatom exudates—a factor in the development of mucus aggregates. Mar. Ecol. Progr. Ser. 96: 33–42.
Malone, B. J., and D. J. McQueen. 1983. Horizontal patchiness in zooplankton populations in two Ontario kettle lakes. Hydrobiologia 99: 101–124.
Malusa, J. R. 1985. Attack mode in a predatory gastropod: Labial spine length and method of prey capture in *Acanthina angelica* Oldroyd. Veliger 28: 1–5.
Marinone, M. C., and H. E. Zagarese. 1991. A field and laboratory study on factors affecting polymorphism in the rotifer *Keratella tropica*. Oecologia 86: 372–377.
Marquis, R. J. 1992. A bite is a bite is a bite? Constraints on response to folivory in *Piper arieianum* (Piperaceae). Ecology 73: 143–152.
Marquis, R. J. 1996. Plant architecture, sectoriality, and plant tolerance to herbivores. Vegetatio 127: 85–97.
Matern, U., H. Strasser, H. Wendorff, and D. Hamerski. 1988. Coumarins and furanohcoumarins. Cell Cult. Somatic Cell Genet. Plants 5: 3–21.
Mathis, A., and R.J.F. Smith. 1993a. Fathead minnows, *Pimephales promelas*, learn to recognize northern pike *Esox lucius*, as predators on the basis of chemical stimuli from minnows in the pike's diet. Anim. Behav. 46: 645–656.
Mathis, A., and R.J.F. Smith. 1993b. Chemical labeling of northern pike (*Esox lucius*) by alarm pheromone of fathead minnows (*Pimephales promelas*). J. Chem. Ecol. 19: 1967–1978.
Mathis, A., D. G. Chivers, and R.J.F. Smith. 1993. Population differences in responses of fathead minnows (*Pimephales promelas*) to visual and chemical stimuli from predators. Ethology 93: 31–40.
Matsuda, H., P. A. Abrams, and M. Hori. 1995. The effect of adaptive anti-predator behavior on exploitative competition and mutualism between predators. Oikos 68: 549–559.
Matsuki, M., and S. F. MacLean, Jr. 1994. Effects of different leaf traits on growth rates of insect herbivores on willows. Oecologia 100: 141–152.
Mattiacci, L., M. Dicke, and M. A. Posthumus. 1994. Induction of parasitoid attracting synomone in Brussels sprouts plants by feeding of *Pieris brassicae* larvae: Role of mechanical damage and herbivore elicitor. J. Chem. Ecol. 20: 2229–2247.
Mattiacci, L., M. Dicke, and M. A. Posthumus. 1995. β-glucosidase: An elicitor of herbivore-induced plant odor that attracts host-searching parasitic wasps. Proc. Natl. Acad. Sci. USA 92: 2036–2040.
Matzinger, P. 1994. Tolerance, danger and the extended family. Ann. Rev. Immunol. 12: 991–1045.
Matzinger, P., and R. Zamoyska. 1982. A beginner's guide to major histocompatibility complex function. Nature 297: 628.

Mauricio, R., M. D. Bowers, and F. A. Bazzaz. 1993. Pattern of leaf damage affects fitness of the annual plant *Raphanus sativus* (Brassicaceae). Ecology 74: 2066–2071.

May, R. M. 1981. Models for single populations. Pages 7–29 in R. M. May, ed., Theoretical Ecology: Principles and Applications. 2d ed. Sinauer Associates, Sunderland, Mass.

May, R. M., and R. M. Anderson. 1978. Regulation and stability of host-parasite interactions: II. Destabilising processes. J. Anim. Ecol. 47: 249–267.

Maynard Smith, J. 1982. Evolution and the theory of games. Cambridge University Press, New York.

Maynard Smith, J. 1989. Evolutionary Genetics. Oxford University Press, Oxford.

Maynard Smith, J., and R. Hoekstra. 1980. Polymorphism in a varied environment: How robust are the models? Genet. Res. 35: 46–57.

McArdle, B. H., and J. H. Lawton. 1979. Effect of prey size and predator instar on the predation of *Daphnia* by *Notonecta*. Ecol. Entomol. 4: 267–275.

McAuslane, H. J., S. B. Vinson, and H. J. Williams. 1991a. Influence of adult experience on host microhabitat location by the generalist parasitoid, *Campoletis sonorensis* (Hymenoptera: Ichneumonidae). J. Insect Behav. 4: 101–113.

McAuslane, H. J., S. B. Vinson, and H. J. Williams. 1991b. Stimuli influencing host microhabitat location in the parasitoid *Campoletis sonorensis*. Entomol. Exp. Appl. 58: 267–277.

McCall, P. J., T.C.J. Turlings, W. J. Lewis, and J. H. Tumlinson. 1993. Role of plant volatiles in host location by the specialist parasitoid *Microplitis croceipes* Cresson (Braconidae: Hymenoptera). J. Insect Behav. 6: 625–639.

McCall, P. J., T.C.J. Turlings, J. H. Loughrin, A. T. Proveaux, and J. H. Tumlinson. 1994. Herbivore-induced volatile emissions from cotton (*Gossypium hirsutum* L.) seedlings. J. Chem. Ecol. 20: 3039–3050.

McCollum, S. A. 1993. Ecological consequences of predator-induced polyphenism in larval hylid frogs. Ph.D. diss., Duke University, Durham, N.C.

McCollum, S. A., and J. Van Buskirk. 1996. Costs and benefits of a predator-induced polyphenism in the gray treefrog *Hyla chrysoscelis*. Evolution 50: 583–593.

McFadden, C. S., M. McFarland, and L. Buss. 1984. Biology of Hydractiniid hydroids: 1. Colony ontogeny in *Hydractinia echinata* (Flemming). Biol. Bull. 166: 54–67.

McKenzie, J. A. 1996. Ecological and Evolutionary Aspects of Insecticide Resistance. R. G. Landes Company, Austin, Texas.

McKey, D. 1974. Adaptive patterns in alkaloid physiology. Am. Nat. 108: 305–320.

McLean, A. R., and T.B.L. Kirkwood. 1990. A model of human immunodeficiency virus infection in T helper cell clones. J. Theor. Biol. 147: 177–203.

McMullen, M. D., and K. D. Simcox. 1995. Genomic organization of disease and insect resistance genes in maize. Am. Phytopath. Soc. 6: 811–815.

McNamara, J. M., and A. I. Houston. 1994. The effect of a change in foraging options on intake rate and predation rate. Am. Nat. 144: 978–1000.

McQueen, D. J., J. R. Post, and E. L. Mills. 1986. Trophic relationships in freshwater pelagic ecosystems. Can. J. Fish. Aquat. Sci. 43: 1571–1581.

Mellors, W. K. 1975. Selective predation of ephippial *Daphnia* and the resistance of ephippial eggs to digestion. Ecology 56: 974–980.

Menge, B. A. 1978. Predation intensity in a rocky intertidal community: Relation between predator foraging activity and environmental harshness. Oecologia 34: 1–16.

Menge, B. A. 1994. Indirect effects in marine rocky intertidal interaction webs: Pattern and importance. Ecol. Mon. 65: 21–74.

Merkel, S. M., K. Kim, P. D. Kim, G. W. Smith, and C. D. Harvell. 1997. Anti-fungal and anti-bacterial activity in healthy and fungal-diseased sea fans (*Grgonia ventalina*). Abstracts for the 97th American Society for Microbiology General Meetings, Miami.

Metcalf, R. L., and R. L. Lampman. 1989. The chemical ecology of diabroticites and Cucurbitaceae. Experientia 45: 240–247.

Michod, R. E. 1979. Evolution of life histories in response to age-specific mortality factors. Am. Nat. 113: 531–550.

Milinski, M. 1977. Do all members of a swarm suffer the same predation? Z. Tierpsychol. 45: 373–388.

Milinski, M., and R. Heller. 1978. Influence of a predator on the optimal foraging behaviour of sticklebacks (*Gasterosteus aculeatus*). Nature 275: 642–644.

Miracle, M. R. 1977. Migration, patchiness, and distribution in time and space of planktonic rotifers. Arch. Hydrobiol. 8: 19–37.

Mole, S. 1994. Trade-offs and constraints in plant-herbivore defense theory: A life-history perspective. Oikos 71: 3–12.

Mopper, S., M. Beck, D. Simberloff, and P. Stiling. 1995. Local adaptation and agents of selection in a mobile insect. Evolution 49: 810–815.

Moran, M. J. 1985. The timing and significance of sheltering behaviour of the predatory intertidal gastropod *Morula marginalba* Blainville (Muricidae). J. Exp. Mar. Biol. Ecol. 93: 103–114.

Moran, N. A. 1992. The evolutionary maintenance of alternative phenotypes. Am. Nat. 139: 971–989.

Moran, N., and W. D. Hamilton. 1980. Low nutritive quality as defense against herbivores. J. Theor. Biol. 86: 247–254.

Morin, P. J. 1983. Predation, competition and the composition of larval anuran guilds. Ecol. Mon. 53: 119–138.

Morris, W. F., and G. Dwyer. 1997. Population consequences of constitutive and inducible plant resistance: Herbivore spatial spread. Am. Nat. 149: 1071–1090.

Mort, M. A. 1986. *Chaoborus* predation and the function of phenotypic variability in *Daphnia*. Hydrobiologia 133: 39–44.

Mosmann, T. R., H. Cherwinski, M. W. Bond, M. A. Giedlin, and R. L. Coffman. 1986. Two types of murine helper T cell clone: 1. Definition according to profiles of lymphokine activities and secreted proteins. J. Immunol. 136: 2348–2357.

Mosmann, T. R., and S. Sad. 1996. The expanding universe of T-cell subsets: Th1, Th2 and more. Immunol. Today 17: 138–146.

Müller, J., and A. Seitz. 1993. Habitat partitioning and differential vertical migration of some *Daphnia* genotypes in a lake. Arch. Hydrobiol., Beih. Ergebn. Limnol. 39: 167–174.

Murant, A. F. 1972. Parsnip mosaic virus. CMI/AAB Descr. Plant Viruses 91: 1–4.

Murant, A. F., and R. A. Goold. 1968. Purification, properties and transmission of parsnip yellow fleck, a semi-persistent, aphid-borne virus. Ann. Appl. Biol. 62: 123–137.

Murdoch, W. W. 1966. "Community structure, population control, and competition"—a critique. Am. Nat. 100: 219–226.
Murdoch, W. W., and A. Oaten. 1975. Predation and population stability. Adv. Ecol. Res. 9: 1–131.
Murdoch, W. W., and A. Sih. 1978. Age-dependent interference in a predatory insect. J. Anim. Ecol. 47: 581–592.
Murphy, P. M. 1993. Molecular mimicry and the generation of host defense protein diversity. Cell 72: 823–826.
Mutikainen, P., and L. F. Delph. 1996. Effects of herbivory on male reproductive success in plants. Oikos 75: 353–358.
Myers, J. H. 1988. The induced defense hypothesis: Does it apply to the population dynamics of insects. Pages 345–366 in K. C. Spencer, ed., Chemical Mediation of Coevolution. Academic Press, San Diego.
Nachman, G. 1991. An acarine predator-prey metapopulation system inhabiting greenhouse cucumbers. Biol. J. Linn. Soc. 42: 285–303.
Nadel, H., and J.J.M. van Alphen. 1987. The role of host- and host-plant odours in the attraction of a parasitoid, *Epidinocarsis lopezi*, to the habitat of its host, the cassava mealybug, *Phenacoccus manihoti*. Entomol. Exp. Appl. 45: 181–186.
Nagelkerken, I., K. Buchan, G. W. Smith, K. Bonair, P. Bush, J. Garzon-Ferreira, L. Botero, P. Gale, C. Heberer, C. Petrovic, L. Pors, and P. Yoshioka. 1997. Widespread tissue mortality in Caribbean sea fans. Proceedings of the 8th Symposium for Reef Studies. Allan Press, New York.
Nei, M., and A. L. Hughes. 1991. Polymorphism and evolution of the major histocompatibility complex loci in mammals. Pages 249–271 in R. K. Selander, A. G. Clark, and T. G. Whittam, eds., Evolution at the Molecular Level. Sinauer Associates, Sunderland, Mass.
Neill, W. E. 1990. Induced vertical migration in copepods as a defence against invertebrate predation. Nature 345: 524–526.
Neill, W. E. 1992. Population variation in the ontogeny of predator-induced vertical migration of copepods. Nature 356: 54–57.
Ngi-Song, A. J., W. A. Overholt, P.G.N. Njagi, M. Dicke, J. N. Ayertey, and W. Lwande. 1996. Volatile infochemicals used in host and host habitat location by *Cotesia flavipes* Cameron and *C. sesamiae* (Cameron) (Hymenoptera: Braconidae), larval parasitoids of stemborers on graminae. J. Chem. Ecol. 22: 307–323.
Nilsson, P. A., C. Brönmark, and L. B. Pettersson. 1995. Benefits of a predator-induced morphology in crucian carp. Oecologia 104: 291–296.
Nitao, J. K. 1990. Metabolism and excretion of the furanocoumarin xanthotoxin by parsnip webworm, *Depressaria pastinacella*. J. Chem. Ecol. 16: 417–428.
Nordlund, D. A., and W. J. Lewis. 1976. Terminology of chemical releasing stimuli in intraspecific and interspecific interactions. J. Chem. Ecol. 2: 211–220.
Nowlis, J. P. 1994. The cause and consequences of host preferences in marine gastropods. Ph.D. diss., Cornell University, Ithaca, N.Y.
Nowlis, J., J. West, and S. May. In prep. Predator-induced increases in sclerite densities of a Caribbean gorgonian.
Nürnberger, T., C. Colling, K. Hahlbrock, T. Jabs, A. Renelt, W. R. Sacks, and D. Scheel. 1994. Perception and transduction of an elicitor signal in cultured parsley cells. Biochem. Soc. Symp. 60: 173–182.

O'Brien, W. J. 1987. Planktivory by freshwater fish: thrust and parry in the pelagia. Pages 3–17 in W. C. Kerfoot and A. Sih, eds., Predation: Direct and Indirect Impacts on Aquatic Communities. University Press of New England, Hanover, N.H.

O'Brien, W. J., D. Kettle, H. Riessen, D. Schmidt, and D. Wright. 1980. Dimorphic *Daphnia longiremis*: Predation and competitive interactions between the two morphs. Pages 497–505 in W. C. Kerfoot, ed., Evolution and Ecology of Zooplankton Communities. University Press of New England, Hanover, N.H.

Ohman, M. D. 1990. The demographic benefits of diel vertical migration by zooplankton. Ecol. Monogr. 60: 257–281.

Ohman, M. D., B. W. Frost, and E. B. Cohen. 1983. Reverse diel vertical migration: An escape from invertebrate predators. Science 220: 1404–1407.

Ohmori, K., M. Hirose, and M. Ohmori. 1992. Function of cAMP as a mat-forming factor in the cyanobacterium *Spirulina platensis*. Plant Cell Physiol. 33: 21–25.

Ohta, T. 1991. Role of diversifying selection and gene conversion in evolution of major histocompatibility complex loci. Proc. Natl. Acad. Sci. USA 88: 6716–6720.

Oleszek, W. 1987. Allelopathic effects of volatiles from some Cruciferae species on lettuce, barnyard grass, and wheat growth. Plant and Soil 102: 271–273.

Olsén, K. H. 1992. Kin recognition in fish mediated by chemical cues. Pages 229–248 in T. J. Hara, ed., Fish Chemoreception. Chapman and Hall, London.

Olson, M. M., and C. R. Roseland. 1991. Induction of the coumarins scopoletin and ayapin in sunflower by insect-feeding stress and effects of coumarins on the feeding of sunflower beetle (Coleoptera: Chrysomelidae). Environ. Entomol. 20: 1166–1172.

Orians, G. H., and D. H. Janzen. 1974. Why are embryos so tasty? Am. Nat. 108: 581–592.

Osborne, S. 1984. Bryozoan interactions: Observations on stolonal interactions. Austral. Mar. Fresh. Res. 35: 453–462.

Ostrofsky, M. L., F. G. Jacobs, and J. Rowan. 1983. Evidence for the production of extracellular herbivore deterrents by *Anabaena flos-aquae*. Fresh. Biol. 13: 501–506.

Padilla, D. K., and S. C. Adolph. 1996. Plastic inducible morphologies are not always adaptive: The importance of time delays in a stochastic environment. Evol. Ecol. 10: 105–117.

Padilla, D. K., C. D. Harvell, J. Marks, and B. Helmuth. 1996. Inducible aggression and intraspecific competition for space in a marine bryozoan, *Membranipora membranacea*. Limnol. Oceanogr. 41: 505–512.

Paine, R. T. 1966. Function of labial spines, composition of diet, and size of certain marine gastropods. Veliger 9: 17–24.

Palmer, A. R. 1985. Quantum changes in gastropod shell morphology need not reflect speciation. Evolution 39: 699–705.

Palo, R. T., and C. T. Robbins. 1991. Plant defences against mammalian herbivory. CRC Press, Boca Raton, Florida.

Papaj, D. R. 1994. Optimizing learning and its effect on evolutionary change in behavior. Pages 133–153 in L. A. Real, ed., Behavioral Mechanism in Evolutionary Ecology. University of Chicago Press, Chicago.

Papaj, D. R., and L.E.M. Vet. 1990. Odor learning and foraging success in the parasitoid, *Leptopilina heterotoma*. J. Chem. Ecol. 16: 3137–3150.
Parejko, K. 1991. Predation by chaoborids on typical and spined *Daphnia pulex*. Freshw. Biol. 25: 211–217.
Parejko, K. 1992. Embryology of *Chaoborus*-induced spines in *Daphnia pulex*. Hydrobiologia 198: 51–59.
Parejko, K., and S. I. Dodson. 1990. Progress towards characterization of a predator/prey kairomone: *Daphnia pulex* and *Chaoborus americanus*. Hydrobiologia 198: 51–59.
Parejko, K., and S. I. Dodson. 1991. The evolutionary ecology of an antipredator reaction norm: *Daphnia pulex* and *Chaoborus americanus*. Evolution 45: 1665–1674.
Parham, P., and T. Ohta. 1996. Population biology of antigen presentation by MHC class I molecules. Science 272: 67–74.
Parker, M. A. 1992. Constraints on the evolution of resistance to pests and pathogens. Pages 181–197 in P. G. Ayres, ed., Environmental Plant Biology Series—Pests and Pathogens: Plant Responses to Foliar Attack. Lancaster, U.K., meeting, April 9–10, 1992. Bios Scientific Publishers, Oxford.
Pasteels, J. M., and M. Rowell-Rahier. 1992. The chemical ecology of herbivory on willows. Proc. Roy. Soc. Edin. B 98: 63–73.
Pastorok, R. A. 1981. Prey vulnerability and size selection by *Chaoborus* larvae. Ecology 62: 1311–1324.
Paul, V. J. 1992. Chemical defenses of benthic marine invertebrates. Pages 164–188 in V. J. Paul, ed., Ecological Roles of Marine Natural Products. Cornell University Press, Ithaca, N.Y.
Paul, V. J., and K. L. Van Alstyne. 1992. Activation of chemical defenses in the tropical green algae *Halimeda* spp. J. Exp. Mar. Biol. Ecol. 160: 191–203.
Pawlik, J. R., M. T. Burch, and W. Fenical. 1987. Patterns of chemical defense among Caribbean gorgonian corals: A preliminary survey. Exp. Mar. Biol. Ecol. 108: 55–66.
Pawlik, J. R., and W. Fenical. 1989. A re-evaluation of the ichthyodeterrent role of prostaglandins in the Caribbean gorgonian coral *Plexaura homomalla*. Mar. Ecol. Progr. Ser. 52: 95–98.
Peacor, S. D., and E. E. Werner 1997. Trait-mediated indirect interactions in a simple aquatic community. Ecology 78: 1146–1156.
Pease, C. M. 1987. An evolutionary epidemic mechanism, with application to type A influenza. Theor. Pop. Biol. 31: 422–451.
Pena-Cortes, H., T. Albrecht, S. Prat, E. W. Weiler, and L. Willmitzer. 1993. Aspirin prevents wound-induced gene expression in tomato leaves by blocking jasmonic acid biosynthesis. Planta 191:123–128.
Perry, D. M. 1985. Function of the shell spine in the predaceous rocky intertidal snail *Acanthina spirata* (Prosobranchia: Muricacae). Mar. Biol. 88: 51–58.
Peters-Regehr, T., J. Kusch, and K. Heckmann. 1997. Primary structure and origin of a predator released protein that induces defensive morphological changes in Euplotes. Europ. J. Protistol. 33: 389–395.
Petitt, F. L., T.C.J. Turlings, and S. P. Wolf. 1992. Adult experience modifies attrac-

tion of the leafminer parasitoid *Opius dissitus* (Hymenoptera: Braconidae) to volatile semiochemicals. J. Insect Behav. 5: 623–634.

Pettersson, L. B., and C. Brönmark. 1997. Density-dependent costs of an inducible morphological defense in crucian carp. Ecology 78: 1805–1815.

Pfennig, D. 1990. The adaptive significance of an environmentally cued developmental switch in an anuran tadpole. Oecologia 85: 101–107.

Pfister, C. A., and M. E. Hay. 1988. Associational plant refuges: Convergent patterns in marine and terrestrial communities result from differing mechanisms. Oecologia 77: 118–129.

Piironen, J., and I. J. Holopainen. 1988. Length structure and reproductive potential of crucian carp (*Carassius carassius* (L.)) populations in some small forest ponds. Annales Zoologici Fennici 25: 203–208.

Pijanowska, J. 1994. Fish enhancement patchiness in *Daphnia* distribution. Verh. Int. Verein. Limnol. 25: 2366–2368.

Pijanowska, J., and P. Dawidowicz. 1987. The lack of vertical migration in *Daphnia*: The effect of homogeneously distributed food. Hydrobiologia 148: 175–181.

Pijanowska, J., L. J. Weider, and W. Lampert. 1993. Predator-mediated genotypic shifts in a prey population: Experimental evidence. Oecologia 96: 40–42.

Pirot, J.-Y., and D. Pont. 1987. Le canard souchet (*Anas clypeata* L) hivernant en Camargue: Alimentation, comportement et dispersion nocturne. Rev. Ecol. 42: 59–79.

Poleo, A.B.S., S. A. Øxnevad, K. Østbye, E. Heibo, R. A. Andersen, and L. A. Vøllestad. 1995. Body morphology of crucian carp *Carassius carassius* in lakes with and without piscivorous fish. Ecography 18: 225–229.

Porter, K. G. 1975. Viable gut passage of gelatinous green algae ingested by *Daphnia*. Verh. Internat. Verein. Limnol. 19: 2840–2850.

Porter, K. G., J. Gerritsen, and J. D. Orcutt, Jr. 1982. The effect of food concentration on swimming patterns, feeding behavior, ingestion, assimilation, and respiration by *Daphnia*. Limnol. Oceanogr. 27: 935–949.

Post, J. R., L. G. Rudstram, D. M. Scahael, and C. Luecke. 1992. Pelagic planktivory by larval fishes in Lake Mendota. Pages 303–318 in J. Kitchell, ed., Food Web Management: A Case Study of Lake Mendota. Springer-Verlag. New York.

Potting, R.P.J., L.E.M. Vet, and M. Dicke. 1995. Host microhabitat location by stemborer parasitoid *Cotesia flavipes*: The role of herbivore volatiles and locally and systemically induced plant volatiles. J. Chem. Ecol. 21: 525–539.

Potts, W. K., and E. K. Wakeland. 1990. Evolution of diversity at the major histocompatibility complex. TREE 5: 181–187.

Pourriot, R. 1964. Étude experimentale de variations morphologiques chez certaines espéces de rotifres. Bull. Société Zool. de France 89: 555–561.

Pourriot, R. 1974. Relations prédateur-proie chez les rotifres: Influence du prédateur (*Asplanchna brightwelli*) sur la morphologie de la proie (*Brachionus bidentata*). Annales d'Hydrobiologie 5: 43–55.

Preudhomme, J. L., and L. A. Hanson. 1990. IgG subclass deficiency. Immunodef. Rev. 2: 129–149.

Prevost, G., and W. J. Lewis. 1990. Genetic differences in the response of *Microplitis croceipes* to volatile semiochemicals. J. Insect Behav. 3: 277–287.

Price, P. W. 1986. Ecological aspects of host plant resistance and biological control:

Interactions among three trophic levels. Pages 11–30 in D. J. Boethel and R. D. Eikenbary, eds., Interactions of Plant Resistance and Parasitoids and Predators of insects. Ellis Horwood, Chichester, U.K.

Price, P. W. 1991. Evolutionary theory of host and parasitoid interactions. Biol. Contr. 1: 83–93.

Price, P. W. 1992. Plant resources as the mechanistic basis for insect herbivore population dynamics. Pages 139–173 in M. D. Hunter, T. Ohgushi, and P. W. Price, eds., Effects of Resource Distribution on Animal-Plant Interactions. Academic Press, New York.

Price, P. W., C. E. Bouton, P. Gross, B. A. McPheron, J. N. Thompson, and A. E. Weis. 1980. Interactions among three trophic levels: Influence of plant on interactions between insect herbivores and natural enemies. Ann. Rev. Ecol. Syst. 11: 41–65.

Quesada, M., K. Bollman, and A. G. Stephenson. 1995. Leaf damage decreases pollen production and hinders pollen performance in *Cucurbita texana*. Ecology 76: 437–443.

Raat, A.J.P. 1988. Synopsis of biological data on the northern pike, *Esox lucius* Linnaeus, 1758. FAO Fisheries Synopsis, vol. 30, part 2. FAO, Rome.

Raffa, K. F., and A. A. Berryman. 1982. Accumulation of monoterpenes and associated volatiles following inoculation of grand fir with a fungus transmitted by the fir engraver, *Scolytus ventralis* (Coleoptera: Scolytidae). Can. Entomol. 114: 797–810.

Raffa, K. F., and A. A. Berryman. 1983. The role of host plant resistance in the colonization behavior and ecology of bark beetles (Coleoptera: Scolytidae). Ecol. Monogr. 53: 27–49.

Raimondi, P. T. 1988. Settlement cues and determination of the vertical limit of an intertidal barnacle. Ecology 69: 400–407.

Raimondi, P. T. 1990. Patterns, mechanisms, consequences of variability in settlement and recruitment of an intertidal barnacle. Ecol. Monogr. 60: 283–309.

Ramcharan, C. W., S. I. Dodson, and J. Lee. 1992. Predation risk, prey behavior, and feeding rate in *Daphnia pulex*. Can. J. Fish. Aquat. Sci. 49: 159–165.

Rasmussen, J. B., R. Hammerschmidt, and M. N. Zook. 1991. Systemic induction of salicylic acid accumulation in cucumber after inoculation with *Pseudomonas syringae pv syringae*. Plant Physiol. 97: 1342–1347.

Raulet, D. H., and W. Held. 1995. Natural killer cell receptors: The ons and offs of NK cell recognition. Cell 82: 697–700.

Rausher, M. D. 1988. Is coevolution dead? Ecology 69: 898–901.

Rausher, M. D. 1996. Genetic analysis of coevolution between plants and their natural enemies. Trends Genet. 12: 212–217.

Rausher, M. D., K. Iwao, E. L. Simms, N. Ohsaki, and D. Hall. 1993. Induced resistance in *Ipomoea purpurea*. Ecology 74: 20–29.

Read, A. P. 1995. Genetics and evolution of infectious diseases in natural populations: Group report. Pages 450–477 in B. T. Grenfell and A. P. Dobson, eds., Ecology of Infectious Diseases in Natural Populations. Cambridge University Press, Cambridge, U.K.

Read, D. P., P. P. Feeny, and R. B. Root. 1970. Habitat selection by the aphid parasite *Diaeretiella rapae* (Hymenoptera: Braconidae) and hyperparasite *Charips brassicae* (Hymenoptera: Cynipidae). Can. Ent. 102: 1567–1578.

Reede, T. 1995. Life history shifts in response to different levels of fish kairomones in *Daphnia*. J. Plankton Res. 17: 1661–1667.

Reede, T., and J. Ringelberg. 1995. The influence of a fish exudate on two clones of the hybrid *Daphnia galeata x hyalina*. Hydrobiologia 307: 207–212.

Rehnberg, B. G., and C. B. Schreck. 1987. Chemosensory detection of predators by coho salmon (*Oncorhyncus kisutch*): Behavioral reaction and the physiological stress response. Can. J. Zool. 65: 481–485.

Repka, S., M. Ketola, and M. Walls. 1994. Specificity of predator-induced neck spine and alteration in life history traits in *Daphnia pulex*. Hydrobiologia 294: 129–140.

Reynolds, C. S. 1984. The Ecology of Freshwater Phytoplankton. Cambridge University Press, Cambridge, U.K.

Reznick, D. 1985. Costs of reproduction: An evaluation of the empirical evidence. Oikos 44: 257–267.

Reznick, D. A., H. Bryga, and J. A. Endler. 1990. Experimentally induced life-history evolution in a natural population. Nature 346: 357–359.

Rhoades, D. F. 1979. Evolution of plant chemical defense against herbivores. Pages 3–54 in G. A. Rosenthal and D. H. Janzen, eds., Herbivores: Their Interaction with Secondary Metabolites. Academic Press, New York.

Rhoades, D. F. 1979. Evolution of plant chemical defenses. Pp. 3–54 in G. A. Rosenthal and D. H. Janzen eds., Herbivores: Their Interation with Secondary Metabolites. Academic Press, New York.

Rhoades, D. F. 1983. Responses of alder and willow to attack by tent caterpillars and webworms: Evidence for pheromonal sensitivity of willows. Pages 55–68 in P. E. Hedin, ed., Plant Resistance to Insects. American Chemical Society, Washington D.C.

Riessen, H. P. 1984. The other side of cyclomorphosis: Why *Daphnia* lose their helmets. Limnol. Oceanogr. 29: 1123–1127.

Riessen, H. P. 1992. Cost-benefit model for the induction of an antipredator defense. Am. Nat. 140: 349–362.

Riessen, H. P., W. J. O'Brien, and B. Loveless. 1984. An analysis of the components of *Chaoborus* predation on zooplankton and the calculation of relative prey vulnerabilities. Ecology 65: 514–522.

Riessen, H. P., and W. G. Sprules. 1990. Demographic costs of antipredator defenses in *Daphnia pulex*. Ecology 71: 1536–1546.

Ringelberg, J. 1964. The positively phototactic reaction of *Daphnia magna* Straus—a contribution to the understanding of diurnal vertical migration. Neth. J. Sea Res. 2: 319–406.

Ringelberg, J. 1980. Aspects of red pigmentation in zooplankton, especially copepods. Pages 91–97 in W. C. Kerfoot, ed., Evolution and Ecology of Zooplankton Communities. University Press of New England, Hanover, N.H.

Ringelberg, J. 1987. Induced behavior in *Daphnia*. Pages 285–323 in R. H. Peters and B. de Bernardi, eds., *Daphnia*. Memorie dell'Istituto Italiano di Idrobiologia, vol. 45. Verbania, Palanza.

Ringelberg, J. 1991a. Enhancement of the phototactic reaction in *Daphnia hyalina* by a chemical mediated by juvenile perch (*Perca fluviatilis*). J. Plankton Res. 13: 17–25.

Ringelberg, J. 1991b. A mechanism of predator-mediated induction of diel vertical migration in *Daphnia hyalina*. J. Plankton Res. 13: 83–89.

Ringelberg, J. 1995. Changes in light intensity and diel vertical migration: A comparison of marine and freshwater environments. J. Mar. Biol. Assoc. U.K. 75: 15–25.
Ringelberg, J., J. Van Kasteel, and H. Servaas. 1967. The sensitivity of *Daphnia magna* Straus to changes in light intensity at various adaptation levels and its implication in diurnal vertical migration. Z. vergl. Physiol. 56: 397–407.
Ringelberg, J., B.J.G. Flik, D. Lindenaar, and K. Royackers. 1991. Diel vertical migration of *Daphnia hyalina* (sensu latiori) in Lake Maarsseveen: 1. Aspects of seasonal and daily timing. Arch. Hydrobiol. 121: 129–145.
Ringelberg, J., and B.J.G. Flik. 1994. Increased phototaxis in the field leads to enhanced diel vertical migration. Limnol. Oceanogr. 39: 1855–1864.
Ringelberg, J., and E. Van Gool. 1995. Migrating *Daphnia* have a memory for fish kairomones. Mar. Freshw. Behav. Physiol. 26: 249–257.
Robinson, B. W., D. S. Wilson, and G. O. Shea. 1996. Trade-offs of ecological specialization: An intraspecific comparison of pumpkinseed sunfish phenotypes. Ecology 77: 170–178.
Roche, K. 1990a. Prey features affecting ingestion rates by *Acanthocyclops robustus* (Copepoda: Cyclopoida) on zooplankton. Oecologia 83: 76–82.
Roche, K. 1990b. Some aspects of vulnerability to cyclopoid predation of zooplankton prey individuals. Hydrobiologia 198: 153–162.
Rodriguez, A. D. 1995. The natural products chemistry of West Indian gorgonian octocorals. Tetrahedron 51: 4571–4618.
Roermund, H.J.W. van. 1995. Understanding biological control of greenhouse whitefly with the parasitoid *Encarsia formosa*. From individual behaviour to population dynamics. Ph.D. diss., Wageningen Agricultural University, The Netherlands.
Roessingh, P., E. Städler, G. R. Fenwick, J. A. Lewis, J. K. Nielsen, J. Hurter, and T. Ramp. 1992. Oviposition and tarsal chemoreceptors of the cabbage root fly are stimulated by glucosinolates and host plant extracts. Entomol. Exp. Appl. 65: 267–282.
Romagnani, S. 1992. Induction of Th1 and Th2 responses: A key role for the 'natural' immune response. Immunol. Today 13: 379–381.
Rotrosen, D., and J. I. Gallin. 1987. Disorders of phagocyte function. Ann. Rev. Immun. 5: 127–150.
Ruohomaki, K., F. S. Chapin III, E. Haukioja, S. Neuvonen, and J. Suomela. 1996. Delayed inducible resistance in mountain birch in response to fertilization and shade. Ecology 77: 2302–2311.
Rutzler, K., D. L. Santavy, and A. Antonius. 1983. The black band disease of Atlantic reef corals: 3. Distribution, ecology and development. P. S. Z. N. I. Mar. Ecol. 4: 329–358.
Ruxton, G. D. 1995. Short-term refuge use and stability of predator-prey models. Theor. Pop. Biol. 47: 1–17.
Sabelis, M. W. 1981. Biological control of two-spotted spider mites using phytoseiid predators: 1. Modelling the predator-prey interaction at the individual level. Agricultural Research Reports 910, Centre for Agricultural Publishing and Documentation, PUDOC, Wageningen, The Netherlands.
Sabelis, M. W., and H. E. van de Baan. 1983. Location of distant spider mite colonies by phytoseiid predators: Demonstration of specific kairomones emitted by *Tetranychus urticae* and *Panonychus ulmi*. Entomol. Exp. Appl. 33: 303–314.

Sabelis, M. W., F. van Alebeek, A. Bal, J. van Bilsen, T. van Heijningen, P. Kaizer, G. Kramer, H. Snellen, R. Veenenbos, and J. Vogelezang. 1983. Experimental validation of a simulation model of the interaction between *Phytoseiulus persimilis* and *Tetranychus urticae* on cucumber. IOBC/WPRS Bull. 6: 207–229.

Sabelis, M. W., B. P. Afman, and P. J. Slim. 1984a. Location of distant spider mite colonies by *Phytoseiulus persimilis*: Localization and extraction of a kairomone. Acarology VI, 1: 431–440.

Sabelis, M. W., J. E. Vermaat, and A. Groeneveld. 1984b. Arrestment responses of the predatory mite, *Phytoseiulus persimilis*, to steep odour gradients of a kairomone. Physiol. Entomol. 9: 437–446.

Sabelis, M. W., and M. Dicke. 1985. Long-range dispersal and searching behaviour. Pages 141–160 in W. Helle and M. W. Sabelis, eds., Spider Mites: Their Biology, Natural Enemies and Control. World Crop Pests, vol. 1b. Elsevier, Amsterdam.

Sabelis, M. W., and J. van der Meer. 1986. Local dynamics of the interaction between predatory mites and two-spotted spider mites. Pages 322–343 in J.A.J. Metz and O. Diekman, eds., Dynamics of Physiologically Structured Populations. Springer Lecture Notes in Biomathematics, vol. 68. Springer-Verlag, Berlin.

Sabelis, M. W., and M.C.M. de Jong. 1988. Should all plants recruit bodyguards? Conditions for a polymorphic ESS of synomone production in plants. Oikos 53: 247–252.

Sabelis, M. W., and O. Diekmann. 1988. Overall population stability despite local extinction: The stabilizing influence of prey dispersal from predator-invaded patches. Theor. Pop. Biol. 34: 169–176.

Sabelis, M. W., O. Diekmann, and V.A.A. Jansen. 1991. Metapopulation persistence despite local extinction: Predator-prey patch models of the Lotka-Volterra type. Biol. J. Linnean Soc. 42: 267–283.

Sadras, V. O. 1997. Interference among cotton neighbours after differential reproductive damage. Oecologia 109: 427–432.

Salt, G. W., ed. 1984. Ecology and Evolutionary Biology. University of Chicago Press, Chicago.

Sapolsky, R. M. 1994. Why zebras don't get ulcers: A guide to stress, stress-related diseases and coping. Freeman, New York.

Sarma, S.S.S. 1987. Experimental studies on the ecology of *Brachionus patulus* (Müller) (Rotifera) in relation to food, temperature and predation. Doctoral diss., University of Delhi, India.

Sato, Y. 1979. Experimental studies on parasitization by *Apanteles glomeratus*: 4. Factors leading a female to the host. Physiol. Entomol. 4: 63–70.

Schluter, D. 1995. Adaptive radiation in sticklebacks: Trade-offs in feeding performance and growth. Ecology 76: 82–90.

Schneider, P. 1937. Sur la variabilité de *Brachionus pala* Ehrenberg dans les conditions expérimentales. Comptes rendus des séances de la Société Biologie et de ses filiales 125: 450–452.

Schoener, T. W. 1993. On the relative importance of direct versus indirect effects in ecological communities. Pages 365–411 in H. Kawanabe, J. E. Cohen, and K. Iwasaka, eds., Mutualism and Community Organisation: Behavioural, Theoretical, and Food Web Approaches. Oxford University Press, Oxford.

Schrag, S. J., and V. Perrot. 1996. Reducing antibiotic resistance. Nature 381: 120–121.

Schultz, J. C. 1983. Habitat selection and foraging tactics of caterpillars in heterogeneous environments. Pages 61–90 in R. F. Denno and M. S. McClure, eds., Variable Plants and Herbivores in Natural and Managed Systems. Academic Press, New York.

Schultz, J. C. 1988. Plant responses induced by herbivores. Trends Ecol. Evol. 3: 45–49.

Schultz, J. C. 1992. Factoring natural enemies into plant tissue availability to herbivores. Pages 175–197 in M. D. Hunter, T. Ohgushi and P. W. Price, eds., Effects of Resource Distribution on Animal-Plant Interactions. Academic Press, New York.

Schultz, J. C., M. D. Hunter, and H. M. Appel. 1992. Antimicrobal activity of polyphenols mediates plant-herbivore interactions. Pages 621–637 in R. W. Hemmingway and P. E. Laks, eds., Plant Polyphenols. Plenum Press, New York

Schwartz, K., T. E. Hausen-Hagge, C. Knobloch, W. Friedrich, E. Kleihauer, and K. Bartram. 1991. Scid in man: B-cell negative SCID patients exhibit an irregular recombination at the JH locus. J. Exp. Med. 174: 1039–1048.

Schwartz, S. S. 1991. Predator-induced alterations in *Daphnia* morphology. J. Plankton Res. 13: 1151–1161.

Scott, M. A., and W. W. Murdoch. 1983. Selective predation by the backswimmer, *Notonecta*. Limnol. Oceanogr. 28: 352–366.

Sebens, K. P., and J. S. Miles. 1988. Sweeper tentacles in a gorgonian octocoral: Morphological modifications for interference competition. Biol. Bull. 175: 378–387.

Seldahl, T., K.-J. Andersson, and G. Högstedt. 1994. Grazing-induced proteinase inhibitors: A possible cause for lemming population cycles. Oikos 70: 3–11

Semlitsch, R. D., and H.-U. Reyer. 1992. Performance of tadpoles from the hybridogenetic *Rana esculenta* complex: Interactions with pond drying and interspecific competition. Evolution 46: 665–676.

Shapiro, A. M., and J. E. DeVay. 1987. Hypersensitivity reaction of *Brassica nigra* L. (Cruciferae) kills eggs of *Pieris* butterflies (Lepidoptera: Pieridae). Oecologia 71: 631–632.

Shapiro, J. 1980. The importance of trophic-level interactions to the abundance and species composition of algae in lakes. Develop. Hydrobiol. 2: 105–116.

Shimoda, T., J. Takabayashi, W. Ashihara, and A. Takafuji. 1997. Response of the predatory insect *Scolothrips takahashii* toward herbivore-induced plant synomone under both laboratory and field conditions. J. Chem. Ecol. 23: 2033–2048.

Shoule, I., and J. Bergelson. 1995. Interplant communication revisited. Ecology 76: 2660–2663.

Shuster, S. M., and M. J. Wade. 1992. Equal mating success among male reproductive strategies in a marine isopod. Nature 350: 608–610.

Sih, A. 1987a. Predators and prey lifestyles: An evolutionary and ecological overview. Pages 203–224 in W. C Kerfoot and A. Sih, eds., Predation: Direct and Indirect Impacts on Aquatic Communities. University of New England Press, Hanover, N.H.

Sih, A. 1987b. Prey refuges and predator-prey stability. Theor. Pop. Biol. 31: 1–12.

Sih, A., and R. D. Moore. 1993. Delayed hatching of salamander eggs in response to enhanced larval predation risk. Am. Nat. 142: 947–960.
Sima, P., and V. Vetvicka. 1990. Evolution of immune reactions. CRC Press, Boca Raton, Florida.
Simms, E. L. 1992. Costs of plant resistance to herbivory. Pages 392–425 in R. S. Fritz and E. L. Simms, eds., Plant Resistance to Herbivores and Pathogens: Ecology, Evolution and Genetics. University of Chicago Press, Chicago.
Simms, E. L., and M. D. Rausher. 1987. Costs and benefits of plant resistance to herbivory. Am. Nat. 130: 570–581.
Sinervo, B., and C. M. Lively. 1996. The rock-paper-scissors game and the evolution of alternative male strategies. Nature 380: 240–243.
Skelly, D. K. 1994. Activity level and the susceptibility of anuran larvae to predation. Anim. Behav. 48: 465–468.
Skelly, D. K. 1995. A behavioral trade-off and its consequences for the distribution of *Pseudacris* treefrog larvae. Ecology 76: 150–164.
Skelly, D. K., and E. E. Werner. 1990. Behavioral and life-historical responses of larval American toads to an odonate predator. Ecology 71: 2313–2322.
Slusarczyk, M. 1995. Predator-induced diapause in *Daphnia*. Ecology 76: 1008–1013.
Smith, D. C., and J. Van Buskirk. 1995. Phenotypic design, plasticity, and ecological performance in two tadpole species. Am. Nat. 145: 211–233.
Smith, G., L. Ives, I. Nagelkerken, and K. Ritchie. 1996. Caribbean sea fan mortalities. Nature 383: 487.
Smith, R.J.F. 1992. Alarm signals in fishes. Rev. Fish Biol. and Fisheries 2: 33–63.
Sommer, U., Z. M. Gliwicz, W. Lampert, and A. Duncan. 1986. The PEG-model of seasonal succession of planktonic events in fresh waters. Arch. Hydrobiol. 106: 433–471.
Spaak, P., and R. Hoekstra. 1993. Clonal structure of the *Daphnia* population in Lake Maarsseveen: Its implications for diel vertical migration. Arch. Hydrobiol., Beih. Ergebn. Limnol. 39: 157–165.
Spitze, K. 1985. Functional response of an ambush predator: *Chaoborus americanus* predation on *Daphnia pulex*. Ecology 66: 938–949.
Spitze, K. 1992. Predator-mediated plasticity of prey life history and morphology: *Chaoborus americanus* predation on *Daphnia pulex*. Am. Nat. 139: 229–247.
Spitze, K., and T. D. Sadler. 1996. Evolution of a generalist genotype: Multivariate analysis of the adaptiveness of phenotypic palsticity. Am. Nat. 148: 108–123.
Stabell, O. B. 1992. Olfactory control of homing behaviour in salmonids. Pages 249–270 in T. J. Hara, ed., Fish Chemoreception. Chapman and Hall, London.
Stabell, O. B., and M. S. Lwin. 1997. Predator-induced phenotypic changes in crucian carp are caused by chemical signals from conspecifics. Environ. Biol. Fish. 49: 145–149.
Stamp, N. E. 1993. A temperate region view of the interaction of temperature food quality and predators on caterpillar foraging. Pages 478–508 in N. E. Stamp and T. M. Casey, eds., Caterpillars: Ecological and Evolutionary Constraints on Foraging. Chapman and Hall, London.
Stamp, N. E., Y. Yang, and T. P. Osier. 1997. Response of an insect predator to prey fed multiple allelochemicals under representative thermal regimes. Ecology 78: 203–214.

Stauffer, H-P., and R. D. Semlitsch. 1993. Effects of visual, chemical and tactile cues of fish on the behavioural responses of tadpoles. Anim. Behav. 46: 355–364.

Stearns, S. C. 1992. The Evolution of Life Histories. Oxford University Press, New York.

Steenis, M. van. 1995. Evaluation and application of parasitoids for biological control of *Aphis gossypii* in glasshouse cucumber crops. Ph.D. diss., Wageningen Agricultural University, The Netherlands.

Steinberg, S., M. Dicke, L.E.M. Vet, and R. Wanningen, 1992. Response of the braconid parasitoid *Cotesia* (= *Apanteles*) *glomerata* to volatile infochemicals: Effects of bioassay set-up, parasitoid age and experience and barometric flux. Entomol. Exp. Appl. 63: 163–175.

Steinberg, S., M. Dicke, and L.E.M. Vet. 1993. Relative importance of infochemicals from first and second trophic levels in long-range host location by the larval parasitoid *Cotesia glomerata*. J. Chem. Ecol. 19: 47–59.

Stemberger, R. S. 1985. Prey selection by the copepod *Diacyclops thomasi*. Oecologia 65: 492–497.

Stemberger, R. S. 1988. Reproductive costs and hydrodynamic benefits of chemically induced defenses in *Keratella testudo*. Limnol. Oceanogr. 33: 593–606.

Stemberger, R. S. 1990. Food limitation, spination, and reproduction in *Brachionus calyciflorus*. Limnol. Oceanogr. 35: 33–44.

Stemberger, R. S., and J. J. Gilbert. 1984. Spine development in the rotifer *Keratella cochlearis*: Induction by cyclopoid copepods and *Asplanchna*. Freshw. Biol. 14: 639–647.

Stemberger, R. S., and J. J. Gilbert. 1985. Body size, food concentration, and population growth in planktonic rotifers. Ecology 66: 1151–1159.

Stemberger, R. S., and J. J. Gilbert. 1987a. Defenses of planktonic rotifers against predators. Pages 227–239 in W. C. Kerfoot and A. Sih, eds. Direct and Indirect Impacts on Aquatic Communities. University Press of New England, Hanover, N.H.

Stemberger, R. S., and J. J. Gilbert. 1987b. Multiple species induction of morphological defenses in the rotifer *Keratella testudo*. Ecology 68: 370–378.

Stemberger, R. S., and J. J. Gilbert. 1987c. Rotifer threshold food concentrations and the size-efficiency hypothesis. Ecology 68: 181–187.

Stenson, J.A.E. 1980. Predation pressure from fish on two *Chaoborus* species as related to their visibility. Pages 618–622 in W. C. Kerfoot, ed., Evolution and Ecology of Zooplankton Communities. University Press of New England, Hanover, N.H.

Stenson, J.A.E. 1987. Variation in the capsule size of *Holopedium gibberum* (Zaddach): A response to invertebrate predation. Ecology 68: 928–934.

Stephens, D. W., and J. R. Krebs. 1986. Foraging Theory. Princeton University Press, Princeton, N.J.

Stephenson, A. G. 1982. Iridoid glycosides in the nectar of *Catalpa speciosa* are unpalatable to nectar thieves. J. Chem. Ecol. 8: 1025–1034.

Sterner, R. W., and R. F. Smith. 1993. Clearance, ingestion and release of N and P by *Daphnia pulex* feeding on *Scenedesmus acutus* of varying quality. Bull. Mar. Sci. 53: 228–239.

Stewart, J. 1994. The primordial VRM system and the evolution of vertebrate immunity. R. G. Landes Austin, Texas.

Stewart, L. J., and D. G. George. 1987. Environmental factors influencing the vertical migration of planktonic rotifers in a hyper-eutrophic tarn. Hydrobiologia 147: 203–208.

Stibor, H. 1992. Predator induced life-history shifts in a freshwater cladoceran. Oecologia 92: 162–165.

Stibor, H. 1995. Chemische Information in limnischen Räuber—Beute Systemen: Der Effekt von Räubersignalen auf den Lebenszyklus von *Daphnia* spp. (Crustacea: Cladocera). Ph.D. diss., Christian-Albrechts University Kiel, Germany.

Stibor, H., and J. Lüning. 1994. Predator-induced phenotypic variability in the pattern of growth and reproduction in *Daphnia hyalina* (Crustacea: Cladocera). Funct. Ecol. 8: 97–101.

Stich, H. B., and W. Lampert. 1981. Predator evasion as an explanation of diurnal vertical migration by zooplankton. Nature 293: 396–398.

Stich, H. B., and W. Lampert. 1984. Growth and reproduction of migrating and nonmigrating *Daphnia* species under simulated food and temperature conditions of diurnal vertical migration. Oecologia 61: 192–196.

Stockhoff, B. A. 1993. Diet heterogeneity: Implications for growth of a generalist herbivore, the gypsy moth. Ecology 74: 1939–1949.

Stoner, A. 1970. Plant feeding by a predaceous insect, *Geocoris punctipes*. J. Econ. Entomol. 63: 1911–1915.

Stout, M. J., J. Workman, and S. S. Duffey. 1994. Differential induction of tomato foliar proteins by arthropod herbivores. J. Chem. Ecol. 20: 2575–2594.

Stout, M. J., K. V. Workman, and S. S. Duffey. 1996a. Identity, spatiality, and variability of induced chemical responses in tomato plants. Entomol. Exp. Appl. 79: 255–271.

Stout, M. J., K. V. Workman, J. S. Workman, and S. S. Duffey. 1996b. Temporal and ontogenetic aspects of protein induction in foliage of the tomato, *Lycopersicon esculentum*. Biochem. Ecol. Syst. 24: 611–625.

Strauss, S. Y., J. K. Conner, and S. L. Rush. 1996. Foliar herbivory affects floral characters and plant attractiveness to pollinators: Implications for male and female plant fitness. Am. Nat. 147: 1098–1107.

Strong, D. R. 1980. Null hypotheses in ecology. Synthesé 43: 271–285.

Strong, D. R., and S. Larsson. 1994. Is the evolution of herbivore resistance influenced by parasitoids? Pages 261–278 in B. A. Hawkins and W. Sheehan, eds., Parasitoid Community Ecology. Oxford University Press, Oxford.

Stutzman, P. 1995. Food quality of gelatinous colonial chlorophytes to the freshwater zooplankters *Daphnia pulicaria* and *Diaptomus oregonensis*. Fresh. Biol. 34: 149–153.

Surico, G., L. Varvaro, and M. Solfrizzo. 1987. Linear furocoumarin accumulation in celery plants infected with *Erwinia carotovora* pv *carotovora*. J. Agri. Food Chem. 35: 406–409.

Swaffar, S. M., and W. J. O'Brien. 1996. Spines of *D. lumholtzi* create feeding difficulties for juvenile bluegill sunfish (*Lepomis macrochirus*). J. Plankton Res. 18: 1055–1061.

Swift, M. C. 1992. Prey capture by the four larval instars of *Chaoborus crystallinus*. Limnol. Oceanogr. 37: 14–24.

Szlauer, L. 1968. Investigations upon ability in plankton crustacea to escape the net. Pol. Arch. Hydrobiol. 15: 79–86.

Takabayashi, J., and M. Dicke. 1996. Plant-carnivore mutualism through herbivore-induced carnivore attractants. Trends Plant Sci. 1: 109–113.

Takabayashi, J., M. Dicke, and M. A. Posthumus. 1991a. Variation in composition of predator-attracting allelochemicals emitted by herbivore-infested plants: Relative influence of plant and herbivore. Chemoecology 2: 1–6.

Takabayashi, J., M. Dicke, and M. A. Posthumus. 1991b. Induction of indirect defence against spider-mites in uninfested lima bean leaves. Phytochemistry 30: 1459–1462.

Takabayashi, J., M. Dicke, S. Takahashi, M. A. Posthumus, and T. A. van Beek. 1994a. Leaf age affects composition of herbivore-induced synomones and attraction of predatory mites. J. Chem. Ecol. 20: 373–386.

Takabayashi, J., M. Dicke, and M. A. Posthumus. 1994b. Volatile herbivore-induced terpenoids in plant-mite interactions: Variation caused by biotic and abiotic factors. J. Chem. Ecol. 20: 1329–1354.

Takabayashi, J., S. Takahashi, M. Dicke, and M. A. Posthumus. 1995. Developmental stage of herbivore *Pseudaletia separata* affects production of herbivore-induced synomone by corn plants. J. Chem. Ecol. 21: 273–287.

Takabayashi, J., Y. Sato, M. Horikoshi, R. Yamaoka, S. Yano, N. Ohsaki, and M. Dicke. 1998. Plant effects on parasitoid foraging: Differences among two tritrophic systems. Biol. Control. 11: 97–103.

Tallamy, D. W. 1985. Squash beetle feeding behavior: An adaptation against induced cucurbit defenses? Ecology 66: 1574–1579.

Tallamy, D. W., and E. S. McCloud. 1991. Squash beetles, cucumber beetles, and inducible cucurbit responses. Pages 155–181 in D. W. Tallamy and M. J. Raupp, eds., Phytochemical Induction by Herbivores. John Wiley, New York.

Tallamy, D. W., and M. J. Raupp. 1991. Phytochemical Induction by Herbivores. John Wiley, New York.

Tanaka, T., S. Yagi, and Y. Nakamatsu. 1992. Regulation of parasitoid sex allocation and host growth by *Cotesia (=Apanteles) kariyai* (Braconidae: Hymenoptera). Ann. Entomol. Soc. Am. 85: 310–316.

Taylor, B. E., and W. Gabriel. 1992. To grow or not to grow: Optimal resource allocation for *Daphnia*. Am. Nat. 139: 248–266.

Taylor, B. E., and W. Gabriel. 1993. Optimal adult growth of *Daphnia* in a seasonal environment. Funct. Ecol. 7: 513–521.

Taylor, P. D. 1996. The selection differential in quantitative genetics and ESS models. Evolution 50: 2106–2110.

Tejedo, M. 1993. Size-dependent vulnerability and behavioral responses of tadpoles of two anuran species to beetle larvae predators. Herpetologica 49: 287–294.

Thaler, J. S., and R. Karban. 1997. A phylogenetic reconstruction of constitutive and induced resistance in *Gossypium*. Am. Nat. 149: 1139–1146.

Thompson, J. N. 1988. Coevolution and alternative hypotheses on insect/plant interactions. Ecology 69: 893–895.

Tietjen, K. G., D. Hunkler, and U. Matern. 1983. Differential response of cultured parsley cells to elicitors from two non-pathogenic strains of fungi: 1. Identification of induced products as coumarin derivatives. Eur. J. Biochem. 131: 401–407.

Tietz, H. M. 1972. An Index to the Described Life Histories, Early Stages, and Hosts of the Macrolepidoptera of the Continental United States and Canada. A. C. Allyn, Sarasota, Florida.

Tjossem, S. F. 1990. Effects of fish chemical cues on vertical migration behavior in *Chaoborus*. Limnol. Oceanogr. 35: 1456–1468.

Toft, C. A. 1991. An ecological perspective: The population and community consequences of parasitism. Pages 319–343 in C. A. Toft, A. Aeschlimann, and L. Bolis, eds., Parasite-Host Associations: Coexistence or Conflict. Oxford University Press, Oxford.

Tollrian, R. 1990. Predator-induced helmet formation in *Daphnia cucullata* (Sars). Arch. Hydrobiol. 119: 191–196.

Tollrian, R. 1991. Some aspects in the costs of cyclomorphosis in *Daphnia cucullata*. Verh. Int. Verein. Limnol. 24: 2802–2803.

Tollrian, R. 1993. Neckteeth formation in *Daphnia pulex* as an example of continuous phenotypic plasticity: Morphological effects of *Chaoborus* kairomone concentration and their quantification. J. Plankton Res. 15: 1309–1318.

Tollrian, R. 1994. Fish-kairomone induced morphological changes in *Daphnia lumholtzi* (Sars). Arch. Hydrobiol. 130: 69–75.

Tollrian, R. 1995a. *Chaoborus crystallinus* predation on *Daphnia pulex*: Can induced morphological changes balance effects of body size on vulnerability? Oecologia 101: 151–155.

Tollrian, R. 1995b. Predator-induced morphological defenses: Costs, life-history shifts, and maternal effects in *Daphnia pulex*. Ecology 76: 1691–1705.

Tollrian, R. Submitted. Infochemicals in aquatic systems: Release and degradation of the *Chaoborus* kairomone.

Tollrian, R., and E. von Elert. 1994. Enrichment and purification of *Chaoborus* kairomone from water: Further steps towards its chemical characterization. Limnol. Oceanogr. 39: 788–796.

Tonn, W. M., C. A. Paszkowski, and I. J. Holopainen. 1989. Responses of crucian carp populations to differential predation pressure in a manipulated pond. Can. J. Zool. 67: 2841–2849.

Tonn, W. M., C. A. Paszkowski, and I. J. Holopainen. 1991. Selective piscivory by perch: Effects of predator size, prey size, and prey species. Verh. Int. Verein. Limnol. 24: 2406–2411.

Tonn, W. M., I. J. Holopainen, and C. A. Paszkowski. 1994. Density-dependent effects and the regulation of crucian carp populations in single-species ponds. Ecology 75: 824–834.

Trainor, F. R. 1991. Discovering the various ecomorphs of *Scenedesmus*. The end of a taxonomic era. Arch. Protistenkd. 139: 125–132.

Trainor, F. R. 1992. Cyclomorphosis in *Scenedesmus armatus* (Chlorophyta): An ordered sequence of ecomorph development. J. Phycol. 28: 553–558.

Trainor, F. R. 1993. Cyclomorphosis in *Scenedesmus subspicatus* (Chlorococcales, Chlorophyta): Stimulation of colony development at low temperature. Phycologia 32: 429–433.

Travis, J. 1994. Evaluating the adaptive role of morphological plasticity. Pages 99–122 in P. C. Wainwright and S. M. Reilly, eds., Ecological Morphology: Integrative Organismal Biology. University of Chicago Press, Chicago.

Tuomi, J., P. Niemela, M. Rousi, S. Siren, and T. Vuorisalo. 1988. Induced accumulation of foliage phenols in mountain birch branch response to defoliation. Am. Nat. 132: 602–608.

Tuomi, J., T. Fagerstrom, and P. Niemala. 1991. Carbon allocation, phenotypic plasticity, and induced defenses. Pages 85–104 in D. W. Tallamy and M. J. Raupp, eds., Phytochemical Induction by Herbivores. John Wiley, New York.

Tuomi, J., and M. Augner. 1993. Synergistic selection of unpalatability in plants. Evolution 47: 668–672.

Tuomi, J., M. Augner, and P. Nilsson. 1994. A dilemma of plant defences: Is it really worth killing the herbivore? J. Theor. Biol. 170: 427–430

Turlings, T.C.J., J. H. Tumlinson, and W. J. Lewis. 1990. Exploitation of herbivore-induced plant odors by host-seeking parasitic wasps. Science 250: 1251–1253.

Turlings, T.C.J. and J. H. Tumlinson. 1991. Do parasitoids use herbivore-induced plant chemical defenses to locate hosts? Florida Entomol. 74: 42–50.

Turlings, T.C.J., J. H. Tumlinson, F. J. Eller, and W. J. Lewis. 1991a. Larval-damaged plants: Source of volatile synomones that guide the parasitoid *Cotesia marginiventris* to the micro-habitat of its hosts. Entomol. Exp. Appl. 58: 75–82.

Turlings, T.C.J., J. H. Tumlinson, R. R. Heath, A. T. Proveaux, and R. E. Doolittle. 1991b. Isolation and identification of allelochemicals that attract the larval parasitoid, *Cotesia marginiventris* (Cresson), to the microhabitat of one of its hosts. J. Chem. Ecol. 17: 2235–2251.

Turlings, T.C.J., and J. H. Tumlinson. 1992. Systemic release of chemical signals by herbivore-injured corn. Proc. Natl. Acad. Sci. USA 89: 8399–8402.

Turlings, T.C.J., F. L. Wäckers, L.E.M. Vet, W. J. Lewis, and J. H. Tumlinson. 1993a. Learning of host-finding cues by Hymenopterous parasitoids. Pages 51–78 in D. R. Papaj and A. C. Lewis, eds., Insect Learning. Chapman and Hall, New York.

Turlings, T.C.J., P. McCall, H. T. Alborn, and J. H. Tumlinson. 1993b. An elicitor in caterpillar oral secretions that induces corn seedlings to emit chemical signals attractive to parasitic wasps. J. Chem. Ecol. 19: 411–425.

Turlings, T.C.J., J. H. Loughrin, P. J. McCall, U.S.R. Rose, W. J. Lewis, and J. H. Tumlinson. 1995. How caterpillar-damaged plants protect themselves by attracting parasitic wasps. Proc. Natl. Acad. Sci. USA 92: 4169–4174.

Valbonesi, A., and P. Luporini. 1990. A new marine species of *Euplotes* (Ciliophora, Hypotrichida) from Antarctica. Bull. Brit. Mus. Nat. Hist. (Zool.) 56: 57–61.

Valbonesi, A., F. Apone, and P. Luporini. In press. Intraclonal polymorphism in the Antarctic ciliate *Euplotes focardii*. Europ. J. Protistol.

Van Alstyne, K. L. 1988. Herbivore grazing increases polyphenolic defenses in the intertidal brown alga *Fucus distichus*. Ecology 69: 655–663.

Van Dam, N. M., and K. Vrieling. 1994. Genetic variation in constitutive and inducible pyrrolizidine alkaloid levels in *Cynoglossum officinale* l. Oecologia 99: 374–378.

Van Donk, E., and D. O. Hessen. 1993. Grazing resistance in nutrient stressed phytoplankton. Oecologia 93: 508–511.

Van Donk, E., and R. D. Gulati. 1995. Transition of a lake to turbid state six years after biomanipulation: Mechanisms and pathways. Wat. Sci. Tech. 32: 197–206.

Van Donk, E., and D. O. Hessen. 1995. Reduced digestibility of UV-B stressed and nutrient-limited algae by *Daphnia magna*. Hydrobiologia 307: 147–151.

Van Gool, E., and J. Ringelberg. 1995. Swimming of *Daphnia galeata x hyalina* in response to changing light intensities: Influence of food availability and predator kairomone. Mar. Freshw. Behav. Physiol. 26: 259–265.

Vanni, M. J. 1987. Effects of food availability and fish predation on a zooplankton community. Ecol. Monogr. 57: 61–88.
Vanni, M. J., and W. Lampert. 1992. Food quality effects on life history traits and fitness in the generalist herbivore *Daphnia*. Oecologia 92: 48–57.
Van Tienderen, P. H. 1991. Evolution of generalists and specialists in spatially heterogeneous environments. Evolution 45: 1317–1331.
Versteeg, R. 1992. NK cells and T cells: Mirror images? Immunol. Today 13: 244–247.
Vet, L.E.M. 1996. Parasitoid foraging: The importance of variation in individual behaviour for population dynamics. Pages 245–256 in R. B. Floyd and A. W. Sheppard, eds., Frontiers and Applications of Population Ecology. CSIRO, Melbourne.
Vet, L.E.M., and A. W. Groenewold. 1990. Semiochemicals and learning in parasitoids. J. Chem. Ecol. 16: 3119–3135.
Vet, L.E.M., W. J. Lewis, D. R. Papaj, and J. C. van Lenteren. 1990. A variable-response model for parasitoid foraging behavior. J. Insect Behav. 3: 471–490.
Vet, L.E.M. and M. Dicke. 1992. Ecology of infochemical use by natural enemies in a tritrophic context. Ann. Rev. Entomol. 37: 141–172.
Vet, L.E.M., W. J. Lewis, and R. T. Carde. 1995. Parasitoid foraging and learning. Pages 65–101 in R. T. Carde and W. J. Bell, eds., Chemical Ecology of Insects, vol. 2. Chapman and Hall, New York.
Via, S., and R. Lande. 1985. Genotype-environment interaction and the evolution of phenotypic plasticity. Evolution 39: 505–522.
Via S., R. Gomulkiewicz, G. De Jong, S. M. Scheiner, C. D. Schlichting, and P. H. Van Tienderen. 1995. Adaptive phenotypic plasticity: Consensus and controversy. Trends Ecol. Evol. 10: 212–217.
Vicari, M., and D. R. Bazely. 1993. Do grasses fight back? The case for antiherbivore defences. Tends Ecol. Evol. 8: 137–141
Volanakis, J. E., Z. B. Zhu, F. M. Schaffer, K. J. Macon, J. Palermos, B. O. Barger, R. Go, R. D. Campbell, H. W. Schroeder, and M. D. Cooper. 1992. Major histocompatibility complex class III genes and susceptibility to immunoglobulin A deficiency and common variable immunodeficiency. J. Clin. Invest. 89: 1914–1922.
Von Elert, E., and C. J. Loose. 1996. Predator-induced diel vertical migration in *Daphnia*: Enrichment and preliminary chemical characterization of a kairomone exuded by fish. J. Chem. Ecol. 22: 885–895.
Vrieling, K., W. Smit, and E. van der Meijden. 1991. Tritrophic interactions between aphids (*Aphis jacobaeae* Schrank), ant species, *Tyria jacobaeae* L. and *Senecio jacobaea* L. lead to maintenance of genetic variation in pyrrolizidine alkaloid concentration. Oecologia 86: 177–182.
Vuorinen, I., M. Rajasilta, and J. Salo. 1983. Selective predation and habitat shift in a copepod species—support for the predation hypthesis. Oecologia 59: 62–64.
Waddington, C. H. 1957. The Strategy of Genes. Allen and Unwin, London.
Wahle, C. M. 1980. Perseus and Medusa revisited. Science 209: 689–691.
Waldbauer, G. P., and S. Friedman. 1991. Self-selection of optimal diets by insects. Ann. Rev. Entomol. 36: 43–63.
Walde, S. J. 1995. Internal dynamics and metapopulations: Experimental tests with predator-prey systems. Pages 173–193 in N. Cappuccino and P. W. Price, eds.,

Population Dynamics: New Approaches and Synthesis. Academic Press, San Diego.

Walls, M., and M. Ketola. 1989. Effects of predator-induced spines on individual fitness in *Daphnia pulex*. Limnol. Oceanogr. 34: 390–396.

Walls, M., H. Caswell, and M. Ketola. 1991. Demographic costs of *Chaoborus*-induced defenses in *Daphnia pulex*: A sensitivity analysis. Oecologia 87: 15–24.

Walter, D. E., and D. J., O'Dowd. 1992. Leaf morphology and predators: Effect of leaf domatia on the abundance of predatory mites (Acari:Phytoseiidae). Environ. Entomol. 21: 478–484.

Walter, M. H. 1989. The induction of phenylpropanoid biosynthetic enzymes by ultraviolet or fungal elicitor in cultured parsley cells is overridden by a heat-shock treatment. Planta 177: 1–8.

Ward, E. R., S. J. Uknes, S. C. Williams, S. S. Dincher, D. L. Wiederhold, D. C. Alexander, P. Ahl-Goy, J. Métraux, and J. A. Ryals. 1991. Coordinate gene activity in response to agents that induce systemic acquired resistance. Plant Cell 3: 1085–1094.

Washburn, J. O., M. E. Gross, D. R. Mercer, and J. R. Anderson. 1988. Predator-induced trophic shift of a free-living ciliate: Parasitism of mosquito larvae by their prey. Science 240: 1193–1195.

Watt, P. J., and S. Young. 1994. Effect of predator chemical cues on *Daphnia* behaviour in both horizontal and vertical planes. Anim. Behav. 48: 861–869.

Webb, P. W. 1975. Hydrodynamics and energetics of fish propulsion. Bull. Fish. Res. Board Can. 190: 1–159.

Webb, P. W. 1984. Body form, locomotion and foraging in aquatic vertebrates. Am. Zool. 24: 107–120.

Webb, P. W. 1986. Effect of body form and response threshold on the vulnerability of four species of teleost prey attacked by largemouthed bass (*Micropterus salmoides*). Can. J. Fish. Aquat. Sci. 43: 763–771.

Webb, P. W., and J. M. Skadsen. 1980. Strike tactics of *Esox*. Can. J. Zool. 58: 1462–1469

Weider, L. J. 1984. Spatial heterogeneity of *Daphnia* genotypes: Vertical migration and habitat partitioning. Limnol. Oceanogr. 29: 225–235.

Weider, L. J., and J. Pijanowska. 1993. Plasticity of *Daphnia* life histories in response to chemical cues from predators. Oikos 67: 385–392.

Wellington, G. M. 1980 Reversal of digestive interactions between Pacific reef corals: Mediation by sweeper tentacles. Oecologia 47: 340–343.

Werner, E. E. 1991. Nonlethal effects of a predator on competitive interactions between two anuran larvae. Ecology 72: 1709–1720.

Werner, E. E. 1992a. Competitive interactions between woodfrog and northern leopard frog larvae: The influence of size and activity. Copeia 1992: 26–35.

Werner, E. E. 1992b. Individual behavior and higher-order interactions. Am. Nat. 140: 5–32.

Werner, E. E. 1994. Individual behavior and higher-order species interactions. Pages 297–324 in L. A. Real, ed., Behavioral Mechanisms in Evolutionary Ecology. University of Chicago Press, Chicago.

Werner, E. E., and B. R. Anholt. 1993. Ecological consequences of the tradeoff be-

tween growth and mortality rates mediated by foraging activity. Am. Nat. 142: 242–272.
Werner, E. E., and B. R. Anholt. 1996. Predator-induced behavioral indirect effects: Consequences to competitive interactions in anuran larvae. Ecology 77: 157–169.
Werner, E. E., and M. A. McPeek. 1994. Direct and indirect effects of predators on two anuran species along an environmental gradient. Ecology 75: 1368–1382.
Wesenberg-Lund, C. 1900. Von dem Abhängigkeitsverhältnis zwischen dem Bau der Planktonorganismen und dem spezifischen Gewicht des Süßwassers. Biol. Zentralblatt 19: 644–656.
West, J. M. 1997. Plasticity in the sclerites of a gorgonian coral: Tests of water motion, light level and damage cues. Biol. Bull. 192: 279–289.
Westphal, E., F. Dreger, and R. Bronner. 1991. Induced resistance in *Solanum dulcamara* triggered by the gall mite *Aceria cladophthirus* (Acari: Eriophyoidea). Exp. Appl. Acarol. 12: 111–118.
Westphal, E., M. J. Perrot-Minnot, S. Kreiter, and J. Gutierrez. 1992. Hypersensitive reaction of *Solanum dulcamara* to the gall mite *Aceria cladophthirus* causes an increased susceptibility to *Tetranychus urticae*. Exp. Appl. Acarol. 15: 15–26.
Whitham, T. G. 1983. Host manipulation of parasites: Within-plant variation as a defense against rapidly evolving pests. Pages 15–41 in R. F. Denno and M. S. McClure, eds., Variable Plants and Herbivores in Natural and Managed Systems. Academic Press, New York.
Whitham, T. G., and C. N. Slobodchikoff. 1981. Evolution by individuals, plant-herbivore interactions, and mosaics of genetic variability: The adaptive significance of somatic mutations in plants. Oecologia 49: 287–292.
Whitham, T. G., A. G. Williams, and A. M. Robinson. 1984. The variation principle: Individual plants as temporal and spatial mosaics of resistance to rapidly evolving pests. Pages 15–51 in P. W. Price, C. N. Slobodchikoff, and W. S. Gaud, eds., A New Ecology: Novel Approaches to Interactive Systems. John Wiley, New York.
Whitman, D. W., and F. J. Eller. 1990. Parasitic wasps orient to green leaf volatiles. Chemoecology 1: 69–75.
Whitman, D. W., and D. A. Nordlund. 1994. Plant chemicals and the location of herbivorous arthropods by their natural enemies. Pages 133–159 in T. N. Ananthakrishnan, ed., Functional Dynamics of Phytophagous Insects. Oxford University Press and IBH, New Delhi.
Wiąckowski, K., and M. Szkarłat. 1996. Effects of food availability on predator-induced morphological defence in the ciliate *Euplotes octocarinatus* (Protista). Hydrobiologia 321: 7–52.
Wicklow, B. J. 1988. Developmental polymorphism induced by intraspecific predation in the ciliated protozoon *Onychodromus quadricornutus*. J. Protozool. 35: 137–141.
Wilbur, H. M. 1987. Regulation of structure in complex systems: Experimental temporary pond communities. Ecology 68: 1437–1452.
Williamson, C. E. 1983. Invertebrate predation on planktonic rotifers. Hydrobiologia 104: 385–396.
Williamson, C. E. 1984. Laboratory and field experiments on the feeding ecology of the cyclopoid copepod, *Mesocyclops edax*. Freshw. Biol. 14: 575–585.

Williamson, C. E. 1987. Predator-prey interactions between omnivorous diaptomid copepods and rotifers: The role of prey morphology and behavior. Limnol. Oceanogr. 32: 167–177.

Williamson, C. E., and N. M. Butler. 1986. Predation on rotifers by the suspension-feeding calanoid copepod *Diaptomus pallidus*. Limnol. Oceanogr. 31: 393–402.

Wilson, D. S. 1989. The diversification of single gene pools by density- and frequency-dependent selection. Pages 366–385 in D. Otte and J. A. Endler, eds., Speciation and Its Consequences. Sinauer Associates, Sunderland, Mass.

Wilson, D. S., and J. Yoshimura. 1994. On the coexistence of specialists and generalists. Am. Nat. 144: 692–707.

Wink, M. 1988. Plant breeding importance of plant secondary metabolites for protection against pathogens and herbivores. Theor. Appl. Gen. 75: 225–233.

Wittmann, J., and F. Schoenbeck. 1996. Studies of tolerance induction in wheat infested with powdery mildew or aphids. Z. Pflanzenkrankheiten und Pflanzenschutz 103: 300–309

Wolfe, G., M. Steinke, and G. O. Kirst. 1997. Grazer-activated chemical defense in a unicellular marine alga. Nature 387: 894–897.

Woltereck, R. 1909. Weitere experimentelle Untersuchungen über Artveränderung, speziell über das Wesen quantitativer Artunterschiede bei Daphnien. Verh. Deutsch. Zool. Ges. 1909: 110–172.

Wood, A. M., and T. Leatham. 1992. The species concept in phytoplankton ecology. J. Phycol. 28: 723–729.

Woodward, B. D. 1982. Tadpole competition in a desert anuran community. Oecologia 54: 96–100.

Woodward, B. D. 1983. Predator-prey interactions and breeding-pond use of temporary-pond species in a desert anuran community. Ecology 64: 1549–1555.

Wooton, J. T. 1994. The nature and consequences of indirect effects in ecological communities. Ann. Rev. Ecol. Syst. 25: 443–466.

Wootton, R. J. 1985. Energetics of reproduction. Pages 231–254 in P. Tytler and P. Calow, eds., Fish Energetics: New Perspectives. Croom Helm, London.

Wratten, S. D., P. J. Edwards, and A. M. Barker. 1990. Consequences of rapid feeding-induced changes in trees for the plant and the insect: Individuals and populations. Pages 137–145 in A. D. Watt, S. R. Leather, M. D. Hunter, and N.A.C. Kidd, eds., Population Dynamics of Forest Insects. Intercept Ltd., Andover, U.K.

Wu, S.-C., and K. Hahlbrock. 1992. In situ localization of phenylpropanoid-related gene expression in different tissues of light- and dark-grown parsley seedlings. Z. Naturforsch. 47: 591–600.

Yoshioka, P. M. 1982. Predator-induced polymorphism in the bryozoan *Membranipora membranacea* (L.). J. Exp. Mar. Biol. Ecol. 62: 233–242.

Young, S., P. J. Watt, J. P. Grover, and D. Thomas. 1994. The unselfish swarm? J. Anim. Ecol. 63: 611–618.

Zagarese, H. E., and M. C. Marinone. 1992. Induction and inhibition of spine development in the rotifer *Keratella tropica*: Evidence from field observations and laboratory experiments. Freshw. Biol. 28: 289–300.

Zangerl, A. R. 1990. Furanocoumarin induction in wild parsnip: Evidence for an induced defense against herbivores. Ecology 71: 1926–1932.

Zangerl, A. R., and M. R. Berenbaum. 1987. Furanocoumarins in wild parsnip: Effects of photosynthetically active radiation, ultraviolet light, and nutrients. Ecology 68: 516–520.

Zangerl, A. R., M. R. Berenbaum, and E. Levine. 1989. Genetic control of seed chemistry and morphology in wild parsnip (*Pastinaca sativa*). J. Hered. 80: 404–407.

Zangerl, A. R., and M. R. Berenbaum. 1990. Furanocoumarin induction in wild parsnip: Genetics and populational variation. Ecology 71: 1933–1940.

Zangerl, A. R., and F. A. Bazzaz. 1992. Theory and pattern in plant defense allocation. Pages 363–391 in R. S. Fritz and E. L. Simms, eds., Plant Resistance to Herbivores and Pathogens: Ecology, Evolution, and Genetics. University of Chicago Press, Chicago.

Zangerl, A. R., and M. R. Berenbaum. 1993. Plant chemistry, insect adaptation to plant chemistry, and host plant utilization patterns. Ecology 74: 47–54.

Zangerl, A. R., and M. R. Berenbaum. 1994/95. Spatial, temporal, and environmental limits on xanthotoxin induction in wild parsnip foliage. Chemoecology 5/6: 37–42.

Zangerl, A. R., and C. E. Rutledge. 1996. The probability of attack and patterns of constitutive and induced defense: A test of optimal defense theory. Am. Nat. 147: 599–608.

Zangerl, A. R., A. M. Arntz, and M. R. Berenbaum. 1997. Physiological price of an induced chemical defense: Photosynthesis, respiration, biosynthesis, and growth. Oecologia 109: 433–441.

Zaret, T. M. 1972. Predators, invisible prey, and the nature of polymorphism in cladocera (class Crustacea). Limnol. Oceanogr. 17: 171–184.

Zaret, T. M. 1975. Strategies for existence of zooplankton prey in homogeneous environments. Verh. Internat. Verein. Limnol. 19: 1484–1489.

Zaret, T. M. 1980. Predation and Freshwater Communities. Yale University Press, New Haven and London.

Zeringue, H. J., and S. P. McCormick. 1989. Relationships between cotton leaf-derived volatiles and growth of *Aspergillus flavus*. J. Am. Org. Chem. Soc. 66: 581–585.

Zinkernagel, R. M., E. Haenseler, T. P. Leist, A. Cerny, H. Hengartner, and A. Althage. 1986. T-cell mediated hepatitis in mice infected with lymphocytic choriomeningitis virus. J. Exp. Med. 164: 1075–1092.

Index

adenosylmethionine:bergaptol O-methyl transferase, 28
absisic acid, 31
Acanthina angelica, 246
Acanthocyclops, 129
Acartia hudsonica, 165
Aceria cladophthirus, 52
activity of prey, 219, 221
Aculus schlechtendali, 69
Acyrthosiphon pisum, 72
adaptation of herbivores, 56. *See also* coevolution
adaptive significance. *See* benefit
Aedes aegypti, 16
A-factor, 153
Agaricia agaricites, 234
aggregation of prey, 189. *See also* swarm formation
Agonopterix clemensella, 19
Agrobacterium tumefaciens, 18, 20, 31
alarm substance, 209. *See also* cue
alertness, 182, 189
alkaloids, 60, 64
allelopathy, 48
Allium, 16
allocation costs, 238, 314–315. *See also* costs; costs: construction; operation
Alternaria, 18, 20; *carthami*, 28
Amblyseius: *andersoni*, 66; *anonymus*, 68; *californicus*, 68; *limonicus*, 68; *swirskii*, 66
Amoeba proteus, 143
Amphipyra tragopogonis, 19
amplification of defenses, 318
angelicin, 24
Ankistrodesmus: *bibraianus*, 94; *falcatus*, 94
Anthocoris, 68
Anthriscus yellows, 18
antibiotics, 48, 96
antibody response, 115. *See also* immune system
Antipathes fiordensis, 234
anti-predator behavior: in anuran larvae, 219–221; in ciliates, 147–148; in zooplankton, 160–177, 177–203. *See also* avoidance; defenses

anuran larvae, 218–231; model system, 220–221
Aphanizomenon, 90; *flos-aquae*, 94
Aphidius colemani, 67
Aphidius ervi, 72
Aphis gossypii, 67
Apoanagyrus lopez, 68
Aporia crataegi, 71
apple, 69
Archips purpurana, 19
Aspergillus niger, 16
Aspidisca costata, 145, 150
Asplanchna, 128–141; *brightwelli*, 129, 133; *girodi*, 129, 133; kairomone, 130, 133; kairomone concentration of, 131; *priodonta*, 133; *sieboldi*, 131
assimilation: efficiency of, 90
Asterionella formosa, 94
attack: frequency of, 306, 309; oscillations of, 306; predictability of, 306; variability in, 306, 307–309. *See also* herbivore; parasites; pathogens; predator
autotoxicity. *See* self-damage
avoidance: of predators, 145, 160–176, 182. *See also* anti-predator behavior, defenses

Bacillus subtilis, 16
Bacillus thuringiensis, 56
bacteria, 48
barnacles, 245–258
B cells, 111
bean, 67, 84
behavioral defenses. *See also* anti-predator behavior; avoidance; defenses
behavioral plasticity, 218
benefit, 34, 312. *See also* ciliates, defenses in; colony formation in algae; crucian carp, defenses in; rotifers, spines in; zooplankton, defenses in; ———, predator avoidance in
bergapten, 24, 26
bet-hedging, 310
bioassay: for *Asplanchna* kairomone, 131; for colony formation in algae, 93
biological control, 59
Biomphalaria mansoni, 16

birch, 57
black swallowtail, 27
Blattella germanica, 16
body size in zooplankton, 186-187
Bos taurus, 19
Bosmina, 129; *longirostris*, 95, 178
bottom up control, 63, 87
Brachionus: *bidentata*, 129; *calyciflorus*, 95, 128; *patulus*, 129; *sericus*, 129; *urceolaris*, 129
Brassica: *juncea*, 50; *napus*, 50; *nigra*, 50; *rapa*, 50
Brassicaceae 50
Briareum asbestinum, 234
bryozoans, 231-244
bullfrog, 222-223, 227-229

cabbage, 71, 78
Campoletis sonorensis, 70
canalization, 251. *See also* defenses, fixed vs. inducible
Candida albicans, 16
cannibal giants, 148, 149
cannibalism, 148, 149, 213
Carassius carassius, 203-217
carnivores, 63-88; response of, 83
carrot thin leaf potyvirus, 18
cassava, 68
Cavariella: *aegopodii*, 18; *pastinacae*, 18
celery: mosaic potyvirus, 18; yellow spot luteovirus,18
cell division, 158
Celleporaria, 234
cell-to-cell communication, 185
Centrospora acerina, 18
Ceratium, 90
Ceratocystis fimbriata, 20
Chaetogaster diastrophus, 144
chalcone synthase, 29
Chaoborus, 129, 164, 180; *flavicans* 165; kairomone, 196-197; larvae, 180; *punctipennis*, 165
Chilo: *partellus*, 69; *orichalcociliellus*, 70
Chlorella vulgaris, 94
Chromatomyia syngenesiae, 72
Chtamalus anisopoma, 246
Chydorus sphaericus, 95
ciliates, 142-159; behavior in, 147-148; defenses in, 155-158
cinnamic acid hydroxylase, 14
cladocera, 177-202, 293

Cleidochasma bassleri, 234
clonal organisms, 307
Coelastrum: *microporum*, 94; *sphaericum*, 94
coenobia, 90
coevolution, 260, 274, 309; diffuse, 308; pairwise, 308; slow down of, 312
Colletotrichum lindemuthianum, 20
colonial marine invertebrates, 231-244
colony formation of algae 89-103
Colpidium: *colpoda*, 145; *kleini*, 145, 150
communication between plants, 40, 75
community structure, 219, 227-230
competition, 227-229; interference, 129, 133-134
conditional defense. *See* defenses, inducible
constraints, 311-312; cost, 311; developmental, 311; ecological, 27; environmental, 311; genetic, 12; phylogenetic, 312. *See also* costs
cooperation, 43
corals, 240-241
coriander feathery red vein nucleorhabdovirus, 18
corn, 69
costs, 29, 34, 49, 82-83, 99-101, 134-140, 197-201, 312-317; allocation, 11, 21, 46, 82, 139, 199, 314-315; in bryozoans, 238-239; in ciliates, 158-159; classification of, 314; construction, 314; in crucian carp, 214; direct, 34; environmental, 11, 26-28, 200-201, 315-316; genetic, 11, 25, 27, 313; of herbivory, 34; of hydrodynamics, 214; indirect, 34, 62-88; infrastructure, 313; operation, 314; opportunity, 200, 238, 315; physiological, 21; plasticity, 253, 256, 313-314; in rotifers, 134-140; self-damage, 315; of vertical migration, 167, 171
Cotesia, 69; *flavipes*, 69; *glomerata*, 71, 75; *kariyai*, 69; *marginiventris*, 69; *platellae*, 71; *rubecula*, 71, 72; *sesamiae*, 70
cotton, 70
coumarate 4 CoA ligase, 14, 29
crests in cladocerans, 184
crucian carp, 203-217; defenses in, 211-214
cucumber, 60, 67, 77, 79; beetle, 52
Cucurbitaceae, 51, 52
Cucurbita moschata, 51
cue, 119-120, 194-197, 263, 272, 309; chemical, 194, 208, 310; mechanical, 310; nonspecific, 79-80; oviposition, 50; re-

INDEX 379

liability of, 257, 278–279, 309; selection against release of, 274; specificity of, 76–79, 308; suitability of, 310; types of, 310; visual, 310; volatility of, 40
Culex pipiens pallens, 16
Curvularia lunata, 16
cyclomorphosis, 90
Cyclops agilis, 95
Cydnodromella pilo, 68
Cylindrocarpon destructans, 18
Cyphoma: *gibbosum*, 234; *signatum*, 234
Cypridopsis vidua, 95
cytochrome P450 monooxygenases, 17
cytokines, 116
cytosceleton: of *Euplotes*, 143

Dacnusa sibirica, 67
damage: concentration of, 47; dispersal of, 47; herbivore induction of, 74; mechanical, 20, 74
dandelion yellow mosaic sequivirus, 18
Daphnia, 90, 129, 160–176, 177–202; *ambigua*, 165, 178, 183; *carinata*, 165, 178, 183; *catawba*, 183; *cephalata*, 178; *cucullata*, 95, 178, 183; *galeata*, 94, 95, 183; *galeata mendotae*, 165, 183; *galeata x hyalina*, 95, 165, 166, 172, 183; *longispina*, 165; *lumholtzi*, 178, 183, 184, 192; *magna*, 92, 94, 95, 165, 172; *obtusa*, 165, 183; *parvula*, 165, 183; *pulex*, 94, 165, 178–184; *pulicaria*, 165, 183; *retrocurva*, 165, 178, 183; *rosea*, 183; *shodleri*, 184
defenses: classification of, 182; constitutive, 45; delayed inducible, 39; fixed vs. inducible, 253; general, 308; indirect, 62–88; inducible, 1–321; life-history, 185–187; localized, 33; morphological, 181–185; multiple types of, 318; post-encounter, 182; pre-encounter, 182; rapid inducible, 33; reversibility of, 286–299, 309, 317–318; specific, 308; systemic, 33. *See also* anti-predator behavior; resistance; response, behavioral
defoliators. *See* herbivores
Delia radicum, 50
density dependence, 215, 224–226
Depressaria pastinacella, 16, 19–20, 26, 32
design of experiments, 136
development: strategies of, 249–256; switch of, 249, 251
Diabrotica undecimpunctata howardi, 32

Diacrisia virginica, 19
Diacyclops, 129
diapause, 187
Diaptomus, 129; *kenai*, 165
dichloroisonicotinic acid, 31
Diglyphus isaea, 67, 72
Dileptus anser, 144
Doridella steinbergae, 324
drag of swimming, 139, 214
dragonfly larvae, 222–224, 227–229
DVM, 160–176

egg size, 138
elicitor, 75; fungal, 28
environment: heterogeneity of, 245
environmental costs. *See* costs
environmental tolerance, 286, 287–299; breadth of adaptation, 292–294; mode of the tolerance curve, 289, 292, 295–296
environmental variance: between-generation, 287, 299; spatial, 287; temporal, 287; within-generation, 287, 299
Epischura, 129
Equus caballus, 19
Erwinia carotovora, 18, 20, 32
Erysiphe sp., 18
Erythropodium caribaeorum, 234
escape reaction. *See also* ciliates, behavior in, 147–148
Escherichia coli, 16
Esox lucius, 204
Eudiaptomus gracilis, 95
Euleia fratria, 19
Euplotes: *aediculatus*, 144; *crenosus*, 154; *daidaleos*, 144, 154; *eurystomus*, 144; *focardii*, 144, 149; *octocarinatus*, 144, 146; *patella*, 144; *plumipes*, 144; *woodruffi*, 154
Euxoa tessellata, 19
evolutionary stable strategy, 255, 261

feeding cycle of predators, 190
ferns, 311
fever, 123
field studies, 131–132
Filinia mystacina, 129
fish, 165; kairomone, 197; larvae, 180–181; piscivorous, 207–217; planktivorous, 180–181
flea beetles, 50
flower, 47, 60
foliar phenolics, 39

food: concentration, 138, 171; quality, 55; threshold of, 138; variability in, 53; webs, 219, 227, 230
foraging behavior, 219
Frankliniella occidentalis, 66
furanocoumarin, 13,14,15,16; angular, 15, 24; inducibility, 28; linear, 15, 24; synthesis, 14, 29
Fusarium: oxysporum, 18, 20, 32; *sambucinum,* 18; *solani,* 18, 20; *sporotrichioides,* 18, 20

Galaxea fascicularis, 234
gall, 51
game theory, 250
generalist genotypes, 175
generalists, 287
genetic assimilation, 258
genetic polymorphism, 168, 172, 194, 238, 251–252; maintenance of, 174–175, 318
genotypes of plants, 80
Gerbera jamesonii, 67
Gibberella pulicaris, 20
gigantism, 148
Glechoma hederacea, 68
glucosides: cyanogenic, 58
glucosinulate, 50, 60
Gorgonacea, 234, 240
Gossypium hirsutum, 70
grazing rate, 97
green frog, 227–229
ground ivy, 68
gypsy moth, 53, 60

handling time, 212
hawthorn, 71
Helicoverpa zea, 70
Heliothis virescens, 16, 70
helmets in cladocerans, 183–185
Helminthosporium carbonum, 20
heracleum latent trichovirus, 18
herbivore: adaptation of, 47, 56; attraction of, 82; dispersal of, 47; insect, 36, 37; movement of, 47, 58; specialist, 47, 49; ungulate, 36
heritability: in bryozoans, 236, in cladocerans, 168; of plant defenses, 21, 25
Herpetocypris reptans, 95
Hipopodina feegeensis, 234
histone, 30
Holopedium gibberum, 178, 184

Homo sapiens, 17, 19
host finding, 47, 49
Hydractinia echinata, 234
hydrogen cyanide, 58
Hydrozoa, 233, 234
Hymenoptera, 65

immune response: time course of, 112–114. *See also* immune system
immune system, 104–126; components of, 107–109; costs of, 123–124; effectiveness of, 120–122; memory of, 111; recognition mechanisms of, 111–114; response to parasites, 110; specificity of, 124; of vertebrates, 106
imperatorin, 24
indirect effects, 229
indol glucosinolate, 50
infection: fungal, 54
infochemicals, 63, 97. *See also* cue
information: reliability of, 85; specificity of, 84–85
ingestion rate, 90
insecticides, 197
interactions: plant-plant, 40, 75
interference, 226; mechanical, 134
interplant communication, 40, 75
isopimpinellin, 24
Itersonilia pastinacae, 18

jasmonic acid, 31

kairomone, 90; activity of, 152; application in experiments, 135; of *Asplanchna,* 130, 133; of *Chaoborus,* 196–197; characterization of, 152–154; of crustaceans, 133–134; of *Lembadion,* 151–154; in natural conditions, 131; specificity of, 133
Keratella: cochlearis, 95, 129, 133; *quadrata,* 129; *slacki,* 129, 133; *testudo,* 129, 133, 139; *tropica,* 129, 133
kleptoparasitism, 213

Lactuca sativa, 16
Lagopus lagopus, 118
Lambornella clarki, 145, 151
landscape: importance of, 207–208
learning: in parasitoids, 81–82
Lebistes reticulatus, 16
Lembadion: bullinum, 143–144; *magnum,* 150

Lepidium virginicum, 50
Leptinotarsa decemlineata, 16
Leucania unipuncta, 19
L-factor, 152
life-table experiments, 135–138
life history, 101; defenses, 185–187; theory of, 185
light, 54; intensity of, 162; limitation of, 22, 24; ultraviolet, 28, 54; UV-A, 20
Lima bean, 66, 76
local adaptation, 173
Lygus lineolaris, 19
Lyriomyza; *bryoniae*, 67; *sativae*, 66; *trifolii*, 67
Lysiphlebus testaceipes, 70

major histocompatibility complex, 112
Malus domestica, 69
Mamestra brassicae, 72
Manduca sexta, 16
Manihot esculenta, 68
marginal value theorem, 263, 271
marmesin synthase, 14
mechanism of spine induction in rotifers, 130
Membranipora membranacea, 233
memory, 111
Mesocyclops, 129
metapopulations, 87
Micractinium pusillum, 94
Microcystis, 90; *aeruginosa*, 94
Microplitis croceipes, 70
migration: nocturnal, 161; reverse, 161, 201
Millepora, 234
mineral nutritients, 39
mite, 65; eriophyid, 51
mixed strategies, 250
modular organisms, 243
Mononychellus progresivus, 68
Montastrea cavernosa, 234
morphological defenses. See defenses
movement: dependence on food, 219–224; speed of, 219, 221
moving target theory, 260, 309
multiple fitness optima, 175
multiple-predator environments, 171, 200, 318
mutualism, 62
Mycobacterium: *avium intracellulare*, 121; *tuberculosis*, 120, 122
myristicin, 17

natural enemies, 48
natural killer cells, 112
neckteeth, 181
nectar: extrafloral, 88
nematode, 258
Nicotiana: *glauca*, 58; *sylvestris*, 58
nicotine, 58
Notonectidae, 133, 166, 181
nutrient, 54; limitation, 22–23

Onychodromus quadricornutus, 145, 149
Opius dissitus, 66
opportunity costs. See costs
optimal defense theory, 313
optimization models, 220
Orgilus lepidus, 72
Orthops scutellata, 19
Oscillatoria agardhii, 93, 94
oviposition cues, 50

Panonychus ulmi, 68
Papaipema marginidens, 19
Papilio: *bairdii*, 19; *brevicauda*, 19; *polyxenes*, 16, 19, 27; *zelicaon*, 19
Paramecium caudatum, 16
parasites: kinds of, 110
parasitism, 151
parasitoids, 59, 74
parsnip: leafcurl virus, 18; mosaic potyvirus, 18; potexvirus,18; webworm, 16, 19–20, 26; wild, 13, 16; yellow fleck sequivirus, 18
Pastinaca sativa, 13, 16, 18–26
pathogens, 51
pear, 68
Pediastrum duplex, 94
peptides, 152
Perca fluviatilis, 170, 210
perch, 210
Periplaneta americana, 16
Petroselinum sativum, 14, 28
phantommidge. See *Chaoborus*
Phaseolus: *lunatus*, 66; *vulgaris*, 67, 84;
phenotype: general purpose, 200
phenotypic plasticity, 13–21; irreversible, 286–299; reversible, 286–299. See also constraints
phenylalanine ammonia lyase, 14, 29
Philaenus spumarius, 18
Phoma complanata, 18, 20, 25
photosynthesis, 22

phototaxis, 162, 164
Phthorimaea operculella, 72
Phyllonorycter blancardella, 69
Phythophthora: *megasperma*, 28; *megasperma f sp. glycinea*, 30
phytoecdysones, 311
Phytomyza: *angelicae*, 19; *pastinacae*, 19
Phytoseiulus persimilis, 66, 67, 76
Pieris: *brassicae*, 71, 75, 78, 79; *napi*, 71; *rapae*, 50, 71
pigeons, 50
pike, 204
Plagiognathus: *obscurus*, 19; *politus*, 19
Planktosphaeria maxima, 94
plant: defense theory, 34; density, 54; volatiles, 74
plasticity costs. *See* costs; costs: genetic; infrastructure
plasticity, morphological, in Amphibia, 229. *See also* defense, morphological
pleiotropy, 26
Plutella xylostella, 72
Pneumocystis carinii, 121
Pocillopora damicornis, 234
Podocoryne, 234
pollinator: deterrence of, 47, 60
population: dynamics, 63, 219; effect of induced indirect defenses on, 86–88; growth rate, 136–137, 158, 187
population biology: effects of defenses on attacker, 319–320
potato, 72
predation pressure: variability in, 207
predation risk, 219. *See also* attack
predation: gape limitation, 211; spacial heterogeneity of, 248; tactile, 182; temporal heterogeneity of, 249; visual, 182; on zooplankton 179–181
predator: avoidance, 145,160–176; behavior, 261, 271; foraging strategy, 260; freshwater, 179–181; invertebrate, 179–180; vertebrate, 180–181
prey choice, 212
protection. *See* benefit
proteinase inhibitor, 86
Proteus vulgaris, 16
proximate factors, 162
Pseudaletia separata, 69, 84
Pseudomonas aeruginosa, 16
Psila rosae, 19
psoralen, 24; synthase, 14

Psylla pyricola, 68
Pyrus communis, 68

Quercus, 57

radiation. *See* light
Rana: *catesbeiana*, 222–223, 227–229; *clamitans*, 227–229; *sylvatica*, 223–224
Raphanus: *raphanistrum*, 50; *sativus*, 16, 50
Raphidocelis subcapitata, 94
reaction time, 292
receptor, 155
refuges: associational, 43
resistance: delayed inducible, 33; rapidly inducible, 33. *See also* defense
resource 220, 222–224; allocation of, 47, 186; depression of, 226; limitation of, 22; partitioning of, 185–187; trade-offs, 12
respiration, 22
response: behavioral, 140; localized, 33; rapid inducible, 33; systemic, 33, 74; time, 143
resting eggs, 187
reversibility of defenses, 286–299
rodents, 36
root flies, 50
rose, 67
rotifers, 258; spines in, 132
rust mite, 52

s-adenosylmethionine: bergaptol O methyltransferase, 14
s-adenosylmethionine: xanthtoxol O methyltransferase, 14
salamander, 258
salicylic acid, 31, 51, 59
scales of defense formation: spatial, 188; temporal, 188
Scenedesmus, 90; *acuminatus*, 94; *acutus*, 92, 94; *obliquus*, 94; *quadricauda*, 94; *subspicatus*, 94
Schizaphis graminum, 70
Scleractinia, 234, 240–241
sclerites, 234
Scolothrips takahashii, 67
self-damage, 47, 58. *See also* costs, self-damage
semiochemicals. *See* infochemicals
Senecio jacobae, 64
S-factor, 152
shading, 24
shoaling. *See* swarm formation

signal. *See* cue
Simocephalus vetulus, 95
Sinapis alba, 50
sinking: loss, 101; rate, 138
size efficiency hypothesis, 180
size: at first reproduction, 186–187; refuge, 212
slugs, 50
Solanum dulcamara, 51
spatial autocorrelation, 249
spatial scales, 319
specialists, 47, 287
sphondin, 23–24, 26
spider mite, 52, 79
spine, 128; in *Daphnia*, 185; development in rotifers, 129; growth in ciliates, 149
Spirulina platensis, 93
Spodoptera: exigua, 16, 69, 86; *litura*, 16
Staphylococcus aureus, 16
Staurastrum, 90
Stenostomum sphagnetorum, 143–144
sterol biosynthesis, 30
stolons, 234
strawberry latent ringspot nepovirus, 18
Streptococcus faecalis, 16
stress, 31
Stylonychia mytilus, 144
Stylopoma duboisii, 234
sunflower, 57
Sus scrophula, 19
susceptibility, 47, 49
swarm formation, 167, 189
sweeper tentacles, 234
swimming: drag of, 139, 214; maneuverability, 214; performance, 214; speed, 188–189
Synchaeta, 129
Synedra tenuis, 94
synergism, 43
synomones, 40, 102

tadpoles, anti-predator behavior of, 221. *See also* anuran larvae
talking trees. *See* interplant communication
tangled bank hypothesis, 257
T cells, 111; helper, 116; inflammatory, 116; killer, 112

Tegella robertsonae, 234, 239–240
terpenoids, 74
Tetrahymena pyriformis, 16
Tetranychidae, 66
Tetranychus urticae, 52, 66, 76, 77
Thalamoporella tubifera, 234
Thamnacus solani, 52
thrips, 59
time lag, 170, 208
time scales, 319
tobacco, 51; necrosis virus, 52; ringspot nepovirus,18
tomato, 51, 57, 67
top down control, 63, 87
trade-off, 175, 219–220. *See also* costs
trichomes, 59
Trichoplusia ni, 16, 19–21
Trichostrongylus tenuis, 118
Triticum aestivum, 70
tritrophic system, 62–88, 317
trophic shift, 151
Tropocyclops, 129, 133
Typhlodromus pyri, 69

ultimate factors, 162
Urostyla grandis, 144
UV light, 20, 28, 54

variability, 47
vertical migration, 160–176; reverse, 200
viruses, 119
visual predators, 164
volatiles: nonspecific, 79–80; specific, 76–79. *See also* cue

water fleas. *See* cladocera, *Daphnia*
wheat, 70
wood frog, 223–224
wounding, 39

xanthotoxin, 20–21, 23–24, 26

yellow cress, 72

Zea mays, 69
zooplankton: defenses in, 190–194; predator avoidance in, 167, 193–194